# Environmental Footprints and Eco-design of Products and Processes

**Series Editor**

Subramanian Senthilkannan Muthu, Head of Sustainability - SgT Group and API, Hong Kong, Kowloon, Hong Kong

W0007712

Indexed by Scopus

This series aims to broadly cover all the aspects related to environmental assessment of products, development of environmental and ecological indicators and eco-design of various products and processes. Below are the areas fall under the aims and scope of this series, but not limited to: Environmental Life Cycle Assessment; Social Life Cycle Assessment; Organizational and Product Carbon Footprints; Ecological, Energy and Water Footprints; Life cycle costing; Environmental and sustainable indicators; Environmental impact assessment methods and tools; Eco-design (sustainable design) aspects and tools; Biodegradation studies; Recycling; Solid waste management; Environmental and social audits; Green Purchasing and tools; Product environmental footprints; Environmental management standards and regulations; Eco-labels; Green Claims and green washing; Assessment of sustainability aspects.

More information about this series at http://www.springer.com/series/13340

Rahul S Mor · Anupama Panghal · Vikas Kumar
Editors

# Challenges and Opportunities of Circular Economy in Agri-Food Sector

Rethinking Waste

 Springer

*Editors*
Rahul S Mor 🄳
Department of Food Engineering
National Institute of Food Technology
Entrepreneurship and Management
(NIFTEM)
Kundli, Sonepat, Haryana, India

Vikas Kumar
Bristol Business School
University of the West of England
Bristol, UK

Anupama Panghal
Department of Food Business Management
and Entrepreneurship Development
National Institute of Food Technology
Entrepreneurship and Management
(NIFTEM)
Kundli, Sonepat, Haryana, India

ISSN 2345-7651                    ISSN 2345-766X  (electronic)
Environmental Footprints and Eco-design of Products and Processes
ISBN 978-981-16-3793-3           ISBN 978-981-16-3791-9   (eBook)
https://doi.org/10.1007/978-981-16-3791-9

This Springer imprint is published by the registered company Springer Nature Singapore Pte Ltd.
The registered company address is: 152 Beach Road, #21-01/04 Gateway East, Singapore 189721,
Singapore

# Foreword

The Agri-Food sector throughout the globe is facing the challenges of resources getting exhausted and raw material depletion. Therefore, industry participants are looking for solutions towards sustainability and resource optimization methods, and circular economy (CE) is one of the steps towards that direction. In this context, the book *Challenges and Opportunities of Circular Economy in Agri-Food Sector: Rethinking Waste* is a scholarly work. In this book, researchers from various domains have contributed their work aligning with the theme. Readers of this book will be advantaged due to immense literature on the primary focus areas of the circular economy towards mitigating the Agri-Food sector challenges. Researchers have highlighted concepts like 6Rs, closed-loop supply chains, waste management, system redesigns, carbon footprints, and Agri-Food business innovations. This book is exceptionally well thought-of-theme based and carefully edited through domain experts. Academicians and researchers will surely benefit from this book towards converting the challenges into opportunities for a resourceful and sustainable Agri-Food sector.

Dr. Chindi Vasudevappa
Vice-Chancellor
National Institute of Food Technology
Entrepreneurship and Management (NIFTEM)
Kundli, Sonepat, Haryana, India

# Preface

The food industry participants realize various types of challenges at different levels of the food supply chain. Issues like depleting resources, excess food wastage, food security concerns, consumer trust, food frauds and recalls and technological advancements add to this. Food industries are looking for sustainable and flexible ways to deal with such issues through circular economy concepts. Researchers and academicians are also working in this direction to comprehend a workable strategic plan. This book is a reflection on all such aspects.

Chapter "Circular Economy in the Agri-Food Sector: An Introduction" presents the concepts of the circular economy, specifically focusing on rethinking waste management. This chapter discusses various innovative business models and strategies towards a circular economy in the Agri-Food sector context. In Chapter "Mapping Facets of Circularity: Going Beyond Reduce, Reuse, Recycle in Agri-Food Supply Chains", the concept of '8Rs' is discussed along with their applicability for enabling and enhancing circularity in the bio- and technosphere of the Agri-Food sector. Extending on the same lines, Chapter "A Conceptual Framework for Food Loss and Waste in Agri-Food Supply Chains: Circular Economy Perspective" presents a conceptual framework for circularity in the Agri-Food supply chains considering the 6Rs and proposing the support towards the sustainable development goals. Further, the perspective of food waste management in the Agri-Food sector through the circular economy is discussed in Chapter "Circular Economy in Agri-Food Sector: Food Waste Management Perspective". It is realized that integrating the circular economy principles into the food supply chain can make the food industry more sustainable. The same is highlighted in Chapter "Sustainable Food Value Chains and Circular Economy", which focuses on the need for a sustainable food value chain to avoid loss during processing and its role in the circular economy. Chapter "Rethinking the Physical Losses Definition in Agri-Food Chains from Eco-Efficiency to Circular Economy" defines physical losses in the Agri-Food chain context and discusses strategies for reducing, eliminating, or transforming physical losses and achieving a more sustainable business strategy.

Further on these lines, Chapter "Can CE Reduce Food Wastage? A Proposed Framework" presents a theoretical framework wherein circular economy (CE) is tested as a moderator leading to pro-environmental behaviour by managing wastes towards achieving sustainability. Adding to this, Chapter "Modeling the Causes of Post-harvest Loss in the Agri-Food Supply Chain to Achieve Sustainable Development Goals: An ISM Approach" highlights the leading causes of food wastage, including poor logistics infrastructure, large numbers of intermediates, and lack of innovative technology, in the Indian Agri-Food supply chain. Reducing wastages may boost the country's economy and support sustainable development goals like zero hunger, sustainable cities, etc. The applications of circularity principles in the wine supply chain having constraints and opportunities are presented in Chapter "From the Vine to the Bottle: How Circular is the Wine Sector? A Glance Over Waste". It further needs developing local arrangements integrating different agents and the alternative valorization routes for by-products to work resources optimally, thus contributing to more deeply incorporated circular practices. Chapter "Carbon Footprint of Raw Milk and Other Dairy Products" studies and presents carbon footprinting or raw milk and dairy products, while Chapter "Valorization of By-Products from Food Processing Through Sustainable Green Approaches" discusses the valorization of by-products from food processing through the adoption of sustainable green approaches. Chapter "Applying Circular Economy Principles to Agriculture: Selected Case Studies from the Indian Context" presents few cases where the circular economy principles are applied in agriculture. Finally, Chapter "Transition Toward a Circular Economy Through Surplus Food Management" describes case studies to demonstrate the importance of local food systems and reducing food waste at different points.

Sonepat, India                                                                          Rahul S Mor
Sonepat, India                                                                      Anupama Panghal
Bristol, UK                                                                              Vikas Kumar

# Acknowledgements

We acknowledge all those people who were involved and helped in completing this book project. Firstly, we would like to thank the authors for contributing their valued time and expertise. Special thanks are due to the reviewers' valuable contributions regarding the improvement of quality, coherence, and content demonstration of the chapters. We also appreciate the referees for reviewing the manuscripts and scholars for editing and organizing the chapters. Finally, the editors are grateful to their parent institutes (National Institute of Food Technology Entrepreneurship and Management, Kundli, Sonipat—131028, India, and Bristol Business School, University of the West of England, Bristol, UK) for providing essential facilities to conduct this research work smoothly.

Rahul S Mor
Anupama Panghal
Vikas Kumar

# Contents

# Editors and Contributors

## About the Editors

**Dr. Rahul S Mor** is a researcher of operational design and development, Agri-Food supply chain, manufacturing systems, etc., and works as Assistant Professor at the Department of Food Engineering, National Institute of Food Technology Entrepreneurship and Management, Kundli, Sonepat, India. He holds a Ph.D. in *Industrial and Production Engineering*, specializing in Supply Chain and Operations Analytics. He has authored/co-authored over 50 publications in international refereed SCI, ABDC, CABS ranked journals of high impact factor & Chapters and Conference papers with Elsevier, Springer, Taylor & Francis, IGI Global, etc. His recent research focuses on 'Managing supply chain disruptions, Benchmarking, Performance measurement, Digital technologies, etc. He has the exposure of executing R&C projects in the manufacturing and food sector.

Dr. Mor is the *Area Editor and Managing Guest Editor*: *Operations Management Research* (Springer Nature), *Int. J. Logistics Research and Applications* (Taylor & Francis); Guest Editor: *IEEE Trans. on Engg. Mgmt.* (IEEE), *Int. J. Transp. Eco.*; Associate Editor: *Supply Chain Forum: An Int. J.* (Taylor & Francis); Editor: *Int. J. Supply and Opr. Mgmt.* (KU); Editorial Board: *J. Dairy Sc.* (Elsevier), *FSFS, IJCEWM, IJLT*, etc., and reviewer of many int. journals. Currently, Dr. Mor is editing various books on *Agri-Food 4.0, Circular Economy, OSCM in Food Industry, Industry 4.0* with Springer, Emerald, EAI-Springer. He is a member of many professional societies.

**Dr. Anupama Panghal** is Assistant Professor at the Department of Food Business Management and Entrepreneurship Development, National Institute of Food Technology Entrepreneurship and Management, Kundli, Sonipat, India. Prior to that she has worked with Thapar University, Patiala, India, and National Agricultural Cooperative Marketing Federation of India, New Delhi. She was also selected by GoI, under the Faculty Secondment Scheme of MHRD, as a visiting professor for the Asian Institute of Technology (AIT), Bangkok, Thailand, for the semester of January 2020. Her academic specialisation is agriculture business management and currently her research interests are in the domain of Agri-Food supply chain management, technology inclusion in Agri-Food value chains and Agri-Food entrepreneurship management. She has authored books, more than 20 research papers in reputed journals, which include Scopus and ABDC indexed ones. Also, she has contributed chapters in the edited books. She has deep interest in carrying out research and consultancy projects in the areas of Agri-Food supply chain and Agri-Food entrepreneurship. Currently, she is working on one project funded by MoFPI on "Development of blockchain based traceability solution for Agri-Food supply chains" and another one on "Promotion of local food entrepreneurship and value chain development". She is also a master trainer for trainers under the PMFME scheme of MoFPI for the domain of Intellectual Property Rights (IPRs) for Agri-Food entrepreneurs. For her work in past 15 years, she has traveled to Australia, Cambodia, Netherlands, Belgium, and Thailand.

**Prof. Vikas Kumar** is a Director of Research and Professor of Operations and Supply Chain Management at Bristol Business School, University of the West of England, Bristol, UK. He is also a visiting Professor at Tong Duc Trang University, Ho Chi Minh, Vietnam. He holds a Ph.D. in Management Studies from the University of Exeter, UK. He serves on the editorial board of several international journals including *Journal of Business Logistics* (Q1), *Sustainability* (Q1), and *International Journal of Mathematical, Engineering and Management* (Q3). He has guest-edited a number of special issues in high impact journals including *International Journal of Production Research*, *Journal of Information Technology*, *Supply Chain Management: An International Journal*, *Production, Planning and Control*, *International Journal of Productivity and Performance Management*, and *Journal of Manufacturing Technology Management*. He is also a reviewer of more than 15 international journals including ABS 4/4* and ABS 3* journals such as *International Journal of Production Research*, *International Journal of Production Economics*, *Production Planning and Control*, *Supply Chain Management. An International Journal*, *Expert Systems with Application*, *International Journal of Production and Operations Management*, *Journal of Intelligent Manufacturing Systems and Computers and Industrial Engineering*. He has published 5 books and more than 200 articles in leading international journals and international conferences.

He has generated research income in the excess of £1 million from research agencies such as Innovate UK, EPSRC, British Council, British Academy, Newton Fund, and Science Foundation of Ireland. He is currently working on two research projects funded by the British Academy/Newton Fund and Royal Academy of Engineering.

# Contributors

**Tripti Agarwal** Department of Agriculture and Environmental Sciences, National Institute of Food Technology Entrepreneurship and Management (NIFTEM), Kundli, Haryana, India

**Vaibhav Aggarwal** Center for Research and Implementation of Sustainable Practices, Jaipur, India

**Santosh K. Arya** R&D Division, Nextnode Bioscience Pvt. Ltd., Kadi, Gujarat, India

**Patricia Calicchio Berardi** LEPABE, Department of Metallurgical and Materials Engineering, Faculty of Engineering, University of Porto, Porto, Portugal; CELOG Centro de Excelência e Logística e Supply Chain, São Paulo School of Business Administration EAESP/FGV, São Paulo, SP, Brazil

**Luciana Stocco Betiol** CELOG Centro de Excelência e Logística e Supply Chain, São Paulo School of Business Administration EAESP/FGV, São Paulo, SP, Brazil; FGVethics, Centro de Estudos em Ética, Transparência, Integridade e Compliance, São Paulo School of Business Administration, EAESP/FGV, São Paulo, SP, Brazil

**María de las Mercedes Capobianco-Uriarte** Department of Economics and Business, University of Almería, Almería, Spain

**María del Pilar Casado-Belmonte** Department of Economics and Business, University of Almería, Almería, Spain

**Vikas Kumar Choubey** National Institute of Technology Patna, Patna, India

**Ricardo Alberto Cravero** National Technological University - Qinnova, Santa Fe, Argentina

**Eoin Cunningham** School of Mechanical and Aerospace Engineering, Queen's University Belfast, Belfast, UK

**Joana Maia Dias** LEPABE, Department of Metallurgical and Materials Engineering, Faculty of Engineering, University of Porto, Porto, Portugal

**Andrew Gough** University of Northampton, Northampton, England

**Jason Grimm** Iowa Valley Resource Conservation and Development, Amana, IA, USA

**Nazlıcan Gözaçan** International Logistics Management Department, Yasar University, Izmir, Turkey

**Hsin-I Hsiao** Department of Food Science, National Taiwan Ocean University, Keelung, Taiwan, ROC

**Shamaila Ishaq** University of Derby, Derby, England

**Yaşanur Kayıkcı** Department of Industrial Engineering, Turkish-German University, Istanbul, Turkey

**Yiğit Kazançoğlu** International Logistics Management Department, Yasar University, Izmir, Turkey

**Caroline C. Krejci** Department of Industrial, Manufacturing, and Systems Engineering, The University of Texas at Arlington, Arlington, TX, USA

**Deepak Kumar** R&D Division, Nextnode Bioscience Pvt. Ltd., Kadi, Gujarat, India

**Mukesh Kumar** National Institute of Technology Patna, Patna, India

**Simmi Ranjan Kumar** Asian Institute of Technology, Khlong Luang District, Thailand

**Vikas Kumar** Bristol Business School, University of the West of England, Bristol, UK

**Jake Kundert** Iowa Valley Resource Conservation and Development, Amana, IA, USA

**Çisem Lafcı** International Logistics Management Department, Yasar University, Izmir, Turkey

**Michi Lopez** Iowa Valley Resource Conservation and Development, Amana, IA, USA

**Ritika Mahajan** Department of Management Studies, MNIT Jaipur, Jaipur, India

**Anuj Mittal** Department of Industrial Engineering Technology, School of Engineering, Dunwoody College of Technology, Minneapolis, MN, USA

**Rahul S Mor** Department of Food Engineering, National Institute of Food Technology Entrepreneurship and Management (NIFTEM), Kundli, Sonepat, India

**Carter Oswood** Feed Iowa First, Cedar Rapids, IA, USA

**Anupama Panghal** Department of FBM & ED, National Institute of Food Technology Entrepreneurship and Management (NIFTEM), Kundli, Sonepat, India

**Jose V. Parambil** Chemical and Biochemical Engineering, Indian Institute of Technology, Patna, Bihta, Bihar, India

**Saugat Prajapati** College of Applied Food and Dairy Technology, Kathmandu, Nepal

**Rishabh Sahu** Department of Agriculture and Environmental Sciences, National Institute of Food Technology Entrepreneurship and Management (NIFTEM), Kundli, Haryana, India

**Md. Shamim** Department of Molecular Biology and Genetic Engineering (PBG), Dr. Kalam Agricultural College, Kishanganj, Bihar Agricultural University, Sabour, Bhagalpur, Bihar, India

**Rohini Sharma** Azim Premji University, Bengaluru, India

**Rohit Sharma** Jaipuria Institute of Management, Noida, Noida, Uttar Pradesh, India

**Anjali Shishodia** LM Thapar School of Management Dera Bassi Campus, Chandigarh, Punjab, India

**Mohammad Wasim Siddiqui** Department of Food Science and Post-Harvest Technology, Bihar Agricultural University, Sabour, Bhagalpur, Bihar, India

**Shilpa Sindhu** Schools of Management, The Northcap University, Gurgaon, Haryana, India

**Beatrice Smyth** School of Mechanical and Aerospace Engineering, Queen's University Belfast, Belfast, UK

**Deepti Srivastava** Integral Institute of Agricultural Science and Technology, Integral University, Lucknow, Uttar Pradesh, India

**Umair Tanveer** University of Bristol, Bristol, England

**Tavishi Tewary** Jaipuria Institute of Management, Noida, Noida, Uttar Pradesh, India

**Jelena V. Vlajic** Queen's Management School, Queen's University Belfast, Belfast, UK

**Tim Walker** arc21, Belfast, UK

# Circular Economy in the Agri-Food Sector: An Introduction

**Rahul S Mor, Anupama Panghal, and Vikas Kumar**

**Abstract** The Agri-Food sector globally faces food resource scarcity and food waste challenges throughout the food supply chain. There are substantial food losses and food wastes in supply chains globally, from farm gate to table, mainly due to lack of required market infrastructure, ignorance and lack of safe transport, improper packaging, lack of standardization and grading, underdeveloped cold chains, and insufficient processing.

## 1 Introduction

The Agri-Food sector globally faces food resource scarcity and food waste challenges throughout the food supply chain. There are substantial food losses and food wastes in supply chains globally, from farm gate to table, mainly due to lack of required market infrastructure, ignorance and lack of safe transport, improper packaging, lack of standardization and grading, underdeveloped cold chains, and insufficient processing [37]. The increasing population demands more food and, in several ways, adds to more food waste, too [28, 24]. Adding to this is the scenario where agriculture is facing resource scarcity, degradation of natural resources, and human health hazards due to the adoption of unsustainable agricultural practices. All this is exerting adverse effects on the ecosystem [22]. Different types of food wastes cause various degrees of adverse externalities to the society, economy, and the environment. While circular agriculture supports the complete recycling of materials

R. S. Mor
Department of Food Engineering, National Institute of Food Technology Entrepreneurship and Management (NIFTEM), Kundli, Sonepat 131028, India

A. Panghal (✉)
Department of FBM & ED, National Institute of Food Technology Entrepreneurship and Management (NIFTEM), Kundli, Sonepat 131028, India
e-mail: anupama@niftem.ac.in

V. Kumar
Bristol Business School, University of the West of England, Bristol BS16 1QY, UK

© The Author(s), under exclusive license to Springer Nature Singapore Pte Ltd. 2021
R. S. Mor et al. (eds.), *Challenges and Opportunities of Circular Economy in Agri-Food Sector*, Environmental Footprints and Eco-design of Products and Processes,
https://doi.org/10.1007/978-981-16-3791-9_1

and implies a complete reorganization of the supply chain system, immediate steps are needed to mitigate waste and losses. A circular economy is based on industrial ecology's theories and principles [9]. Circular economy looks for solutions towards food traceability, transparency, integration, and, eventually, sustainability.

Eradicating waste by applying circular models of business may lead to enormous benefits. Although circular agriculture is a new concept yet, there is an excellent opportunity for the Agri-Food sector in developing and underdeveloped economies to encourage the sustainable practices of the food chain. Such an approach would help in attaining several sustainable goals globally. Circular Economy looks to remove reasonable waste within the market. If we look through the angle of a circular economy, waste doesn't discuss the same old meaning of *"junk"*, but it describes any insufficient use of resources or assets [14, 39]. There are different types of waste that circular models designed to remove, viz.,

*Wasted resources*—Material and energy that is unutilized efficiently.

*Wasted capacities*—Products and assets that cannot be utilized effectively.

*Wasted lifecycles*—Products reaching service period untimely or early without a second lease of scope.

*Wasted embedded values*—Constituents, material, and energy unrecovered from the residual current of products.

In a circular economy, food products are derived from a healthy production system to provide healthy nutrition-rich foodstuffs. Food loss and waste can be designed in a way in the entire food supply chain to minimize food wastes and losses. The concepts of 6R, i.e., Reuse, Recycle, Reduce, Recover, Remanufacture, Redesign, are essentially associated with the circular economy [8]. Building upon the foundational framework of 6R, the critical elements in a circular economy include *designing for the future, the inclusion of digital technology, preservation, and extension of existing resources, prioritizing regenerative resource use, utilizing waste as a resource/raw material, revisiting business models, and collaborative efforts to create value.* The majority of the population does not realize that food waste is a significant problem globally. About one-third of all the food produced worldwide is wasted for various reasons, equivalent to 1.3 billion tons annually. Developed countries waste food more than underdeveloped or emerging countries. United States of America (USA) generates considerable food waste among all the nations. USA wastes 30% of all the food produced, estimated to be around USD 48.3 billion annually. The hospitality industry is also guilty of generating a considerable amount of wastage where overbuying, spoilage and excess food production are the most common causes. Consumers from developed countries (Europe and North America) generate higher waste of food products with 95–115 kg per capita annually compared to 6–11 kg in developing and underdeveloped countries (Africa, South, and Southeast Asia).

Developing countries generate 40% of food waste at post-harvesting and processing stages, and the developed countries generate more than 40% of waste at retail and consumption stages. Bulk quantities of food products at the retail stage are unused, which causes unnecessary wastage because of quality standards and enhance appearance. Approx. 870 million people can be fed if about one-fourth of the food wasted globally is saved [25, 36]. Globally it has been estimated that waste generated from cereals is around 30%, root crops, fruits and vegetables generate 40–50%, oilseeds, meat, dairy generate 20%, and the rest 30% is caused by fish [35]. Waste generated from food also leads to wastage of various resources like water, land, labor, energy, and capital. Apart from natural resource loss, greenhouse gas emissions lead to global warming and climate change, thereby causing environmental and ecological degradation.

Additionally, an essential factor of this concept is the need to lower resource consumption and waste discharge into the environment. Basic principles of the circular economy remain the same in the agriculture sector too. The majority of the studies on principles focus on the environmental factors of the sustainable system. Still, the other factors like social and economic are not discussed, following a rather technical approach. In many theories, different social system elements like equity, inclusiveness, and gender remain implicit. Many sectors of the agricultural industry can be described under the circular agriculture concept, which involves agro-ecology. The innovative aspect of circularity is its application to the entire food framework, including handling and utilization. This perspective offers numerous choices for reusing supplements, components, and natural waste on a basic level. In this context, the current book presents the circular economy towards managing waste in the Agri-Food sector. The challenges faced, and opportunities perceived towards transitioning to the circular economy for food waste management are explored.

## 2 Opportunities and Challenges

Several challenges surface on various facets in the food sector, including resource scarcity, uncertainty, high wastages, product quality, unremunerated prices to farmers, exploitation and environmental impact, no information sharing, and extended channels. In the same lines, circular economy adoption faces several challenges at various food production stages, food consumption, and food waste &surplus management [9]. Simultaneously, there is a paucity of reliable parameters for measuring and documenting progress in the acquisition of circular economy principles [3, 13]. There exists a gap in understanding and perceiving the potential benefits of technological solutions towards food waste prevention. Lack of understanding further leads to a lack of guidance towards an advanced technological approach towards circular economy solutions [6]. The current global food production and consumption system have necessitated for change towards circularity in the economy and the agriculture sector. There is a huge wastage of resources and raw materials to produce food globally.

The circular agriculture system's primary purpose is to judiciously use the resources along the lines of controlled measures to reduce waste by closing the resources' loops. In circular agriculture, waste can be used as organic material to enhance and develop newer and better products, including food crops, vegetables, fruits, feed, energy, and livestock. Circular economy adoption at the industry level may find an opportunity in green product design, business model innovations, digitalization, and reverse logistics. New technologies and supply chain methods define the future of Agri-Food systems to be termed Agri-Food 4.0, in line with Industry 4.0 [17, 20].

## 2.1 Green Product Designing

The increasing legal, social, and economic pressures for adopting sustainability measures in the business make industries focus on designing sustainable and eco-friendly products. The majority of product attributes fixed at the time of product design itself decide its contribution towards waste. That is the reason companies are focusing on product design features to be such that they should not violate the requirements of reuse, recycle, return, remanufacturing, etc., at the end of the life cycle [21]. Designing the products for a long life cycle and designing for the product-life extension can help develop sustainable products towards a circular economy [5].

The fundamental idea behind adopting green design for products is to have a critical consideration at the designing stage itself for environmental impacts of the product in later phases of its life cycle. The product, once manufactured, sold, used, and scrapped, should not contribute towards wastages. The parameters of resource conservation while procuring, production, and distribution stages of the product are considered along with eco-friendliness [45].

## 2.2 The Digitalization of Agri-Food Supply Chains

The long supply chains from farm to fork have resulted in making the consumers ignorant about the origins of their food and the impact of their production and consumption choices on the overall environment and society [19]. Digitalization enables food supply chains to be highly integrated, effective, and responsive to consumer demands and regulatory requirements. Digitally driven supply chains are powerful and more successful. They are enhanced with a new information technology infrastructure that substantially affects supply chains and firm performance. The application of digital technology, i.e., the IoT [42], blockchain [42, 47], robotics, cloud acquisition, and artificial intelligence [11], enables the supply chain to become more responsive and proactive by the implementation of embedded chips, sensors, and software [40]. Transparency, traceability, environmental and social impacts, legal culpability, and e-market accessibility lead to a high-value proposition in the Agri-Food supply chain [12]. The blockchain, IoT, and big data technologies shorten agricultural supply chains making them more efficient, offering social benefits to the rural and deprived section of the world. It provides visibility, traceability of food products, transparency, authenticity, and quality of agri-products [10, 43]. The diffusion of digital technologies in Agri-Food supply chains results in enhanced efficiency and productivity as the technologies like big data, IoT, and blockchain bring transparency to food supply chains [2]. Also, it results in increased yield, reduced food wastage, taps changes in consumer patterns, and fulfilling the sustainable development goals [1, 34]. The IoT-enabled and cloud data acquisition-based supply chains are transformed from fragmented and ambiguous ones to integrated and transparent ones [31]. At the same time implementation of digital technologies is quite challenging as well. Different implementation challenges for digital technologies faced by Agri-Food leaders and companies as mentioned in the literature are:

1. Training of existing staff and recruitment of a skilled workforce [7, 18, 26, 38]
2. Financial constraints [15, 23, 29]
3. Lack of government support and policies [15, 18, 33]
4. Lack of digital culture [32]
5. Lack of infrastructure and internet-based networks [23, 30, 41]
6. Lack of integration of technology platforms [48]
7. Lack of coordination and collaboration [23, 4, 16, 30, 44, 46]
8. Unclear economic benefits by the implementation of digital technologies [18]

## 2.3   Innovative Business Models

Innovative business models are emerging in the Agri-Food sector to fulfil the existing gap towards closed-loop food supply chains. Broadly the emerging models may be categorized as in Table 1.

## 2.4   Reverse Logistics

Reverse logistics-related activities are crucial for creating closed-loop supply chains for the industry. Recall, return, recycling, review, and repair included in reverse logistics have significant environmental and economic impacts. Due to the dynamic and complex nature of reverse logistics activities, they are actively considered while calculating circular economy initiatives' impact. Reverse logistics activities strongly correlate to green performance management by reducing food waste and loss, thereby necessitating close examination.

## 3   System Transition: A Way Forward

To begin with the circular agriculture system, more emphasis is needed on environmental and health concerns. More farmers and consumers are to be convinced about

**Table 1** Emerging business models towards circular economy

| S. No. | Strategy/focus | Motives | Few initiatives |
|---|---|---|---|
| 1 | Agri-tech startups based on emerging digital technologies | Bringing automation in the Agri-Food supply chains by employing various IoT and AI-based applications | 1. Aquaconnect<br>They connect farmers with the upstream and downstream supply chain through AI and remote-sensing technologies to help shrimp and fish farmers improve efficiency and farming revenue. They use Farm MOJO, its AI-enabled mobile app, for this purpose (https://indiaai.gov.in/article/providing-ai-driven-solutions-for-agriculture-and-food-commodity-value-chains)<br>2. BharatAgri<br>They are using their algorithm to tell farmers what, when, and how to grow. They are monetizing an information-based service in agriculture<br>3. NEERx Technovation<br>They have developed a sensor called 'SHOOL: Smart sensor for Hydrology and Land application', for farm microclimate information using dielectric technology. It can help prevent pest infestation, improve water and fertilizer retention, drought prevention, soil degradation, water harvesting, reduce agri-input cost, and improve productivity (https://yourstory.com/herstory/2020/03/women-entrepreneurs-agri-tech-startups?utm_pageloadtype=scroll)<br>4. Qzense, Bengaluru<br>They are using IoT solutions to minimize food loss due to internal spoilage, ripeness, sweetness, and shelf life. They are using a unique combination of near-infrared spectral sensors and olfactory sensors. They are presently used for procurement and inward quality checks and grading the Commodities based on spoilage and sweetness levels<br>5. Intello Lab, Delhi<br>They are working towards gauging product quality using AI tools (computer vision and deep learning). They work on a B2B model with food growers, processors, retailers, food service companies, and other stakeholders in the food supply chain (https://yourstory.com/herstory/2020/06/women-entrepreneurs-agritech-product-supply-chain)<br>6. Agnext<br>They are focused on improving the quality of agricultural produce, rapid commodity assessment, etc., using a combination of AI-driven hardware, software, and analytics (Spectometry + Computer Vision + IOT sensing solutions) (https://indiaai.gov.in/article/providing-ai-driven-solutions-for-agriculture-and-food-commodity-value-chains) |

(continued)

**Table 1** (continued)

| S. No. | Strategy/focus | Motives | Few initiatives |
|---|---|---|---|
| 2 | Blockchain-based business models/startups | Providing a distributed digital database, which is tamper-resistant and supports product tracking and tracing, and is accessible to all supply chain participants<br><br>The critical product information such as origin and expiration dates, batch numbers, processing data, storage temperatures, and shipping details get digitized and entered into the blockchain at every step along the chain and cannot be tempered | 1. The food companies, including Dole, Unilever, Walmart, Golden State Foods, Kroger, Nestle, Tyson Foods, McLane Company, and McCormick & Co., making Blockchain Collaboration to identify where the global supply chain can benefit from Blockchain. (*Source*: http://fortune.com/2017/08/22/walmart-blockchain-ibm-food-nestle-unilever-tyson-dole/)<br>2. *Provenance* serves over 200 food businesses with product traceability software. Provenance works with grocers to bring shoppers to produce data and with seafood companies to highlight their workers' fair treatment and more. On December 12, 2017, they announced a new partnership with Unilever, Sainsbury's, and others to track social sustainability and financial efficiency across supply chains<br>3. *Arc-Net* teamed up with PwC Netherlands in August 2017 to target food fraud (the intentional substitution, addition, tampering, or misrepresentation of food, ingredients, or packaging)<br>4. Other startup focuses on supporting small farmers by improving product traceability. These are relatively small so far and include:<br>• Brazil-based *Bart.Digital*, which provides secure financial documentation for small farmers<br>• Colorado-based *Bext360* targets explicitly fair-trade coffee (got Ireland's Moyee Coffee, complete Blockchain)<br>• Ambrosus is also working on Blockchain technology that leverages intelligent packaging sensors to track food and medicine (https://www.cbinsights.com/research/blockchain-grocery-supply-chain/) |

(continued)

**Table 1** (continued)

| S. No. | Strategy/focus | Motives | Few initiatives |
|---|---|---|---|
| | Local Food Supply Chain Based Models | Helping the farmers to connect to the buyers directly | 1. Farmers Markets<br>2. Direct Online Marketing models like Farmer Uncle, Ninzacart<br>3. Farm Pick |
| | Additive Manufacturing based (3D Food Printing) | Permits to digitize and personalize the nutrition and energy requirements Customized food near to consumer Short Supply Chains Less food wastage hence optimum utilization and management of resources | 1. Several big companies are actively adopting 3D food printing like Nestle, Mondelez, Hershey's, Barilla<br>2. Presently, 3D food printing has its applications in confectionery and bakery food products like cake, pastry, ice cream, biscuits, chocolates, pizza, burger, coffee, etc.<br>3. It has huge scope in preparing food products for astronauts, adults with special food requirements, food products for sportspersons and allergic people, etc. |

*Source* Authors

the possible health and environmental benefits of circular agriculture. For making circular agriculture a successful venture, there is a need to encourage the production of exportable agricultural commodities and liberal Exim policies globally. For proper execution and business development, more emphasis has to put on better marketing strategy, managerial expertise, access to a worldwide learned organization, and being in the position to receive funds for investment. More investment needs to be done in the public–private partnership (PPP) model to develop the necessary market infrastructure. Significant constraints and risks identifying with various new circular agricultural businesses are specified at the production and system level. There are tedious and expensive registration measures for new products at the production level, an absence of information about new products among new customers, and tedious cycles to get the new circular model adopted. Winding up not so important cycle will help achieve minimum economic or environmental advantages at the system level. It may be a cumbersome task to bring the required changes in the mindset of producers and consumers.

In contrast, other linear processes involving waste streams and others in the food system can be continued to be used as it doesn't include external factors. Risk factors at the system level consist of organic waste, which contains toxic substances or pathogens and circulates in the food system, which needs to be addressed through suitable research and extension interventions. Along with these, it is to be noted that only technical or economic aspects such as recycling of nutrients or building the business can affect and create negative implications for other vulnerable groups to promote circularity. It is suggested for governments globally in general and developing countries in particular and their public and private accomplices to advance circular agriculture to improve various purposes, including better environmental monitoring, climate moderation, public health, and revenue generation at the same time. The social, ecological, and financial measurements should be considered for designing circular initiatives, assuming a proper observing framework. Those social perspectives, such as comprehensiveness, value, youth, and gender, are not adequately integrated into many circular ideas. Besides, it is prescribed to incorporate the private sector for the advancement of new activities. Likewise, it is prescribed to encourage circular initiatives to gain from existing activities and extra pilots. This is expected to additionally investigate the capability of the promising idea of circularity in farming and food in low and medium countries in particular. In developing economies, food wastage of food can be observed mainly during the food value chain's early levels, primarily responsible for financial, management, and technical constraints in harvesting, transportation, packaging, storage, cold chain, and processing facilities. Food waste issues can be tackled by making a massive investment in the early stages of the supply chain at the level of farmers through direct support in terms of technology, machinery, etc., and in the creation of lacking transportation, marketing, cold chain, modern storage, and processing infrastructure.

On the other hand, the wastage of food can be observed at later stages of the supply chain in developed countries. Consumer behavior plays a vital role in developed countries, which is not similar to developing countries. Many studies recognized failure to coordinate among various parties concerned in the supply chain as major contributory factors. Farmer-buyer agreements are often helpful to extend the extent of coordination. Furthermore, increasing perception among industries, retailers, and consumers and finding valuable benefits for food that gets wasted may reduce the number of losses and wastes in developed countries.

The unutilized fund, especially at the consumer level, is one of the most significant contributors to food waste generation [27]. There are possibilities for minimizing and reusing the food waste at the consumer level by smart shopping in the sense of planning meals for the week and taking into account the food items in the refrigerator or in-store to avoid overbuying. This will reduce food waste and spoilage. Various old preserving techniques to preserve fresh foods have been developed over ancient times as there was no technology like refrigeration system. Pickling and sun-drying techniques have helped to preserve foods over a long time. The fresh food may be preserved to reduce spoilage. Around 40% of waste is generated from peeling the outer layers or removing the inside seeds. Food wastage can be avoided if the apples or cucumbers peels containing lots of nutrition are utilized. Even seeds from melon, watermelon, or pumpkin can be sun-dried and roasted to have these as nuts that contain various minerals and vitamins.

Perishable items like milk, fruits, or vegetables with a different life cycle need to be used smartly. A deep freezer is mainly for fish, meat, and other non-veg items. Keeping in other areas will spoil these items and cause more food waste. In most households, people throw away previous meals leftovers as they think it is harmful to consume previous meals causing infection. Right ways to preserve and use the previous meal leftovers for the next meal can save much food from going to the dustbins. Restaurants and hotels mostly throw away their leftovers, causing a huge pile of food wastes. Recent trends to give away leftover food to homeless people and beggars who cannot afford to buy food can reduce food waste. Health drinks from fruits help to prepare quick recipes saving time, effort and have a nutritious diet. The serving sizes on the plate need to be checked to avoid food waste. Customers generally order more than their capacity in restaurants and hotels and throw away food. Governments in different countries like Singapore have put taxes on food waste, helping the hospitality industry keep a check on food waste. Local food banks can be promoted by different private agencies or government funding to establish different localities where supervisors can redirect avoidable food waste for distribution among various institutions or apply other solutions. Before buying in stores and markets, the date of expiry must be noted by customers for packaged foods. It is essential to avoid spoiled food from buying. Even if the food is purchased with a brief period of the lifecycle, it must be consumed to prevent spoilage and food waste. Developing and underdeveloped countries have inferior food product maintenance after harvesting, and rules and regulations must be developed to reduce this food waste. Though most food companies stock out those lots that are nearing expiring dates, malpractices can be observed to resell the expired products with different labels and expiry dates. Strict measures by government agencies can keep a check on these malpractices.

A kitchen garden on a terrace is a prevalent scenario in metropolitan cities where the urban population is developing their small gardens over a terrace or small patch of land. Composting of food waste which is organic manure for plants provides natural nutrients for their growth. This reduces the waste load in landfills and is utilized very efficiently within the cycle for food production. Various Non-Governmental Organizations provide various terrace farming solutions for composting at their home. From plate leftovers like fish or meat, bones can be feed to stray dogs. Commercial and household food waste can be used to generate energy, which generally fills up the landfills. Municipal wastewater generated from organic waste in industrial areas such as food and beverage manufacturing, biodiesel production, electricity generation has found immense utilization of these resources. Non-eatable food waste may produce methane gas for kitchens in biogas plants that do not require complex setups and can be beneficial and cheap utilization of resources.

# 4 Conclusion

There is a high potential for circular agriculture globally and low and high-income countries (LMICs) to generate additional income and gainful employment. There are endless ways to reduce, reuse and recycle huge food wastes globally. Circular agriculture can save natural resources as well as the ecosystem and reduces wasteful expenditures on foods. Unnecessary health expenditures can be curtailed by improving the environment, thereby lesser health hazards to humankind. Considering more about our food wastes generated from each household, we can help make a specific transformation to preserve a portion of the world's most significant assets. Indeed, even minimal changes to how shopping, cooking, and consuming food help reducing the adverse effects on the environment and ecosystem in a big way. With little effort, food wastage may be reduced significantly. This would help accomplish different objectives simultaneously, viz. less ecological contamination, environment relief, improved public health, better revenue for farmers and other stakeholders, and finding new work opportunities in the circular Agri-Food sector.

Circular agriculture being a new concept, has many local practices and innovations followed by local people for generations globally. These practices and innovations are found to be economically feasible as most of this created additional income and gainful employment for the local population. The social benefits of circular agriculture can be observed in improved living conditions and new openings. The ecological advantages incorporate better waste administration, a diminished utilization of regular sources, lower fossil fuel byproducts, less natural contamination. Additionally, due to the utilization of natural composts, soil quality and soil biodiversity can improve.

**Acknowledgment** The authors are thankful to their parent institutes (National Institute of Food Technology Entrepreneurship and Management, Kundli, Sonepat – 131028, India, and Bristol Business School, University of the West of England, Bristol, UK) for providing support and essential facilities to conduct this research work smoothly.

# References

1. Anastasiadis F, Tsolakis N, Srai JS (2018) Digital technologies towards resource efficiency in the agrifood sector: key challenges in developing countries. Sustainability (Switzerland) 10(12). https://doi.org/10.3390/su10124850
2. Astill J, Dara RA, Campbell M, Farber JM, Fraser ED, Sharif S, Yada RY (2019) Transparency in food supply chains: a review of enabling technology solutions. Trends Food Sci Technol 91:240–247
3. Aznar-Sánchez JA, Mendoza JMF, Ingrao C, Failla S, Bezama A, Nemecek T, Gallego-Schmid A (2020) Indicators for circular economy in the agri-food sector. Resour Conserv Recycl 163:105028
4. Bag S, Wood LC, Xu L, Dhamija P, Kayikci Y (2020) Big data analytics as an operational excellence approach to enhance sustainable supply chain performance. Res, Conser Recycling 153:104559
5. Bocken NMP, de Pauw I, Bakker C, van der Grinten B (2016) Product design and business model strategies for a circular economy. J Ind Prod Eng 33(5):308–320. https://doi.org/10.1080/21681015.2016.1172124
6. Ciccullo F, Cagliano R, Bartezzaghi G, Perego A (2021) Implementing the circular economy paradigm in the Agri-food supply chain: the role of food waste prevention technologies. Resour Conserv Recycl 164:105114
7. Hecklau F, Galeitzke M, Flachs S, Kohl H (2016) Holistic approach for human resource management in Industry 4.0. Procedia Cirp 54:1–6
8. Jawahir IS, Bradley R (2016) Technological elements of circular economy and the principles of 6R-based closed-loop material flow in sustainable manufacturing. Procedia Cirp 40:103–108
9. Jurgilevich A, Birge T, Kentala-Lehtonen J, Korhonen-Kurki K, Pietikäinen J, Saikku L, Schösler H (2016) Transition towards circular economy in the food system. Sustainability 8(1):69
10. Kamble SS, Gunasekaran A, Gawankar SA (2020) Achieving sustainable performance in a data-driven agriculture supply chain: a review for research and applications. Int J Prod Econ 219:179–194
11. Kaur H (2019) Modelling internet of things driven sustainable food security system. Benchmarking. https://doi.org/10.1108/BIJ-12-2018-0431
12. Kittipanya-ngam P, Tan KH (2020) A framework for food supply chain digitalization: lessons from Thailand. Prod Plan Control 31(2–3):158–172. https://doi.org/10.1080/09537287.2019.1631462
13. Kristensen HS, Mosgaard MA (2020) A review of micro-level indicators for a circular economy–moving away from the three dimensions of sustainability? J Clean Prod 243:118531
14. Kumar V, Sezersan I, Garza-Reyes JA, Gonzalez ED, Moh'd Anwer AS (2019) Circular economy in the manufacturing sector: benefits, opportunities and barriers. Manage Decis 57(4):1067–1086. https://doi.org/10.1108/MD-09-2018-1070
15. Kumar R, Singh RK, Dwivedi YK (2020) Application of industry 4.0 technologies in SMEs for ethical and sustainable operations: analysis of challenges. J Clean Product 275:124063. https://doi.org/10.1016/j.jclepro.2020.124063
16. Kumar R (2020) Sustainable supply chain management in the era of digitalization: issues and challenges. In: Idemudia EC (eds) Handbook of Research on Social and Organizational Dynamics in the Digital Era, pp. 446–460, IGI Global. https://doi.org/:10.4018/978-1-5225-8933-4.ch021
17. Lezoche M, Hernandez JE, Díaz MDMEA, Panetto H, Kacprzyk J (2020) Agri-food 4.0: a survey of the supply chains and technologies for the future agriculture. Comput Ind 117:103187
18. Luthra S, Mangla SK (2018) Evaluating challenges to Industry 4.0 initiatives for supply chain sustainability in emerging economies. Process Saf Environ Prot 117:168–179. https://doi.org/10.1016/j.psep.2018.04.018

19. Marsden T, Smith E (2005) Ecological entrepreneurship: sustainable development in local communities through quality food production and local branding. Geoforum 36(4):440–451. https://doi.org/10.1016/j.geoforum.2004.07.008
20. Masi D, Kumar V, Garza-Reyes JA, Godsell J (2018) Towards a more circular economy: exploring the awareness, practices, and barriers from a focal firm perspective. Prod Plan Control 29(6):539–550
21. Mesa JA, Esparragoza I, Maury H (2019) Trends and perspectives of sustainable product design for open-architecture products: facing the circular economy model. Int J Precis Eng Manuf-Green Technol 6(2):377–391
22. Mishra B, Varjani S, Pradhan I, Ekambaram N, Teixeira JA, Ngo HH, Guo W (2020) Insights into interdisciplinary approaches for bioremediation of organic pollutants: innovations, challenges and perspectives. Proc Natl Acad Sci India Sect B: Biol Sci 1–8
23. Moktadir MA, Ali SM, Kusi-Sarpong S, Shaikh MAA (2018) Assessing challenges for implementing industry 4.0: implications for process safety and environmental protection. Process Safety Environ Protect 117:730–741. https://doi.org/10.1016/j.psep.2018.04.020
24. Mor RS, Jaiswal S, Singh S, Bhardwaj A (2019) Demand forecasting of the short-lifecycle dairy products, In: Chahal H, Jyoti J, Wirtz J (eds) Understanding the role of business analytics. Springer, Singapore. https://doi.org/10.1007/978-981-13-1334-9_6
25. Mor RS, Bhardwaj A, Singh S (2018) A structured literature review of the supply chain practices in food processing industry. In: Proceedings of the 2018 international conference on industrial engineering and operations management, Bandung, Indonesia, 6–8 Mar 2018, pp 588–599
26. Müller JM, Kiel D, Voigt KI (2018) What drives the implementation of Industry 4.0? The role of opportunities and challenges in the context of sustainability. Sustainability (Switzerland) 10(1). https://doi.org/10.3390/su10010247
27. Nahman A, De Lange W, Oelofse S, Godfrey L (2012) The costs of household food waste in South Africa. Waste Manage 32(11):2147–2153
28. Ng HS, Kee PE, Yim HS, Chen PT, Wei YH, Lan JCW (2020) Recent advances on the sustainable approaches for conversion and reutilization of food wastes to valuable bioproducts. Bioresour Technol 302:122889
29. Nicoletti B (2017) Agile procurement. Agile Procurement 2. https://doi.org/10.1007/978-3-319-61085-6
30. Pfohl H-C, Yahsi B, Kurnaz T (2017) Concept and diffusion-factors of Industry 4.0 in the supply chain. 381–390. https://doi.org/10.1007/978-3-319-45117-6_33
31. Rai A, Patnayakuni R, Seth N (2006) This content downloaded from 216.227.221.251 on Tue. Manager MIS Q 30(2):226–246
32. Ras E, Wild F, Stahl C, Baudet A (2017) Bridging the skills gap of workers in industry 4.0 by human performance augmentation tools—challenges and roadmap. In: ACM international conference proceeding series, Part F128530, pp 428–432. https://doi.org/10.1145/3056540.3076192
33. Rauch E, Dallasega P, Unterhofer M (2019) Requirements and barriers for introducing smart manufacturing in small and medium-sized enterprises. IEEE Eng Manage Rev 47(3):87–94. https://doi.org/10.1109/EMR.2019.2931564
34. Renda A (2019) The age of foodtech: optimizing the agri-food chain with digital technologies. In: Valentini R, Sievenpiper J, Antonelli M, Dembska K (eds) Achieving the sustainable development goals through sustainable food systems. Springer, Cham. https://doi.org/10.1007/978-3-030-23969-5_10
35. Salman W, Ney Y, Nasim MJ, Bohn T, Jacob C (2020) Turning apparent waste into new value: up-cycling strategies exemplified by Brewer's Spent Grains (BSG). Curr Nutraceuticals
36. Sharma P, Gaur VK, Sirohi R, Varjani S, Kim SH, Wong JW (2021) Sustainable processing of food waste for production of bio-based products for circular bioeconomy. Bioresour Technol 124684
37. Sindhu S, Panghal A (2016) Robust retail supply chains—the driving practices. Int J Adv Oper Manage 8(1):64–78

38. Sommer L (2015) Industrial revolution—Industry 4.0: are German manufacturing SMEs the First victims of this revolution? J Ind Eng Manage 8(5):1512–1532. https://doi.org/10.3926/jiem.1470

39. Upadhyay A, Akter S, Adams L, Kumar V, Varma N (2019) Investigating "circular business models" in the manufacturing and service sectors. J Manuf Technol Manag 30(3):590–606. https://doi.org/10.1108/JMTM-02-2018-0063

40. Van Nunen JAEE, Zuidwijk RA, Moonen HM (2005) Smart and sustainable supply Chains. In: Smart business networks. Springer, Berlin, pp 159–167. https://doi.org/10.1007/3-540-26694-1_11

41. Varghese A, Tandur D (2014) Wireless requirements and challenges in Industry 4.0. In: Proceedings of 2014 international conference on contemporary computing and informatics, IC3I 2014, pp 634–638. https://doi.org/10.1109/IC3I.2014.7019732

42. Venkatesh VG, Kang K, Wang B, Zhong RY, Zhang A (2020) System architecture for blockchain based transparency of supply chain social sustainability. Robot Comput-Integr Manuf 63(March 2019):101896. https://doi.org/10.1016/j.rcim.2019.101896

43. Wang M, Kumar V, Ruan X, Neutzling DM (2019) Farmers' attitudes towards participation in short food supply chains: evidence from a Chinese field research. Revista Ciências Administrativas/J Administ Sci 24(3):1–12. https://doi.org/10.5020/2318-0722.2018.9067

44. Wang S, Wan J, Li D, Zhang C (2016) Implementing smart factory of Industrie 4.0: an outlook. Int J Distrib Sens Netw. https://doi.org/10.1155/2016/3159805

45. Ying J, Li-jun Z (2012) Study on green supply chain management based on circular economy. Phys Procedia 25:1682–1688

46. Zezulka F, Marcon P, Vesely I, Sajdl O (2016) Industry 4.0—an introduction in the phenomenon. IFAC-PapersOnLine 49(25):8–12. https://doi.org/10.1016/j.ifacol.2016.12.002

47. Zhang C, Chen Y (2020) A review of research relevant to the emerging industry trends: Industry 4.0, IoT, blockchain, and business analytics. J Ind Integr Manage 5(1):165–180. https://doi.org/10.1142/S2424862219500192

48. Zhou K, Liu T, Zhou L (2016) Industry 4.0: towards future industrial opportunities and challenges. In: 2015 12th international conference on fuzzy systems and knowledge discovery, FSKD 2015, pp 2147–2152. https://doi.org/10.1109/FSKD.2015.7382284

# Mapping Facets of Circularity: Going Beyond Reduce, Reuse, Recycle in Agri-Food Supply Chains

Jelena V. Vlajic, Eoin Cunningham, Hsin-I Hsiao, Beatrice Smyth, and Tim Walker

**Abstract** The world is facing increasing economic, environmental, and social sustainability challenges. Agri-Food supply chains play an intrinsic part in these as both, heavy users of natural resources and a supplier of essential resources satisfying societal needs. In the context of the circular economy, focusing on Reduce, Reuse and Recycle, (3R waste hierarchy) is no longer sufficient. Here an extension is proposed through rethinking the system, redesigning its infrastructure, processes and products, replacing traditional technologies, materials and energy sources with more effective ones and repurposing or recontextualising the use of products, components and by-products. While principles behind these 'Rs' are generally known, their application is industry and material/product-specific. Within the Agri-Food supply chain it is subject to strict food safety regulations, packaging requirements, ethical considerations, etc. In this paper, we explore 'R' principles discussed in the academic literature, identify '8Rs' concept applicable to biological (e.g., crops, prepared food) and technical flows (e.g., packaging) that appear at Agri-Food producers and discuss their role in enabling and enhancing circularity in bio- and techno-sphere. We present applications of these 'Rs' in four cases relevant to the Agri-Food sector.

J. V. Vlajic (✉)
Queen's Management School, Queen's University Belfast, Belfast BT9 5EE, UK
e-mail: j.vlajic@qub.ac.uk

E. Cunningham · B. Smyth
School of Mechanical and Aerospace Engineering, Queen's University Belfast, Belfast BT9 5AH, UK
e-mail: e.cunningham@qub.ac.uk

B. Smyth
e-mail: beatrice.smyth@qub.ac.uk

H.-I. Hsiao
Department of Food Science, National Taiwan Ocean University, Keelung, Taiwan, ROC
e-mail: hi.hsiao@email.ntou.edu.tw

T. Walker
arc21, 2nd Floor, Belfast Castle, Antrim Road, Belfast BT15 5GR, UK
e-mail: tim.walker@arc21.org.uk

© The Author(s), under exclusive license to Springer Nature Singapore Pte Ltd. 2021
R. S. Mor et al. (eds.), *Challenges and Opportunities of Circular Economy in Agri-Food Sector*, Environmental Footprints and Eco-design of Products and Processes,
https://doi.org/10.1007/978-981-16-3791-9_2

# 1  Introduction

Diminishing availability of non-renewable natural resources, risks of depletion of renewable and regenerative natural resources, rising air, water and soil pollution and contamination jeopardise global ecosystems. Though productivity and survival of the Agri-Food sector depends intrinsically on the health of these ecosystems, the detrimental environmental impact of the sector is noticeable: it contributes around 25% of global emission of GHGs [79], accounts for about 70% of total global water withdrawals, and food production practices adversely affect soil quality and biodiversity [26]. Additionally, overwhelming food losses and waste indicate that both, food production and consumption, are not sustainable: FAO found that in 2016 more than 10% food losses occurred between post-harvest to global distribution and exceeds 20% in less developed parts of the world [26]. At the same time, around 40% of food waste occurs in developed countries, such as the USA, while millions of meals are needed for hunger relief [63]. In the UK, there is an excess of around 3.6 million tonnes of farm output, from which, around 45% enters the waste stream and around 55% is handled as surplus. This is equivalent to 7% of the total food harvest, and a £1.2 billion potential financial loss (wrap.org.uk).

Food packaging has an important role in relation to food safety, preservation of food quality, and safe delivery and use [33, 89], but it also represents a challenge for recovery operations [71], as only around 51% and 67% of packaging waste is recovered in the US and the EU in 2014, respectively [25, 88]. The most challenging type of packaging waste is one made of plastic-based materials. In the UK, only 46.2% of plastic waste enters the recycling stream [74]. The rest is a major source of soil and water pollution.

Detrimental environmental effects of food and packaging waste indicate over-production and over-consumption, interrupted circularity of flows within bio- and techno-sphere (e.g., materials that cannot be recycled), mismatches between type of products, components or materials and recovery streams (e.g. end-of-use products enter recycling stream despite intact functionality) and cross-contamination of flows that are supposed to circle in bio- or techno-sphere exclusively (e.g. microplastic found in food and fresh water). Recent studies suggest that this might be an unintended consequence of using 'Reduce, Reuse, Recycle', i.e., the 3R waste hierarchy model [24] is still a predominantly linear economy based on 'Take-Make-Dispose' philosophy and sporadic circular flows. In line with the development of a circular economy, there are calls for the adoption of process and system thinking which complements sustainability aspirations [22, 50, 57]. There are multiple proposals for an extension of 3R waste hierarchy [75] towards increased circularity by rethinking the system role and functions towards effective regeneration and restoration of bio- and techno-sphere. This can be achieved through redesigning its infrastructure, processes and products, replacing traditional technologies, materials and energy sources with more effective ones and repurposing or recontextualising use of products, components and by-products [6, 19, 35, 42, 57, 58, 68, 96].

This chapter is structured as follows: In Sect. 2 we define circularity and present the *8R framework*, i.e., an overview of those 'R' principles that are extracted from the literature on the circular economy, sustainable production, and consumption, as well as circular supply chains. In Sect. 3, we apply the identified 'R' principles on design, production and use of biological products (crops, food items and final food products). In Sect. 4 we apply it to technical products (packaging) in the Agri-Food sector. Finally, in Sect. 5 we present four typical cases of Agri-Food producers and demonstrate the application of these R principles in the context of managing the production of pre-prepared foods and meals, wastewater and run-offs, agricultural biomass waste and packaging.

## 2 Circularity and R Principles

*Circularity* represents a performance that indicates to what extent raw materials, parts and final products are restored or regenerated into technical or biological flows, respectively. Circularity performance is represented by various circularity metrics, which indicate how well the principles of the circular economy are applied on the product or service level [14], or within so-called mezzo (e.g. industrial parks) and macro systems (e.g. cities, regions, nations). Though in principle more circularity implies more environmental benefits [68], that is not always the case [43]. Thus, there is a need for more research to understand, under which conditions higher circularity results in environmentally beneficial and cost-efficient business operations [29].

Circularity can be achieved by creating (a) closed-loop supply chains, where reverse flows cycle back in the system, to the organisations where they started [91], or (b) open-loop supply chains, where reverse flows cycle and cascade to alternative supply chains in the same sector ('open-loop, same sector') or in a different industry ('open-loop, cross-sector'), [95]. Traditionally, 3R principles are used in both types of supply chains and ordered hierarchically based on their circularity [68].

Next to basic 3R principles, the literature identifies other R frameworks: 4R framework [24], 5R framework [22], 6R framework [40, 45], 9R framework [15, 42], 10R framework [2, 68] or other [76]. Based on these frameworks, we present 8R circularity concept based on R principles that contribute to the development of circular and sustainable systems, strategies, and practices.

- **Rethink (R1)**. As a principle, it has a wider connotation. It includes reconceptualization of ideas, processes, constitutive elements of Circular Economy and other R principles [58]. A functional circular economy requires rethinking of both the production and consumption systems and processes to prevent waste, recover the value of products, components and materials and achieve sustainable circularity of flows. That means, if circular flows enter Biosphere and Technosphere without contaminating either [9], *regenerative* and *restorative* processes in these environments are established. Rethinking of production systems requires initiatives for the transition to the circular economy, such as development of new business models

[49], development of specialised recycling and reprocessing systems [15] and design of innovative, easily recoverable and valuable products [65]. Rethinking of consumption systems requires societal changes, such as *reconsideration* of the importance of ownership (e.g., sharing versus owning) and behavioural changes that lead to less consumption [5], as well as initiatives for increased citizen engagement towards the circular economy. A powerful initiative can lead to *refusal* to accept certain materials as input for production [15] or to use products that have a negative effect on circularity [58].

- **Redesign (R2).** It focuses on improving the design of products, processes, services and infrastructure to enable input from recycled materials, extend the useful life of products and enable easy recovery of products, components and materials and their return to biological and technical cycles [22, 86]. Redesign aims to improve the economic viability of a company, as well as reduce environmental impact and improve social impact within and beyond its supply chain. However, the success of redesign requires consumer acceptance of redesigned products and alignment or adaptation of business entities in the supply chains to changes in supply chain processes and infrastructure [46]. For example, to improve the purchasing process and make it part of circularity, tracing of recovered products and materials must be enabled [15, 47].

- **Reduce (R3).** It refers to a broad range of strategies that aim to reduce inefficient use of resources in the pre-manufacturing phase, inefficient use of energy and the use of *virgin materials* in manufacturing, generation of *toxic by-products* in manufacturing and post-manufacturing phase [30]. These strategies consider product and process design, changes of a layout, working procedures and flows, as well as elements of sharing economy that result in reduced consumption overall [20]. New technologies such as the Internet of Things (IoT), Block-chain, Augmented reality and robotisation can help to achieve eco-efficiency and optimise the use of resources [39].

- **Replace (R4).** It refers to substitution, i.e., it aims to substitute existing technologies, materials and energy sources with more sustainable ones. New technologies are based on eco-design and eco-efficient technological innovation [28], the use of Industry 4.0 hardware and software for efficient data collection and control, as well as increased use of 3D printing which enables replacement of parts printed without the release of residual waste [15, 39]. The development of new materials considers new non-toxic raw materials [50, 77], and durable and reusable smart materials or advanced materials based on nanotechnology [15]. Lastly, there is a wide spectre of renewable energy sources based on the use of sun, wind, waves, tide, or residual heat from industry [5].

- **Reuse (R5).** It refers to the recovery of products and components that result from commercial returns, lease returns, warranty returns, or returns due to overstocking [91]. This recovery type is based on the repeated use of a product or a component for a similar purpose as originally intended [73]. In the literature, different activities, such as re-labelling, repacking, repairing, or refurbishing are considered as variations of reuse [21]. In some studies, reuse is considered via the lens of resale or donations of second-hand or surplus products [95].

- **Repurpose (R6).** It represents using products, discarded products or their parts, or by-products for another function, i.e., their intended purpose is changed [26]. Repurposed products, by-products and components enter a distinct new life cycle [76], while maintaining or improving their value [58]. Application of repurposing principles is challenging: while many parts can be repurposed in a variety of products, not many products can be easily repurposed; moreover, the scale of an operation is small, and traceability of parts and products can be lost [58]. Changes of the context in which a product used are termed as *recontextualising*, and it considers changes in ownership, in the role that product is designed for [19].
- **Recycle (R7).** It refers to a recovery of materials that do not retain the functionality of used parts or products [73]. This depends on the complexity of products and their constitutive parts [7]. Organised collection and disassembly might precede recycling [65, 72] while mixing of recycled and virgin materials determines the quality of resulting products [5, 31].
- **Recovery of other resources (R8).** It considers natural renewable resources used in processes of production and consumption, e.g., energy, water, soil, air.

## 3 Circularity and R Principles Applied to Biological Flows (Food and Organic Material) in Agri-Food Supply Chains

The goal of sustainable Agri-Food production is to preserve the eco-system while producing high-quality nutritious food products in a cost-efficient manner while concurrently recovering food products, by-products and inputs for the benefit of the society [93]. However, conventional agriculture is predominantly a linear process involving the three main steps of make, take, and dispose. Consequently, though highly productive and consumer-oriented, this linear model results in pollution, negative impacts on biodiversity, over fertilisation, run-off to waterbodies, inadequate management of effluent and extensive demands on fossil resources. Additionally, 140 billion tonnes of agricultural biomass waste are generated annually across the globe [85], posing a commercial and environmental challenge in regard to their effective disposal.

In the context of Agri-Food businesses, the 3R hierarchy is focused on food losses and waste reduction, i.e., Prevention, Reuse and Recycling principles [83]. However, these R principles cannot be adequately applied to tackle the aforementioned issues and provide a window into potential solutions, such as the use of agricultural biomass waste as a source of sustainable raw materials for the development of natural polymers. We, therefore, adapt the 8R conception presented in the previous section to biological materials and food products typically present in Agri-Food supply chains.

- **Rethink (R1).** It is associated with a strategic view on balancing Agri-Food production and processing resources, capacities and capabilities with market demand, as well as on environmental and social impacts of food production and consumption [93]. Rethinking represents the *reconsideration* of ways in which food is produced, and what food is. As such, it leads to the creation of new forms of Agri-Food systems, such as technologically enabled precision agriculture, urban farming techniques, vertical agriculture [84], and the creation of new solutions for the regeneration of land and water [15]. Rethinking, related to food consumption considers initiatives to educate consumers and change their behaviour while improving health and nutrition, e.g., solutions that optimise nutritional value, environmental impact, and costs for individual consumers [36]. Consequently, increasing numbers of consumers *refuse* to buy GMO-based food, food products associated with non-sustainable production (e.g., palm oil without sustainable sourcing certification), or food of animal origin [38, 78, 79].

- **Redesign (R2)**. It considers specific strategies, in context of food and organic material, that aim to improve outcomes of Agri-Food production, such as the use of traditional or contemporary techniques to improve the resistance of plants and their fruits to insects and various diseases, as well as harsh climatic conditions (e.g., frost, droughts or humid climates) [8]. Redesign of food production processes refers to changes that enable sustainable intensification [69] or expansion of mixed farming systems that enable easier nutrient exchange, cropping and pasture rotations [66]. This might require upskilling of workers, re-training or even recruitment of professionals that have specific knowledge [1]. Redesign in the context of food services mostly considers remaking or revamping menus. Remaking takes place when food services feature meals made of surplus foodstuffs [27], while revamping considers emphasis on new features including better presentation of food on the plate, or special dining experiences.

- **Reduce (R3)**. It considers, in the context of food and organic material, *prevention practices* (e.g. avoiding surpluses and avoidable food waste and losses) [83]. In some studies, the reduce principle considers all strategies that result in the reduction of food waste and losses, e.g. reuse/food redistribution and recycling [26, 29, 93, 99]. According to Teuber and Jensen [83], prevention targeted at consumers results in the highest economic value per tonne of food waste avoided, and it is often connected to advances in packaging and packaging materials (more about this in the next section). In the context of Agri-Food production, reduction is primarily seen through the lens of *eco-efficient use* of natural resources (e.g. land, water, etc.) and man-made resources (e.g. machinery, logistics), as well as the use of new technologies to track, trace and improve forecasting related to supply, demand and changes in quality and quantity of fresh foods over time [13, 60]. Bio-by-products, often inedible and of low value, are mostly associated with unavoidable waste and out of scope in relation to the reduce principle.

- **Replace (R4)**. It considers, in the context of food and biological flows, the substitution of inputs, technologies or parts of production systems, or the substitution of the ingredients products are made of. New, more productive crop varieties and livestock breeds, precision farming, satellite navigation [69], and vertical farming systems [4] can substitute inputs, technologies or elements of traditional food production systems.

- **Reuse (R5)**. In terms of food products, reuse takes two distinctive forms: redistribution and resale. In either case, it is related to a *surplus* of packed or unpacked food products [82], which results from overproduction or product returns. *Redistribution* typically represents food donations distributed by producers, wholesalers or retailers to charities, food banks [90], and day centres and night shelters for homeless and vulnerable people [12], or internal redistribution to a company's canteens and employees [29]. *Resale* of food products takes place if there is demand from customers/the market. The resale channel depends on the quality of food products: typically, high-quality food material and products are resold to food services or on food markets, while out-of-specification, perishable or deteriorated fresh food material and products are resold to pet-food producers or farmers [91].
- **Repurpose (R6)**. When referring to biological flows, repurpose considers multiple uses of biological materials [49]. Discarded products, by-products or waste materials can be used in the same or other industries (instead of entering waste streams) [56], requiring inter- and cross-industry collaboration or trade [95]. On a smaller scale, repurposing often occurs amongst food processors, where a surplus of high-quality food products, raw materials or by-products is used as an input for the production of other food products [27, 93], e.g., bread is used for the production of bagel chips, breadcrumbs or croutons for salads, and fruit-mix salads are used for the production of cakes. Additionally, agricultural production results in abundant organic waste that is usually designated to waste-to-energy streams. This change from the typical waste-to-energy route, i.e., *recontextualization* as raw materials for production of biopolymer products, results in more value from recovery [37].
- **Recycle (R7)**. In the context of biological flows, recycle considers material recovery related to food products, food waste or by-products. As food products can be of plant, fungi or animal origin, recycling of each type of organic material is regulated. Recycling of plant- or fungi-based materials can take various forms: from simple, on-site land spreading to use of specialised facilities (e.g., anaerobic digesters or in-vessel facilities, which produce digestate and compost, respectively) [67]. In many instances, recovery of material is accompanied by recovery of energy, called thermal [44]. Recycling, which has a high potential for reducing food waste, can also convert materials into valuable products that can serve as raw resources for other industries [83].
- **Recovery of other resources (R8)**. Related to biological flows, this considers the recovery of water, soil, and air from pollution. In the Agri-Food sector, recovery of these resources is especially important as soil quality and water accessibility determine the quantity and quality of crop production. Each of these resources requires specific methods for recovery, that range from 'natural' solutions to specialised treatment facilities for purification of water or air. For example, amendment of polluted soil can be achieved by direct use of fruit and vegetable waste [83], next to the use of fertilisers. Purification of wastewater often caused by agricultural and food production activities requires specialised treatment facilities. However, there are also other solutions for soil and water recovery, such as *bioremediation*. Bioremediation is the managed decomposition of organic contaminants and

is based on the fact that all organic materials are prone to biodegradation [41]. Living organisms, such as bacteria, fungi or plants, can be used to degrade or detoxify pollutants from soil, water or air, and biological processes can be focused so that pollutants are mineralised into biomass, water, carbon dioxide and other non-hazardous products [64]. Micro-organisms, which can be either naturally present (e.g., in soils) or specially selected for the contaminant of concern [81], are commonly used for bioremediation through the underlying processes of anaerobic and/or aerobic degradation [41]. Likewise, fast-growing plants like willow or reeds can also be used for bioremediation (Fig. 1). Whatever organism is used, the correct environmental conditions (e.g. temperature, humidity, nutrients) need to be present to promote its growth and the nutrient level should be such that the organism uses the contaminant of concern as a foodstuff [41]. Although bioremediation is a slow process, it brings with it the advantage of avoiding high capital costs associated with heavy equipment and infrastructure [41], and generally has a lower energy and resource demand than conventional treatment methods. Waste by-products can also be avoided [64], thus presenting a more environmentally friendly and economical alternative.

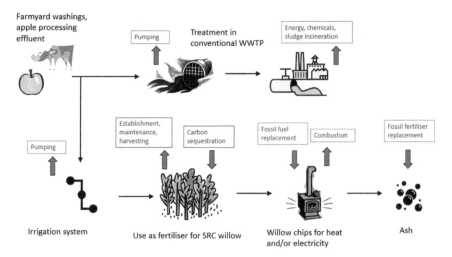

**Fig. 1** Comparison of carbon (greenhouse gas) and energy flows in a conventional and willow-based system used for managing farmyard washings or apple processing effluent. ↑ indicates carbon emissions and energy/resource demands, while ↓ indicates carbon sequestration or carbon/energy/resource savings. *Source* Author

## 4 Circularity and R Principles Applied to Technical Flows (Packaging) Present in Agri-Food Supply Chains

Food packaging has an important function in value-adding and recovery operations [71] and, as such, has an important role in food safety, food quality, preservation, and delivery. Next to this, it functions as a medium for information about packaging content, weight, brand, and packaging recyclability. Additionally, there is increasing pressure to improve the circularity of packaging materials, especially those made from plastics, metals, glass and paper [32]. Though the 3R waste hierarchy can be applied to packaging, consideration of new materials, new technological solutions for packaging and its use, changes in regulations related to design and recovery of these new packaging materials require going beyond the 3Rs. We adapt the 8R concept to food packaging typically present in Agri-Food supply chains.

- **Rethink (R1)**. In relation to packaging, this *(re)considers* the need for and uses of packaging, as well as practical *"design for zero waste"* approaches combined with innovations in material science, e.g., development of smart materials or biomaterials. This requires cross-sectoral collaboration, such as between food producers and wholesalers/retailers and packaging companies, (re)processors, government and environmental agencies and local authorities [94]. Evolution of packaging towards integration with electronics and cloud data solutions is in an increasing trend [11]. Moreover, citizen campaigns are often focused on pollution caused by packaging, and consequently, environmentally conscious consumers *refuse* to use non-recyclable materials, e.g., coffee cups, single-use plastics, etc.
- **Redesign (R2)**. This considers changes in the packaging design to improve its functional characteristics or recyclability. Creation of active packaging extends the freshness of food products, while the creation of intelligent packaging enables tracking and monitoring conditions of packed foods [11]. Moreover, the improvement of packaging by adding sensor elements to bio-based materials creates bio-based smart packaging [34]. Halonen et al. explain that they include packaging that represents a safe 'envelope' for food products, advanced coatings and additives for the preservation of foods, and renewable sensor materials which can detect food deterioration or quality.
- **Reduce (R3)**. In the context of packaging, reduce refers to the extent of the reduction or avoidance of packaging that allows a food product to retain its appearance and quality [71]. This reduction can be achieved incrementally by improving eco-efficiency and aiming to reduce the quantity of material or virgin material used in the production of packaging, as well as minimising scrap and defects in production [28, 61].
- **Replace (R4)**. This considers the substitution of packaging materials with other, less resource-intensive and polluting types of material, resulting in more durable, easier to reuse or recycled packaging. Strategies to reduce waste of food products and extend shelf-life of foods [60] include: replacing packaging that provides a label with a 'best-before' date with intelligent packaging which shows how long packaging has been open; replacement of traditional packaging for 'family

size' portions with single-person portions which have been individually split and packaged for individual consumption (so-called snap-packs); and replacement of conventional single-use food packaging with modified atmosphere packaging, active or resealable packaging [60]. Another example is the use of edible packaging solutions, which have been shown to extend shelf life and reduce environmental impact [11, 16]. However, while some types of innovative packaging solutions enable multiple reuses, not all of them have a better environmental impact from the aspect of material recovery.

- **Reuse (R5)**. Primary packaging is in many cases designed for single-use due to food safety requirements, thus limiting options for reuse. Transport packaging is typically reused, as a part of take-back management [5].
- **Repurpose (R6)**. In technical flows, repurposing takes various forms. There are no examples of repurposing of food packaging in the same industry, but packaging materials can be *recontextualized* and used in other industries [49]. There are some promising uses of packaging waste: plastic waste can be repurposed for production of asphalt pavement materials [98], while waste glass can be repurposed for production of building bricks [18].
- **Recycling (R7)**. Recycling of technical materials such as packaging is regulated via the Extended Producer Responsibility concept and corresponding Producer Responsibility Obligations defined in EU Directive 94/62/EC. In the UK, businesses that handle (import, produce, use) more than 50 tonnes of packaging annually and have a turnover more than £2 million or equivalent must comply with the packaging regulations. They must register with the regional environmental agency (e.g., Northern Ireland Environmental Agency) and manage packaging waste themselves or by using an approved compliance scheme. Reprocessors refer to companies that recycle material to issue Packaging Recovery Notes, which are bought by business or compliance schemes. Material for recycling can be collected by using take-back management [5] or waste collection services. Major issues that determine the quality of recycled materials are related to the level of contamination of collected material.
- **Recovery of other resources (R8)**. In technical flows, this considers various solutions for energy recovery, as well as other elements, such as $CO_2$ generated by production and logistics activities [93].

## 5  Application of R Principles

In this section, we present an overview of the extent to which companies in the Agri-Food sector apply R principles.

## 5.1   Case 1: Application of R Principles in the Food Cold Chain—Examples of a Preventing and Reducing Food Waste

Chilled and frozen food represents a significant part of the food market. In the last decade, due to the rise in employment and the consequential change in customer behaviour in developed countries, prepared frozen food increased its market share in the overall food industry. In 2019 in Taiwan, the frozen food industry was 50.41 billion NTD, an increase of 3.68 billion NTD from 2018, reaching around a 12% share of the overall food industry in the country [55]. Manufacturers of prepared frozen food are in a large number of cases focal companies in supply chains, responsible not only for production but also for distribution and sale. In a tropical climate such as in Taiwan, reduction and prevention of food waste, especially of perishable products is an essential part of all production and logistics activities. We present here a typical example of mid-size manufacturers of prepared frozen food located in Taiwan, which depicts the application of various R principles.

The manufacturer of prepared frozen foods supplies restaurants and coffee shops with more than 70 types of frozen foods made of meat, vegetables, rice, and seafood. Each day, the production of prepared frozen foods results in around 500 kg of organic waste, and there are occasional product returns. Food waste and returns end up in recycling stream due to human errors in handling food items, over-purchasing or errors in managing cold chain conditions, as well as out-of-specification raw materials and technical failures. To prevent food waste and increase circularity, the company considered various R principles.

To prevent food waste along the cold chain, the company is *rethinking* (R1) its cold chain design, operations and management. In particular, they are considering various options for the creation of a smart cold chain to improve prediction, operating control, data accuracy and quality, as well as to make easier communication and collaboration with suppliers and customers. A combination of hardware, such as Internet of Things (IoT) devices, Radio Frequency Identification (RFID) tags and Wireless Sensor Networks (WSN), and Time–Temperature Integrators (TTIs) with user-friendly software for shelf-life modelling such as Cold Chain Predictor, Seafood Spoilage Predictor, and ComBase, enables informed decision making and assessment of food waste risks [59]. With real-time information, decisions about optimal handling or stock control can be better informed, which should prevent food waste, and thus achieve cold chain management with low economic losses and low environmental impacts.

The company *redesigned* (R2) pre-prepared meals by combining a surplus of various finished food products to create a new pre-prepared meal. For example, the surplus of various vegetables, seafood and meat products are combined to form a new meal. Additionally, redesign takes place also within production in terms of re-training of the workforce. As Taiwanese food producers rely on a significant number of foreign workers, who are familiar only with specific operations or equipment, reduction of human errors and improvement of workers skills to fit a wide range of operations in production and logistics requires their re-training.

In situations with only limited quantities of redesigned pre-prepared meals or a lack of customers, the typical practice is internal *redistribution* (R5), e.g., employees receive free meals. Alternatively, some companies are organising tasting sessions, where they try new recipes and new combinations.

Use of out-of-specification raw materials often results in more intensive processing and residual by-products. Instead of ending in the recycling stream, the company *repurposes* (R6) these by-products by transforming them into final components ready to add to various meals. For example, fat removed from pork is transformed into lard and used for the preparation of pork floss, which is used as a topping to add flavour to various pre-prepared meals. Moreover, surplus food that is not sold or used in human consumption, as well as by-products and selected organic wastes, are sent to farms to be converted to animal feed.

Practices and strategies related to prevention of food waste in the cold chain case fit well within the concept of R principles. The principles *reuse* and *repurpose* are used as practices that aim to reduce cost burden due to over-purchasing or over-production, *redesign* is used as a strategy to create new products and increase their value, and *rethinking* represents a vision of how various supply chain actors can be linked to prevent overall waste, recover the value of products, components and materials, and achieve sustainable circularity of flows.

## 5.2   Case 2: Application of R Principles in Agriculture—Examples of Recovering Wastewater and Run-Off

In Northern Ireland, 63% of water bodies do not achieve the 'good or better' status required by the Water Framework Directive, with agricultural run-off a significant polluting contributor [17]. Much of the existing wastewater treatment infrastructure in the region is either operating at maximum capacity or past the end of its design life [62], which presents challenge for managing agricultural effluents even if the logistical difficulties of collection and transport can be overcome. In the following paragraphs, we present two examples that outline the implementation of short rotation coppice (SRC) willow in the context of rethinking and redesigning Agri-Food systems to support circularity across the whole food-energy-water nexus.

*Example 1. Rethinking and redesign of wastewater flows in a typical dairy farm in Northern Ireland*

To tackle challenges for managing agricultural effluents, it is proposed that through *rethinking* (R1) the system, the flow of nutrients could be better managed by the use of SRC willow, which not only provides a circular approach for nutrient flows but also serves as a bioenergy source.

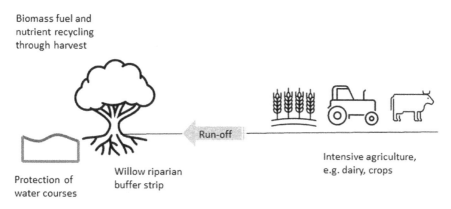

**Fig. 2** SRC willow riparian buffer strip. *Source* Author

Irish dairy production is indicative of intensive agriculture across Europe [3], where agricultural run-off and farmyard washings represent some of the principal causes of water pollution [23]. *Redesign* (R2) of a farm, e.g., adding an SRC willow riparian buffer strip (Fig. 2), shows promise for ameliorating the effects of intensive agriculture on a wider scale. An investigation of SRC willow buffer strips in a typical dairy farm in N. Ireland [48] found that total nitrogen and phosphorous leachate could be reduced by 14% and 9% respectively, with a 16.5% decrease in $CO_2$ emissions from fossil fuel displacement and minimal impact on milk production.

Additionally, the *reduction* (R3) in nutrient run-off to waterways decreases demands on existing water treatment plants. Demands on wastewater treatment plants can also be reduced through management of dirty water, such as farmyard washings, by using SRC willow instead of conventional wastewater treatment. In the research farm at the Agri-Food and Biosciences Institute, Hillsborough, N. Ireland, it was found that for an equivalent nutrient load, a conventional wastewater treatment plant consumes 2.6 $MJ/m^3$ of wastewater, while the net energy production of the willow system, which manages dirty water and produces wood fuel, is 48 $MJ/m^3$ [80].

The outputs of a willow buffer system (wood fuel and ash) can *replace* (R4) both fossil fuels and fossil-based fertilisers, providing a more efficient energy crop. The nutrients taken up by the plants can be extracted from the ash after the combustion of the harvested bioenergy crop, thus avoiding nutrient wastage and associated pollution, and allowing the nutrients to be *recirculated* in another cycle.

*Example 2. Redesign of wastewater flows and energy recovery at a fruit processing factory*

Food processing factories use water intensively and have a high demand for energy. A local apple processing company put in place an adjacent SRC willow plantation [97]. The biomass is used as a heat source to meet the energy demands of the fruit processing plant and the wastewater from the plant being is to the willow for bioremediation (see Fig. 2). Analysis of the lifecycle energy and greenhouse gas balances found that greenhouse gas savings of over 80% were achieved compared to the conventional wastewater management system and that energy and emissions savings of the order of 10% were realised for the company's heating system due to reduced oil demand [97].

These case studies show rethinking, i.e. the benefits of moving beyond conventional pollution prevention practices towards sustainable systems thinking. By *redesigning* (R2) the systems, wastewater with environmental polluting potential is managed so that environmental benefits are achieved through energy savings, fossil fuel displacement and decreased carbon emissions.

## 5.3   Case 3: Application of R-principles in Cross-Industry Exchanges—Examples of Repurposing Organic Waste as an Energy Source to Raw Materials for Novel Biopolymers

The poultry industry is a sector with an abundant and varied supply of biowaste streams with potential applications in biobased polymers (fillers, functional additives, precursors). One interesting example of repurposing of organic waste is Moy Park Ltd. based in Northern Ireland. Moy Park supply 25% of all poultry sold within Western Europe from 12 plants within NI, England, France and the Netherlands.

In recent years the company has been extremely forward-thinking in their efforts to find cost-effective ways to capture waste stream management issues within their core business. Having broken down their production process into key stages (Fig. 3), they identified and quantified 7 key streams (designated waste or low-value co-products) including litter (40,000 tonnes per year), litter ash (25,000 tonnes per year), bone (82,000 tonnes per year), blood (2500 tonnes per year), feather (10,000 tonnes per year), eggshell (2500 tonnes per year) and meal (45,000 tonnes per year) [51].

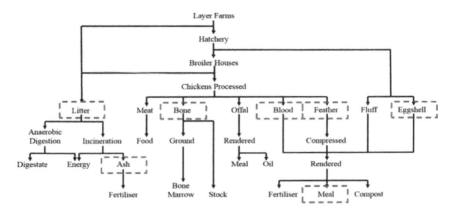

**Fig. 3** Poultry production with identified waste streams. Adapted from [51]

In the past, the waste-to-bioenergy route has been favoured for agricultural wastes, but these can also be repurposed, i.e., utilised for the manufacture of novel biomaterials. While their continuing development and conversion to polymer additives are at various points on the technology readiness level (TRL) spectrum there is ample evidence for utilising these materials within the polymer processing industry.

Coupled with oil savings, *repurposing* (R4), i.e., *recontextualising* poultry wastes and using them as polymer feedstocks will yield financial gains for both the polymer and poultry industries, estimated at over £1,964,000 per day in the UK alone [53]. The research shows the environmental and economic value, attainable via repurposing, and highlights the important role the biomaterials sector could play in tackling environmental issues. The following types of agricultural wastes are found most promising for this:

- Eggshells: Composed predominantly of calcium carbonate they can be readily utilised as a polymer filler or functional additive in loadings up to 55% [52]. Moreover, the eggshell membrane is a potential source for collagen mining for applications including healthcare and cosmetics [70].
- Poultry Litter Ash: Rich in potassium their fine particle size and morphology make them readily usable as a filler material, again in loadings upwards of 55% [52].
- Feathers: Much like poultry meal, feathers contain a high organic content in the form of Keratin, a natural fibrous protein which has potential to be utilised for fibre reinforcement or more excitingly be utilised as a novel biopolymer.

Repurposing, i.e. recontextualising of Agri-Food waste for the development of new materials can mitigate the impending depletion of natural resources. These bio-based resources, at present posing the industry a commercial and environmental challenge in regard to their effective disposal, contain the building blocks for the development of natural polymers and represent a source of sustainable raw materials. At present they are simply underutilised. Their availability across the globe is in stark contrast with the enormous disparities, and resulting turmoil, associated with worldwide distribution of fossil resources.

## 5.4 Case 4: Application of R Principles to Food Packaging—Examples of Reducing and Recovering Packaging Waste

All companies that handle more than 50 tonnes of packaging in the last calendar year, have a turnover of more than £2 million, including any subsidiaries, operate in the UK and produce, use or sell packaging products are required to comply with the Producer Responsibility Obligations (https://www.gov.uk/guidance/packaging-producer-responsibilities). Local authorities, such as Belfast City Council, have a statutory duty to manage the wastes produced by their area and have introduced programmes to educate/inform, guide, and coordinate activities involving primarily households and some local businesses. This has also involved establishing new partnerships, such as arc21 to assist in dealing with the resulting waste management issues. Previously, programmes were developed to promote innovation and new initiatives to improve environmental performance (e.g. the Business Improvements through Environmental Solutions programme, BITES), to engage more widely to determine how agri-business sector wastes could be revalorised as the product (e.g. project funded by the EU Interreg) and to consider how better supply-chain management could improve recycling collection performance, i.e. how to better align waste collection with local reprocessing requirements while enhancing local employment opportunities (e.g. projects funded by Invest NI).

In Northern Ireland, the (re)processors in the Collaborative Circular Economy Network (CCEN) have used funding from Invest NI funds to review their products and determine how and if they can align with developments in local government waste collection operations, local sortation operators and with the market and consumer expectations. The network includes (re)processors supplying food manufacturers with glass and paper products and their focus has been reuse (glass packaging) and redesign (paper packaging material). Various R principles are considered and used by these producers and (re)processors.

The results from the initial study by the CCEN were presented at a Chartered Institution of Wastes Management conference [10] and revealed that over £100 m worth of economic value per year is generated in Northern Ireland from manufacturing new products from paper, plastics, and glass. It also highlighted that further £50 m of economic potential could be realised if additional high-quality recyclables were available locally. One of the major factors limiting the expansion of this manufacturing potential was the lack of availability of high-quality recyclables with each of the companies having to import recyclables to supply their business. This motivated involved stakeholders to *rethink* (R1) how this system can be improved, in particular for plastic and glass packaging.

- Plastics: To tackle these issues, most plastic and food producers in Northern Ireland have signed up to the Plastic Pact as well as the 'On-Pack Recycling Label' (OPRL) guidelines to reduce "bad plastics" and improve the communication around recycling options for end-of-life packaging. By following the Plastic Pact and OPRL guidance, even companies outside of CCEN are seeking to maintain and grow market share while pre-empting proposals on the introduction of the UK Plastic Tax in 2022.
- Glass: While the recycling of glass has no major issues, major food and beverage companies increasing collaboration with glass (re)processors to tackle issues of limited availability of raw materials for glass products on the island of Ireland, the carbon impact of coloured glass and consideration of alternative energy sources for glass (re)processors.

There is an observed trend of food packaging *redesign* (R2) towards improved recyclability. For example, food packaging producers changed the structure of plastic packaging by removing polyvinylidene chloride (PdVC) barriers within flexible films, by using single materials/polymers to increase the recovery of the material and by adding recycled content were applicable/approved for direct food contact. Next to this, the producer light-weighted lines are introduced throughout the business.

In food packaging, there are initiatives for the *reduction* (R3) of

- Packaging weight: for example, one food packaging producer reduced the weight of plastic milk bottles, and glass (re)processors reduced weight of glass bottles
- The pigment use or colourings: for example, the food packaging producer reduced pigment used in its milk bottle caps to decrease contamination within the High-Density Polyethylene (HDPE) stream; glass (re)processors are working to relax colour specification for clear glass which could reduce the carbon impact of jars and bottles.

In line with packaging *redesign* (R4), there are initiatives to increase circularity by replacing materials that are traditionally used by ones that are easier to recycle: for example, food packaging producer replaced polystyrene (PS) product lines of yoghurt pots by polypropylene (PP) and/or Polyethylene terephthalate (PET).

*Recycling* (R7) is a standard practice that takes place for food and beverage packaging (except for certain types of plastics). Agri-businesses are contributing to circularity by increasingly using recycled PET (called rPET), which led packaging producers to launch more products with 30–100% rPET. Similarly, there is a move in beverage companies from collecting and reusing glass bottles to recycling them. Based on data collected in one regional glass (re)processing company in Northern Ireland in 2020, more than 550 million glass bottles collected by councils and other suppliers are recycled annually.

Practices and strategies related to reduction and recovery of food packaging fit well within the concept of R principles. Recycling is increasing within Agri-Food packaging, as it is driven largely by legislative requirements, while Redesign and Reduction are used as strategies to improve recyclability, increase circularity, and increase efficiency in use of resources. Rethinking the packaging design, collection, recovery, and network of companies that can collaborate to enable smooth return flows is increasingly being recognised as essential to improve the sustainability of packaging. And as the circular economy becomes better articulated, and legislative drivers begin to impact upon companies, there is clearer vision that these agendas can be achieved within supply chains while also delivering individual organisation's environmental, social and governance (ESG) commitments.

# 6 Conclusions, Limitations, and Directions for Future Research

The objective of this study was to explore 'R' principles widely discussed in the academic literature and contextualise them in the framework of circularity of flows within bio- and techno-sphere and Agri-Food supply chains.

The transition from linear 'Take-Make-Dispose' approach towards a circular model that enables recovery of product, components and materials, as well as energy and natural renewable resources [87] requires going beyond 3R waste hierarchy model. Based on the literature [6, 19, 35, 42, 57, 58, 68], it requires:

- Setting a clear *vision of* how circular economy leads to sustainable societies, regenerative processes in biosphere and restorative processes in techno-sphere. *Rethinking* related to purpose, goals and ways to increased effectiveness of production and consumption systems is the first principle behind the transition to circular economy and sustainability.
- Defining *strategies* that will lead to the achievement of defined goals and purpose. *Redesign* of production and consumption systems, its infrastructure, processes and products, *replacement* of traditional technologies, materials and energy sources with more effective ones and *reduction* of inefficiencies in the use of resources and polluting inputs and outputs associated with production and consumptions are key principles that drive context-specific strategies towards circularity and sustainability.
- Operationalising and optimising business models associated with circular economy practices. *Reuse, repurposing, recycling* and *recovery* of renewable resources are key principles behind circular economy practices.

Rethinking, redesign, reduction, replacement, reuse, repurpose, recycle and recovery represent the 8R concept (Fig. 4) that considers vision, strategies and practices that help to envisage, plan and establish circularity and make the transition towards a circular economy and a sustainable society. This concept represents a synthesis of previously developed R concepts and frameworks. However, though

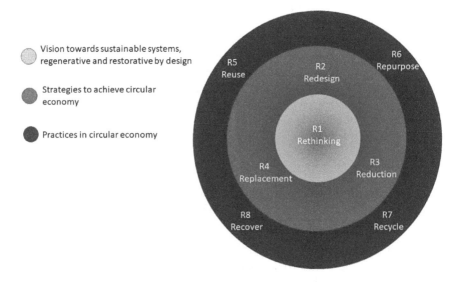

**Fig. 4** Vison, strategies, and circular economy practices in 8R concept. *Source* Author

existing R concepts and frameworks indicate priorities in resource recovery and/or waste management, they do not distinguish difference in nature of Rs. With this study, we contribute to the knowledge in area of resource recovery and/or waste management and sustainability by offering different view of Rs. These are not presented as priorities, but as principles that refer to:

- a vision of the circularity that fits within and enable co-existence of natural and man-made world,
- strategies that lead to circularity and result in long-lasting implications on use and recovery of resources and
- practices based on circular flows that need further organisational and technological enhancement, as well as adoption along supply chains.

It is worth emphasising however that application of 8R principles is context specific. In Table 1, we present a summary of findings from four cases in the Agri-Food sector, by illustrating examples of 8Rs in context of bio- and technical materials.

These examples show how these 8R principles can be applied in the Agri-Food sector. This summary of findings (in Table 1) can be used as a map for companies in Agri-Food sector and provide guidance on how each of these principles can be applied for various Agri-Food issues related to both, Agri-Food materials and products, and packaging.

Limitation of the study and future directions are:

- The list of examples is not exhaustive, as it is based on cases related to Agri-Food production. Inclusion of other actors in the Agri-Food supply chain, such as wholesalers, retailers, food services, food banks, etc. would provide a more comprehensive list of possible applications of 8R concept to prevent food waste and improve its valorisation.
- The cases illustrated in this study are based on Agri-Food companies in developing economies, the UK and Taiwan. As R strategies and practices are context-dependent, its development and applications might be different in the context of developing countries, which face different challenges [92].
- Achieving advances in the Agri-Food sector, in particular regarding Agri-Food waste treatment and its valorisation is possible only with investment in infrastructure, and operational improvements, policy updates and behavioural change [54]. This requires multi-disciplinary approach to develop new conversion technologies for the production of, for example, high value (bio)polymers, (bio)chemicals and (bio)actives, construction materials, packaging, usable water, fertilisers, bioenergy and enable their wide application across different industry sectors.
- Ultimately, the issue of the circular economy is how we move from theory to practice which is increasingly being driven by recognition that our current linear system of 'take-make-waste' is not sustainable and, indeed, is having a catastrophic effect upon biodiversity. Obviously, Agri-Food supply chains are integral to such change. As [9] highlight in their seminal book, "waste is food" and in the context of the circular economy simply focusing on the waste hierarchy is

**Table 1** Examples of the application of R principles in biosphere and techno-sphere of the Agri-Food sector

| Focus | Principle | Examples in the Agri-Food sector | |
|---|---|---|---|
| | | Biosphere | Techno-sphere |
| Vision | Rethinking | • Consideration of new technologies (e.g. intelligent cold food chain) to prevent food waste along the supply chain and improve circularity<br>• Consideration of bioremediation to regenerate waste flows back to nature<br>• Consideration of agri-waste streams for other roles/other contexts of use (recontextualization) | • Consideration of innovative collaborative programmes and initiatives that engage multiple stakeholders and aim to improve environmental performance<br>• Consideration of new functionalities of packaging |
| Strategy | Redesign | • Change of recipes for pre-prepared meals<br>• Redirection of the wastewater flow | • Change of packaging structure and composition in plastic food packaging |
| | Reduction | • Increase of effectiveness and efficiency of working procedures to reduce scrap and defects<br>• Use SRC willow as a buffer to filter wastewater and run-offs and reduce nutrient run-off and the load on the wastewater treatment plant | • Reduction of packaging weight (plastics)<br>• Reduction of contamination within different plastic streams<br>• Reduction of input of PET virgin materials |
| | Replacement | • Use of wood fuel and ash instead fossil fuels and fossil-based fertilisers | • Replacement of input materials and components in food packaging to increase recyclability |
| Practice | Reuse | • Internal re-distribution of newly combined meals<br>• Use of newly combined meals for food marketing purposes | • Reuse of glass bottles in the production of beverages |
| | Repurpose | • Use of by-products as raw materials for other finished food products<br>• Use by-products and organic waste as animal feed | • Recontextualization: use of agricultural wastes for novel bio-materials/polymers |
| | Recycling | • Organic waste to compost<br>• Anaerobic digestion (compost) | • Recovery of plastic, glass, paper material used for packaging |

(continued)

**Table 1** (continued)

| Focus | Principle | Examples in the Agri-Food sector | |
|---|---|---|---|
| | | Biosphere | Techno-sphere |
| | Recovery (of other resources) | • Organic waste to energy<br>• Energy recovery via the use of biomass<br>• Energy recovery (anaerobic digestion and incineration) | • Energy recovery (incineration of packaging material) |

*Source* Author

no longer sufficient. It is however challenging to rethink supply chains, redesign infrastructure, processes and products, replace traditional technologies, materials and energy sources and repurpose products and components. This paper provides a taste of what companies within Agri-Food supply chains are considering as well as highlights areas where further research will be needed to both assist organisations in reboot of their supply chains, and where the additional study would be beneficial.

**Acknowledgements** We would like to thank Shao-Ju Hsu for data collection of food waste prevention and reduction in food production factories in Taiwan.

# References

1. Badurdeen F, Jawahir IS (2017) Strategies for value creation through sustainable manufacturing. In: Procedia manufacturing: 14th global conference on sustainable manufacturing, GCSM, vol 8(October 2016), pp 20–27. https://doi.org/10.1016/j.promfg.2017.02.002
2. Bag S, Gupta S, Kumar S (2021) Industry 4.0 adoption and 10R advance manufacturing capabilities for sustainable development. Int J Prod Econ 231(December 2019):107844. https://doi.org/10.1016/j.ijpe.2020.107844
3. Balaine L, Dillon EJ, Läpple D, Lynch J (2020) Land use policy can technology help achieve sustainable intensification? Evidence from milk recording on Irish dairy farms. Land Use Policy 92:104437
4. Banerjee C, Adenaeuer L (2014) Up, up and away! The economics of vertical farming. J Agric Stud 2(1):40. https://doi.org/10.5296/jas.v2i1.4526
5. Bauwens T, Hekkert M, Kirchherr J (2020) Circular futures: what will they look like? Ecol Econ 175(November 2019):106703. https://doi.org/10.1016/j.ecolecon.2020.106703
6. Blomsma F, Brennan G (2017) The emergence of circular economy: a new framing around prolonging resource productivity. J Ind Ecol 21(3):603–614. https://doi.org/10.1111/jiec.12603
7. Blomsma F, Tennant M (2020) Circular economy: preserving materials or products? Introducing the resource states framework. Resour Conserv Recycl 156(January):104698. https://doi.org/10.1016/j.resconrec.2020.104698
8. Bradford KJ, Dahal P, Van Asbrouck J, Kunusoth K, Bello P, Thompson J, Wu F (2018) The dry chain: reducing postharvest losses and improving food safety in humid climates. Trends Food Sci Technol 71:84–93
9. Braungart M, McDonough W (2009) Cradle to Cradle. Remaking the way we make things. Vintage Books

10. CCEN (2017) Collaborative circular economy network. Scoping Study
11. Chen LH, Hung P, Ma H (2020) Integrating circular business models and development tools in the circular economy transition process: a firm-level framework. Bus Strategy Environ 29(5):1887–1898. https://doi.org/10.1002/bse.2477
12. Cherrett T, Maynard S, McLeod F, Hickford A (2015) Reverse logistics for the management of waste. In: McKinnon A, Browne M, Piecyk M, Whiteing A (eds) Green logistics. Improving environmental sustainability of logistics, 3rd edn. Kogan Page, pp 338–357
13. Ciccullo F, Cagliano R, Bartezzaghi G, Perego A (2021) Implementing the circular economy paradigm in the agri-food supply chain: the role of food waste prevention technologies. Resour Conserv Recycl 164(June 2020):105114. https://doi.org/10.1016/j.resconrec.2020.105114
14. Corona B, Shen L, Reike D, Rosales Carreón J, Worrell E (2019) Towards sustainable development through the circular economy—a review and critical assessment on current circularity metrics. Resour Conserv Recycl 151(May 2019):104498. https://doi.org/10.1016/j.resconrec.2019.104498
15. Cramer J (2014) Milieu. Amsterdam University Press
16. Cruz-Romero MC, Murphy T, Morris M, Cummins E, Kerry JP (2013) Antimicrobial activity of chitosan, organic acids and nano-sized solubilisates for potential use in smart antimicrobially-active packaging for potential food applications. Food Control 34(2):393–397. https://doi.org/10.1016/j.foodcont.2013.04.042
17. DAERA (2016) Delivering our future, valuing our soils: a sustainable agricultural land management strategy for Northern Ireland. https://www.daera-ni.gov.uk/publications/sustainable-agricultural-land-management-strategy-report-and-executive-summary
18. Demir I (2009) Reuse of waste glass in building brick production. Waste Manage Res 27(6):572–577. https://doi.org/10.1177/0734242X08096528
19. den Hollander MC, Bakker CA, Hultink EJ (2017) Product design in a circular economy: development of a typology of key concepts and terms. J Ind Ecol 21(3):517–525. https://doi.org/10.1111/jiec.12610
20. Di Leo A, Michelini L, Principato L (2020) Sharing platform and innovative business models: enablers and barriers in the innovation process. In: Kosseva MR, Webb C (eds) Food industry wastes. Assessment and recuperation of commodities. Academic Press, Elsevier, pp 431–449
21. Difrancesco RM, Huchzermeier A (2016) Closed-loop supply chains: a guide to theory and practice. Int J Log Res Appl 19(5):443–464. https://doi.org/10.1080/13675567.2015.1116503
22. Doppelt B (2010) Leading change toward sustainability: a change management guide for business, Government and Civil Society (Updated 2n). Routledge
23. EEA (2018) European waters—assessment of status and pressures 2018. https://doi.org/10.2800/303664
24. Directive 2008/98/EC of the European Parliament and of the Council of 19 November 2008 on waste and repealing certain Directives (2008)
25. Eurostat (2018) Packaging waste statistics
26. FAO (2019) The state of food and agriculture. Moving forward on food loss and waste reduction. In: Food and Agriculture Organization of the United Nations. https://doi.org/10.4324/9789781315764788
27. Filimonau V, Todorova E (2020) Management of hospitality food waste and the role of consumer behavior. In: Kosseva MR, Webb C (eds) Food industry wastes. Assessment and recuperation of commodities, 2nd edn. Academic Press, Elsevier, pp 451–466
28. Garetti M, Taisch M (2012) Sustainable manufacturing: trends and research challenges. Prod Plan Control 23(2–3):83–104. https://doi.org/10.1080/09537287.2011.591619
29. Garrone P, Melacini M, Perego A, Sert S (2016) Reducing food waste in food manufacturing companies. J Clean Prod 137:1076–1085. https://doi.org/10.1016/j.jclepro.2016.07.145
30. Govindan K, Jha PCC, Garg K (2016) Product recovery optimization in closed-loop supply chain to improve sustainability in manufacturing. Int J Prod Res 54(5):1463–1486. https://doi.org/10.1080/00207543.2015.1083625
31. Gudmestad OT (2019) Concerns to be considered during recycling operations. Resources 8(2):76. https://doi.org/10.3390/resources8020076

32. Hahladakis JN, Iacovidou E (2018) Closing the loop on plastic packaging materials: what is quality and how does it affect their circularity? Sci Total Environ 630:1394–1400. https://doi. org/10.1016/j.scitotenv.2018.02.330

33. Halloran A, Clement J, Kornum N, Bucatariu C, Magid J (2014) Addressing food waste reduction in Denmark. Food Policy 49(P1):294–301. https://doi.org/10.1016/j.foodpol.2014. 09.005

34. Halonen N, Pálvölgyi PS, Bassani A, Fiorentini C, Nair R, Spigno G, Kordas K (2020) Bio-based smart materials for food packaging and sensors—a review. Front Mater 7(April):1–14. https://doi.org/10.3389/fmats.2020.00082

35. Hansen EG, Revellio F (2020) Circular value creation architectures: make, ally, buy, or laissez-faire. J Ind Ecol 24(6):1250–1273. https://doi.org/10.1111/jiec.13016

36. Heller MC, Keoleian GA, Willett WC (2013) Toward a life cycle-based, diet-level framework for food environmental impact and nutritional quality assessment: a critical review. Environ Sci Technol 47(22):12632–12647. https://doi.org/10.1021/es4025113

37. Herrero M, Laca A, Laca A, Diaz M (2020) Application of life cycle assessment to food industry wastes. In: Kosseva MR, Webb C (eds) Food industry wastes. Assessment and recuperation of commodities. Academic Press, Elsevier, pp 331–353

38. Honkanen P, Verplankern B, Olsen SO (2006) Ethical values and motives driving organic. J Consum Behav 5:420–430. https://doi.org/10.1002/cb.190

39. Ivanov D (2020) Viable supply chain model: integrating agility, resilience and sustainability perspectives—lessons from and thinking beyond the COVID-19 pandemic. Ann Oper Res. https://doi.org/10.1007/s10479-020-03640-6

40. Jawahir IS, Dillon OWJ, Rouch KEE, Joshi KJ, Venkatachalam A, Jaafar IH (2006) Total life-cycle considerations in product design for sustainability: a framework for comprehensive evaluation. In: Trends in the development of machinery and associated technology, pp 1–10. http://citeseerx.ist.psu.edu/viewdoc/download?doi=10.1.1.402.3563&rep=rep1&type=pdf

41. Kiely G (1997) Environmental engineering. McGraw-Hill

42. Kirchherr J, Reike D, Hekkert M (2017) Conceptualizing the circular economy: an analysis of 114 definitions. Resour Conserv Recycl 127(April):221–232. https://doi.org/10.1016/j.rescon rec.2017.09.005

43. Klemeš JJ, Fan YV, Jiang P (2020) Plastics: friends or foes? The circularity and plastic waste footprint. Energy Sources, Part A: Recov Util Environ Effects 00(00):1–17. https://doi.org/10. 1080/15567036.2020.1801906

44. Kosseva MR (2020) Sources, characteristics, treatment, and analyses of animal-based food wastes. In: Kosseva MR, Webb C (eds) Food industry wastes. Assessment and recuperation of commodities, 2nd edn. Academic Press, Elsevier, pp 67–85

45. Kuik SS, Nagalingam SV, Amer Y (2011) Sustainable supply chain for collaborative manu-facturing. J Manuf Technol Manag 22(8):984–1001. https://doi.org/10.1108/174103811111 77449

46. Lacasa E, Santolaya JL, Biedermann A (2016) Obtaining sustainable production from the product design analysis. J Clean Prod 139:706–716. https://doi.org/10.1016/j.jclepro.2016. 08.078

47. Leigh M, Li X (2015) Industrial ecology, industrial symbiosis and supply chain environmental sustainability: a case study of a large UK distributor. J Clean Prod 106:632–643. https://doi. org/10.1016/j.jclepro.2014.09.022

48. Livingstone D, Smyth BM, Foley AM, Murray ST, Lyons G, Johnston C (2021) Willow coppice in intensive agricultural applications to reduce strain on the food-energy-water nexus. Biomass Bioenergy 144:105903

49. Lüdeke-Freund F, Gold S, Bocken NMP (2019) A review and typology of circular economy business model patterns. J Ind Ecol 23(1):36–61. https://doi.org/10.1111/jiec.12763

50. McDonough W, Braungart M (2013) The upcycle. Beyond sustainability—designing for abundance (I). North Point Press

51. McGauran T (2019) Development of novel bio-polymers derived from poultry waste streams. Queen's University Belfast

52. McGauran T, Dunne N, Smyth BM, Cunningham E (2020) Incorporation of poultry eggshell and litter ash as high loading polymer fillers in polypropylene. Compos Part C: Open Access 3:100080

53. McGauran T, Smyth B, Dunne N, Cunningham E (2021) Feasibility of the use of poultry waste as polymer additives and implications for energy, cost and carbon. J Clean Prod 291:125948

54. Mehta N, Cunningham E, Roy D, Cathcart A, Dempster M, Berry E, Smyth B (2021) Acceptable or weird? Exploring environmentalists', students' and the general public's perceptions of bio-based plastics

55. Ministry of Economic Affairs R.O.C. (2020) Food industry statistics

56. Mirabella N, Castellani V, Sala S (2014) Current options for the valorization of food manufacturing waste: a review. J Clean Prod 65:28–41. https://doi.org/10.1016/j.jclepro.2013.10.051

57. Morseletto P (2020) Restorative and regenerative: exploring the concepts in the circular economy. J Ind Ecol 24(4):763–773. https://doi.org/10.1111/jiec.12987

58. Morseletto P (2020b) Targets for a circular economy. Resour Conserv Recycl 153(November 2019):104553. https://doi.org/10.1016/j.resconrec.2019.104553

59. Ndraha N, Hsiao H-I, Vlajic J, Yang M-F, Lin H-TV (2018) Time-temperature abuse in the food cold chain: review of issues, challenges, and recommendations. Food Control 89:12–21. https://doi.org/10.1016/j.foodcont.2018.01.027

60. Ndraha N, Vlajic J, Chang C-C, Hsiao H-I (2020) Challenges with food waste management in the food cold chains. In: Kosseva MR, Webb C (eds) Food industry wastes, 2nd edn. Academic Press, Elsevier, pp 467–483. https://doi.org/10.1016/B978-0-12-817121-9.00022-X

61. Niero M, Hauschild MZ, Hoffmeyer SB, Olsen SI (2017) Combining eco-efficiency and eco-effectiveness for continuous loop beverage packaging systems: lessons from the Carlsberg circular community. J Ind Ecol 21(3):742–753. https://doi.org/10.1111/jiec.12554

62. NIW (2019) PC21 Outline Capital Submission

63. NRDC (2017) Wasted: how America is losing up to 40 percent of its food from farm to fork to landfill

64. Nriagu J (2019) Encyclopedia of environmental health, 2nd edn. Elsevier

65. Pagell M, Wu Z, Murthy NN (2007) The supply chain implications of recycling. Bus Horiz 50(2):133–143. https://doi.org/10.1016/j.bushor.2006.08.007

66. Pagotto M, Halog A (2016) Towards a circular economy in Australian agri-food industry: an application of input-output oriented approaches for analyzing resource efficiency and competitiveness potential. J Ind Ecol 20(5):1176–1186. https://doi.org/10.1111/jiec.12373

67. Papargyropoulou E, Lozano R, Steinberger KJ, Wright N, Ujang ZB (2014) The food waste hierarchy as a framework for the management of food surplus and food waste. J Clean Prod 76:106–115. https://doi.org/10.1016/j.jclepro.2014.04.020

68. Potting J, Hekkert M, Worrell E, Hanemaaijer A (2017) Circular economy: measuring innovation in the product chain. In: PBL Netherlands Environmental Assessment Agency (Issue 2544)

69. Pretty J, Benton TG, BhaPretty J, Benton TG, Bharucha ZP, Dicks LV, Flora CB, Godfray HCJ, Goulson D, Hartley S, Lampkin N, Morris C, Pierzynski G (2018) Global assessment of agricultural system redesign for sustainable intensification. Nat Sustain 1(8):441–446. https://doi.org/10.1038/s41893-018-0114-0

70. Radhakrishnan R, Ghosh P, Selvakumar TA (2020) Poultry spent wastes: an emerging trend in collagen mining. Adv Tissue Eng Regen Med Open Access 6(2):26–35

71. Radhakrishnan S (2016) Environmental footprints of packaging. In: Muthu SS (ed) Environmental footprints of packaging. Environmental footprints and eco-design of products and processes. Springer, Berlin, pp 165–192. https://doi.org/10.1007/978-981-287-913-4

72. Rahimifard S, Coates G, Staikos T, Edwards C, Abu-Bakar M (2009) Barriers, drivers and challenges for sustainable product recovery and recycling. Int J Sustain Eng 2(2):80–90. https://doi.org/10.1080/19397030903019766

73. Rahman S (2012) Reverse logistics. In: Mangan J, Lalwani C, Butcher T, Javadpour R (eds) Global logistics and supply chain management, 2nd edn. Wiley, New York, pp 338–354

74. RECOUP (2018) UK Household Plastics Collection Survey

75. Redlingshöfer B, Barles S, Weisz H (2020) Are waste hierarchies effective in reducing environmental impacts from food waste? A systematic review for OECD countries. Resour Conserv Recycl 156(January):104723. https://doi.org/10.1016/j.resconrec.2020.104723

76. Reike D, Vermeulen WJVV, Witjes S (2018) The circular economy: new or Refurbished as CE 3.0?—Exploring controversies in the conceptualization of the circular economy through a focus on history and resource value retention options. Resour Conserv Recycl 135:246–264. https://doi.org/10.1016/j.resconrec.2017.08.027

77. Russo I, Confente I, Scarpi D, Hazen BT (2019) From trash to treasure: the impact of consumer perception of bio-waste products in closed-loop supply chains. J Clean Prod 218:966–974. https://doi.org/10.1016/j.jclepro.2019.02.044

78. Schröder MJA, McEachern MG (2004) Consumer value conflicts surrounding ethical food purchase decisions: a focus on animal welfare. Int J Consum Stud 28(2):168–177. https://doi.org/10.1111/j.1470-6431.2003.00357.x

79. Searchinger T, Waite R, Hanson C, Ranganathan J (2019) Creating a sustainable food future. A menu of solutions to feed nearly 10 billion people by 2050. In: World resources report. www.SustainableFoodFuture.org

80. Smyth B, Fearon T, Olave R, Johnston C, Forbes G (2014) Energy balance of SRC willow used for managing farmyard washings—how does it compare to a conventional wastewater treatment works? In: Water efficiency conference

81. Speight JG (2018) Reaction mechanisms in environmental engineering. Butterworth-Heinemann

82. Teigiserova DA, Hamelin L, Thomsen M (2020) Towards transparent valorization of food surplus, waste and loss: clarifying definitions, food waste hierarchy, and role in the circular economy. Sci Total Environ 706:136033. https://doi.org/10.1016/j.scitotenv.2019.136033

83. Teuber R, Jensen JD (2020) Definitions, measurement, and drivers of food loss and waste. In: Kosseva MR, Webb C (eds) Food industry wastes. Assessment and recuperation of commodities, 2nd edn. Academic Press, Elsevier, pp 3–18

84. Tseng ML, Chiu ASF, Chien CF, Tan RR (2019) Pathways and barriers to circularity in food systems. Resour Conserv Recycl 143:236–237. https://doi.org/10.1016/j.resconrec.2019.01.015

85. UNEP (2009) Converting waste agricultural biomass into a resource compendium of technologies. https://wedocs.unep.org/bitstream/handle/20.500.11822/7614/WasteAgriculturalBiomassEST_Compendium.pdf?sequence=3&isAllowed=y

86. UNEP (2011) Towards a Green economy: pathways to sustainable development and poverty eradication—a synthesis for policy makers. In: United Nations Environment Programme. www.unep.org/greeneconomy%0ADisclaimer

87. Urbinati A, Chiaroni D, Chiesa V (2017) Towards a new taxonomy of circular economy business models. J Clean Prod 168:487–498. https://doi.org/10.1016/j.jclepro.2017.09.047

88. US-EPA (2016) Advancing sustainable materials management: 2014 Fact Sheet

89. Verghese K, Lewis H, Lockrey S, Williams HHH, Verghese BK, Lewis H, Lockrey S, Williams HHH, Verghese K, Lewis H, Lockrey S, Williams HHH, Verghese BK, Lewis H, Lockrey S, Williams HHH (2015) Packaging's role in minimizing food loss and waste across the supply chain. Packag Technol Sci 28(7):603–620. https://doi.org/10.1002/pts

90. Vlajic JV, Bogdanova M, Mijailovic R (2016) Waste not, want not: managing perishables in small and medium retail enterprises. In: The proceedings of 21st international symposium on logistics (ISL 2016), vol 21, pp 330–339. https://pure.qub.ac.uk/portal/files/121421006/Final_proceeding_with_abstract_update_OCT_05_1_.pdf

91. Vlajic JV, Mijailovic R, Bogdanova M (2018) Creating loops with value recovery: empirical study of fresh food supply chains. Prod Plan Control 29(6):522–538. https://doi.org/10.1080/09537287.2018.1449264

92. Vlajic JV (2015) Vulnerability and robustness of SME supply chains: an empirical study of risk and disturbance management of fresh food processors in a developing market. Organ Resil 2015:85–102. https://doi.org/10.1201/b19305-8

93. Vlajic JV (2020) Towards sustainable production and a circular economy: an extension of 3R practices in the agri- food industry. International Society for the Circular Economy, July
94. Vlajic JV, Hsiao H-I (2018) Collaboration in circular supply chains. In: Logistics network conference, Sept, 5–7
95. Weetman C (2017) A circular economy handbook for business and supply chains. Repair, remake, redesign, rethink. Kogan Page. https://www.koganpage.com/product/supply-chains-for-a-circular-economy-9780749476755
96. Wognum N, Trienekens J, Wever M, Vlajic J, van der Vorst J, Omta O, Hermansen J, Nguyen TLT (2009) Organisation, logistics and environmental issues in the European pork chain. In: Trienekens J, Petersen B, Wognum N, Brinkmann D (eds) European pork chains. Diversity and quality challenges in consumer-oriented production and distribution. Wageningen Academic Publishers. https://doi.org/10.3920/978-90-8686-660-1
97. Wolsey M, Smyth B, Johnston C (2018) Biomass for heat generation and wastewater management in the agri-food sector—are circular economy benefits realised? In: 26th European biomass conference and exhibition
98. Wu S, Montalvo L (2021) Repurposing waste plastics into cleaner asphalt pavement materials: a critical literature review. J Clean Prod 280:124355. https://doi.org/10.1016/j.jclepro.2020.124355
99. Zero Waste Europe (2019) Food systems: a recipe for food waste prevention (Policy briefing) (Issue January). https://zerowasteeurope.eu/downloads/food-systems-a-recipe-for-food-waste-prevention/

# A Conceptual Framework for Food Loss and Waste in Agri-Food Supply Chains: Circular Economy Perspective

Yaşanur Kayıkcı⬤, Nazlıcan Gözaçan, Çisem Lafcı, and Yiğit Kazançoğlu

**Abstract** The importance given to Circular Economy (CE) has further increased in the Agri-Food Supply Chain (AFSC) to combat the challenges of food loss and waste which could be caused by various reasons, such as poor stock management, economic behavior and also, the occurrence of COVID-19 outbreak. The transition from linear to circularity can also enable competitive sustainability from farm to fork in AFSC, which consists of different stages: farmers, food processors, food distributors, food retailers, consumers. Food loss mainly occurs in AFSC at near-farm stages (i.e., harvesting, processing) while food waste happens in AFSC at near-fork stages (i.e., retail, post-consumption). Thus, 6Rs (remanufacture, redesign, reduce, recycle, reuse, and recover) of CE principles can offer various benefits to close the loop of the wastages along with the AFSC. In this chapter, a conceptual framework for circularity in the AFSC is proposed considering the 6Rs. Furthermore, this framework also supports the Sustainable Development Goals. The applicability of the proposed framework is examined and discussed in the case of Turkey using SWOT Analysis. Key findings indicate that there is confusion about food loss and waste issues in Turkey. In addition, the solutions and developments for FLW problem are generally focused on food waste.

**Keywords** Circular economy · Agri-Food supply chain · Agri-Food · Circularity · Food loss and waste · SDGs · SWOT analysis

Y. Kayıkcı (✉)
Department of Industrial Engineering, Turkish-German University, Şahinkaya Cad. 106-34820, Istanbul, Turkey
e-mail: yasanur@tau.edu.tr

N. Gözaçan · Ç. Lafcı · Y. Kazançoğlu
International Logistics Management Department, Yasar University, Izmir, Turkey

© The Author(s), under exclusive license to Springer Nature Singapore Pte Ltd. 2021          41
R. S. Mor et al. (eds.), *Challenges and Opportunities of Circular Economy in Agri-Food Sector*, Environmental Footprints and Eco-design of Products and Processes, https://doi.org/10.1007/978-981-16-3791-9_3

# 1 Introduction

The unsustainability of the current food system is caused by the inefficiencies that lead to loss of productivity, energy, natural resources, biodiversity [31], and increase in Greenhouse Gas (GHG) emissions, resource scarcity, climate change, and so on. Also, these inefficiencies cost approximately trillions of dollars in a year according to the Food and Agricultural Organization of the United Nations (FAO) [31]. The mismanagement of natural resources and processes can be classified as one of the reasons that cause these inefficiencies [15].

Moreover, another challenge faced by food retailers worldwide is Food Loss and Waste (FLW) which is considered as a substantial contributor to the overall generation of waste all along the supply chain [8, 21]. Depending upon different estimations about FLW, almost 30–50% of food intended for human consumption is being lost or wasted at some points of Agri-Food Supply Chain (AFSC) [26, 31, 47], and result in serious sustainability issues in terms of economic, environmental, and social pillars of sustainability [16]. Also, corresponding to the latest emerging disease COVID-19, the amount of FLW generated in households, the associated economic cost of FLW generation increased by 12% and 11%, respectively [4]. For these purposes, the current economic model based on the take-make-dispose method has been highly criticized in terms of its unsustainability lately. Therefore, the transition process of the production systems into more sustainable approaches gained momentum [15] and the transition into Circular Economy (CE) can be proposed as a solution to these issues. CE can be defined as a renewable and regenerative system that aims to reduce waste by the effective and sustainable design of materials, products, processes, and business models [7, 31, 47]. Sustainable development goals (SDGs) are necessary to guarantee the wider usage of CE principles throughout the sector and economy [6, 34]

In 2015, the Sustainable Development Goals (SDGs) on sustainable agriculture and food waste were implemented as part of the 2030 Agenda [52]. The SDGs related to AFSC are SDG-2 "Zero Hunger", SDG-6 "Clean Water and Sanitation"; SDG-7 "Affordable and Clean Energy"; SDG-8 "Decent Work and Economic Growth"; SDG-9 "Industry, Innovation and Infrastructure"; SDG-11 "Sustainable Cities and Communities"; SDG-12 "Responsible Consumption and Production"; SDG-13 "Climate Action"; and SDG-15 "Life on Land". In order to meet SDGs, the implementation of CE principles in the AFSC is imperative to reduce FLW and land use; recovering natural capital, biodiversity, and eco-systemic assets; and improving soil quality [15].

There are several studies in the literature focused on FLW. This study differs from other studies by supporting the circular AFSC framework with SWOT analysis of Turkey. The main motivation is the applicability of the CE principles for FLW issue in Turkey's case. In this context, SWOT analysis has been proposed for the adaptation of the CE principles (6Rs) into Turkey to reduce FLW. Thus, the objective of this chapter is to construct a SWOT analysis-based conceptual framework for the circularity of the AFSC because wastages are becoming a crucial problem of the AFSC in terms of sustainability (economic, environmental and social) issues. FLW were examined as an important contributor to these critical problems in AFSCs. For this purpose, 6Rs (remanufacture, redesign, reduce, recycle, reuse, and recover) of CE principles have been examined in detail to close the loop to address the circularity function in the AFSC to eliminate the FLW. Also, the concept was supported by SDGs and a framework has been presented. The applicability of the proposed framework is investigated and discussed for the Turkey case with a SWOT analysis.

## 2 Food Loss and Waste in AFSC

Around one-third of the world's edible food is estimated to be lost, discarded, or wasted in global food supply chains [17]. Food loss and waste are different from each other due to the processes in which they occur. Food loss can be defined as food that is no longer suitable for human consumption due to reasons such as spoilage, spilling, bruising, and wilting, etc. in the process up to the retail process is called food loss [16]. Recently, the shutdown of coffee shops and restaurants led to a sharp decrease in demand for milk and dairy products during the pandemic, resulting in a significant food loss of up to 80% of the milk and dairy products produced [42]. Moreover, many dairy farmers in the UK have to dispose of thousands of liters of fresh milk because of the supply chain deterioration caused by COVID-19 [5]. Food waste can be defined as food that has good quality and suitable for human consumption but is not consumed and being discarded before and after the consumption [16]. In particular, animal-based products and meat have been associated with significant waste of resources [24] that also requires large quantities of agricultural land [19].

There are many underlying reasons behind the occurrence of FLW as presented in Fig. 1. For instance, poor harvesting techniques that are combined with insufficient harvesting equipment may cause crop/grain left behind in the field. Moreover, food products harvested at the wrong time may play a critical role since deformations such as blackening, rotting, and crushing in food items occur on over-maturing food and reduce the consumable of these items. Also, the usage of chemical substances for the

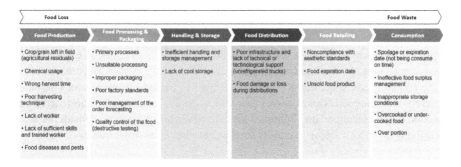

**Fig. 1** Underlying reasons for FLW at different stages of AFSC. *Source* Authors

production process of the food can pose a threat to food security and hygiene, and contamination by changing the structure of the food and affect the perishability of the food which is assumed as a low-quality food product by the consumer. Besides, an insufficient number of workers causes the product to remain in the field. Furthermore, improper harvesting and likewise loss occur when the area is harvested by workers without sufficient skills and training. According to the results of Gustavsson et al. [26] and Lipinski et al. [37], food processing and handling is the stage that has 24% of the global FLW occurred [44]. For instance, milling, cleaning, classifying, drying, winnowing and packaging, etc. are the most frequently used processes during food processing and packaging that cause FLW. In these food processing and packaging stages, FLW can occur in many ways such as damaged perishable foodstuff (e. g. seafood, fruits, and vegetables), spilled milk due to improper packaging [44]. Also, poor factory standards such as temperature, lighting, etc. can pose losses in food. Poor management of order forecasting is another crucial component that has serious effects on food losses because most used forecasting tools are generally not even close to optimal solutions and it ignores the risk of human error which unpredictable.

Another important stage that FLW occurs is the handling and storage process of food. Mismanagement of these processes can cause losses in this stage of the AFSC. Also, a lack of cool storage might be required for perishable foods such as seafood, milk, meat, etc. The lack of these systems leads to food loss as the product will spoil during the stock, and reprocessing of these products was very difficult. FLW can also occur in the food distribution stage which includes poor infrastructure and lack of technical or technological support such as improper coverage of the trucks, failure in cool storage conditions, packaging, insufficient ventilation, delays in transportation [44]. The reasons underneath food waste are determined as follows: noncompliance with aesthetic standards, food expiration date, and unsold food product. The noncompliance with aesthetic standards is usually caused by the insufficient packaging of the products or the deformation of the boxes or packages due to the impacts they are subjected to during distribution. Moreover, the food expiration date is a substantial component for the foodstuffs to be considered as waste. The close expiration dates of products often create a barrier to consumers from purchasing these products and cause products not to be sold.

The consumption stage of the food, which is also known as post-consumer, is among the stages food waste occur. Principally, not being consumed on time is one of the main reasons result in food waste which leads to spoilage and expiry date of the product. Also, ineffective food surplus management causes food waste because food surplus means a removal of the intentionally abandoned edible food, that is appropriate for human consumption, from the sale or otherwise some failures at the consumption stage due to several reasons [31]. In addition, overcooking and undercooking the food is another way that affects the quality of the food and therefore the food is not consumed and becomes a waste. Over portion can be considered as a contributor to food waste as it causes food waste due to leaving leftovers.

## 3  Circular Applications for SDGs in AFSC

SDGs were launched in 2015 to be a "blueprint to achieve a better and more sustainable future for all" as a part of the 2030 Agenda [52]. The concept consists of 17 interlinked global goals. However, the 9 interlinked global goals with a potential for circular AFSC were determined by evaluating [18], which highlighted the SDGs related to AFSC, and Schroeder et al. [45], which studied the relevance of CE principles to SDGs. The interlinked global goals are demonstrated in Fig. 2.

The principles of CE are imperative along the AFSC to achieve SDGs. CE principles can be implemented as a "toolbox" to accomplish a large range of SDGs [45]. Accordingly, FLW is also seen as the key to achieving SDGs within the scope of the farm to fork policy [11]. Thus, the main consideration of this framework is adjusting FLW flow under a circular approach with 6Rs to accomplish SDGs. The framework of circularity in AFSC is displayed in Fig. 3. Considering SDG-13, "Climate Action", the Agri-Food sector is recognized as a major sector in the production of GHG emissions, as it contains about 24% of global GHG emissions [30, 46]. Minimizing the utilization of natural resources would help reduce GHGs caused by FLW, among other pollutants [36]. Thus, CE activities such as chemical leasing and nutrient recycling in agriculture could decrease global carbon dioxide equivalent ($CO_2$-eq.) amount by up to 7.5 billion tones [13].

As the supplying process is performed through resources which are solar, water, soil, and nutrients; food production initiates three categories which are crop production, livestock, and aquaculture. Agricultural water use primarily includes this process to harvest for crop production and numerous other requirements of livestock systems, such as the need for animal water, washing facilities, etc. [3]. Commonly, primary or secondary processes of AFSC create wastewater that can be reused for

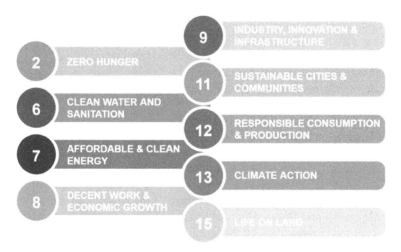

**Fig. 2** The interlinked global goals with a potential for circular AFSC. *Source* [52]

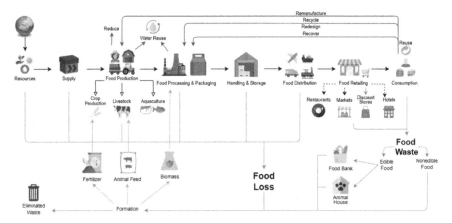

**Fig. 3** Framework of circularity in AFSC. *Source* Authors

crop production improve soil [54] after eliminating any existing pathogens [29]. In this way, following the SDG-6, a large percentage of the irrigation needs will be met by a circular approach [43]. CE practices can assist expand accessibility to clean drinking water and equitable sanitation, decrease toxicity and enhance the quality of water. These practices include such as small-scale water purification, sustainable sanitation, wastewater treatment, water reuse and recycling, nutrient recovery, biogas systems, etc. [45].

CE is considered an important element for the transformation to systems of sustainable consumption and production [53]. Moreover, SDG-12 is defined as "By 2030, halve per capita global food waste at the retail and consumer levels and reduce food losses along production and supply chains, including post-harvest losses." [25]. Less organic assets and components are used in food production by applying to reduce from 6Rs. Also, waste from consumption can be re-included in the production, processing, and packaging processes of the chain by going through remanufacture, recycle, recover, reuse, and redesign processes. Food waste has direct implications for multiple interrelated environmental impacts associated with the use of fertile soils, fresh water, fertilizers, energy, and the release of greenhouse gases for producing food [16, 24, 40]. For the inevitable FLW, the best option is its reuse, primarily for human consumption, and secondly for animal feed [46].

Food waste occurs at the end of the AFSC originating at the retail while food losses occur in the production, post-harvest, and processing stages [9]. Food waste is divided into two as edible and nonedible. Edible food is sent to animal houses for animal feeding which is an option for FLW [46], and food banks according to their types, which corresponds to SDG-2 "Zero Hunger" [25]. Nonedible food, waste from food banks and animal houses, and food loss are recruited into the loop as fertilizer, animal feed, and biomass through the formation process to provide clean and more efficient encouraging growth and helping the environment as in SDG-7. Also, when we refer to SDG-11 as "By 2030, reduce the adverse per capita environmental impact of cities, including by paying special attention to air quality and municipal and other waste management" [25], food waste includes urban waste.

Considering SDG-8, CE approaches also could bring important benefits of cost savings, job creation, innovation, productivity, and resource efficiency in both developed and developing countries [45]. In the modest scenario, European CE job opportunities vary from 634,769; in the optimistic scenario, 747,829 by 2025 [12]. Also, new skilled jobs could be created in recycling up to 85% and remanufacturing up to 50% by 2030 in the UK [38]. The definition of the target SDG-9 is "Develop quality, reliable, sustainable, and resilient infrastructure, including regional and trans-border infrastructure, to support economic development and human well-being, with a focus on affordable and equitable access for all" [25]. Accordingly, the concept of CE is "restorative and regenerative by design" [14]. CE applications redesign this industry to new and more innovative infrastructures such as renewable energy, circular water, and waste/resource management, reverse logistics, support for research, and innovation to make AFSCs more resilient and sustainable. CE practices strive to restore natural capital by utilizing sustainable and regenerative agriculture and agroforestry applications that encompass and guard biodiversity and restoring biological material to soils as nutrients that are crucial for healing terrestrial ecosystems [45]. This situation corresponds to the definition of SDG-15 "Life on Land".

## 4 Case Focus: Turkey

In this section, SWOT analysis, which is a strategic planning technique, is used to identify strengths, weaknesses, opportunities, and threats of Turkey related to circular AFSC in terms of SDGs considering Fig. 3. The SWOT matrix is shown in Fig. 4.

Turkey is a large country with the population which is more than 83 million at the beginning of 2021 [51]. Considering SDG-8, the share of agricultural employment in total employment is about 20% and Turkey ranks 12th in the world with this rate [32]. While the population of the village decreased by 11.4% between 2013 and 2020 [51], current records showing that the migration from the city to the village increased with the occurrence of the COVID-19 outbreak.

The food industry is the major sector in Turkey [33] with 7% of GDP contribution [49]. Also, agricultural production and the food industry play important roles in the economy as 30. 8% of Turkey's land is used for agricultural purposes [1]. Moreover, Turkey ranks as a significant producer and exporter of fruits, vegetables, and nuts [39]. Food supply is therefore not one of Turkey's main problems, however, the large quantity of municipal solid waste produced is a concern as FLW has not been controlled [44]. However, the lacks of the infrastructure of land consolidation and irrigation in Turkey cause high cost as well as the increase in the amount of fuel that the farmers use [10]. When the concentration is on SDG-9 and SDG-11, policy recommendations were presented for the effective implementation of support, incentives, rewards, and sanctions for a sustainable land and irrigation management [32] which corresponds to the definition of SDG-6.

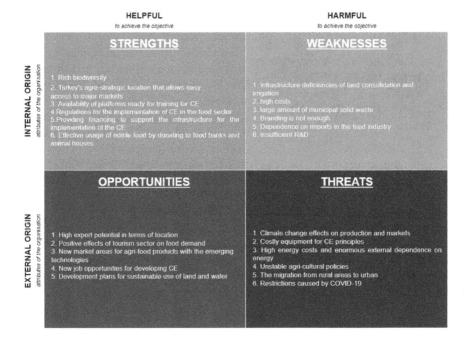

**Fig. 4** SWOT analysis of Turkey related to circular AFSCs in terms of SDGs

Turkey has greatly expanded its poultry production, particularly in the last 20 years, built contemporary facilities using emerging technologies, and has become one of the world's leading producers and exporters of poultry meat [35]. Besides, Turkey has become the second country after France with 14.1 million head of cattle among the countries of the European Union in 2016 [32].

One of the most important industries for Turkey's exports is aquaculture. In tandem with the advances in the cultivation of aquaculture and the export of processing technologies for aquatic products in Turkey, there is also a major growth in the production and export of aquatic products [32]. Thus, aquaculture accounts for approximately 20% of the world's animal protein intake, actively participating in global food security and human nutrition [39]. In addition, eco-friendly production techniques are in the process of increasing traceability in aquaculture products [32].

However, the majority of the Agri-Food industry in Turkey consists of small and medium-sized companies and, in this situation, the transformation of the technical system and the efficiency and capability of production are also adversely affected [27, 35]. Also, Turkey, as a developing country, in the future will be increasingly important in the agricultural and food sectors due to its agro-strategic location that allows easy access to major markets [32]. However, the political challenges and uncertainty observed in recent years have adversely affected the sector's export capability [35].

The agriculture sector includes emissions from enteric fermentation, manure management, rice cultivation, agricultural soils, field burning of agricultural residues, and urea application. In 2017, the agriculture sector accounted for 11.9% of total emissions in Turkey which does not cause a significant amount considering the climate action, namely SDG-13. However, enteric fermentation is by far the largest source of GHG emissions of agriculture in Turkey. Enteric fermentation and agricultural soils dominate the trends in this sector [50].

FAO [16] highlights that the amount of food wasted in Turkey are 26 million tons. 53% of fruit and vegetables are wasted until they reach the consumer from the field. Various applications to prevent FLW have been launched in Turkey. For instance, tax reductions are applied on goods for donations made to food banks [23]. Accordingly focusing on SDG-2, the retailer METRO in Turkey, performed the food donations of 164 tons (350 thousand meals) edible product (should be withdrawn from the shelves but have not passed the expiration date) in 2019 [41]. 15% of the food waste generated was also given to animal houses. For SDG-7, the company also aims to turn nonedible foods into fertilizers by sending them to the formation process which will provide a significant return to Turkey since 9.15 million tons were imported for animal feed [48]. Also, according to TGDF [48], natural gas, which is obtained from organic waste, is estimated to meet 6% of Turkey's energy needs by the definition of SDG-15 and SDG-12. In addition, AA [2] reduced food waste by 50% by donating 1,896 tons of food to feed street and forest animals, while delivering more than 4 million meals to the people who need it.

In this context, Migros and Animal Rights Federation in Turkey (HAYTAP) are conducting projects for shelter, street, and forest animals, setting out to protect the right to life of species, which is one of the basic principles of the concept of sustainability [28]. Within the scope of this project, which started in Istanbul and reached 205 stores in 24 provinces, food products such as fruits, vegetables, meat, charcuterie, and dry food that will be removed from the shelf due to the expiration date are delivered to HAYTAP volunteers. Furthermore, these products are subjected to sorting, cooking, and cleaning processes and are used for feeding shelter, street, and forest animals [28]. Moreover, the national "Protect Food" initiative was initiated by the FAO and the Ministry of Agriculture and Forestry to increase public awareness of the FLW and to unite all AFSC members [20]. Furthermore, logistics support and finance are provided by Fairy company to increase the food donation and distribution capacity of the Food Rescue Association in the regions of the Aegean and Mediterranean where food rescue activities cannot yet be carried out and those in need cannot benefit from food banks [22].

## 5 Discussion

In this work, a conceptual framework for FLW is proposed in Sect. 3. In addition, a SWOT analysis is performed for Turkey considering this framework. Through this analysis, it can be obtained that there is awareness about FLW. However, there is also confusion about the difference between food loss and waste considering the circular applications for FLW in Turkey. With this understanding, some projects are focusing on the consumption stage but somehow aim to solve food loss. Solving this confusion is important for the applicability of the CE principles for the FLW issue since FLW is also seen as the key to achieving SDGs within the scope of the farm to fork policy.

Furthermore, Turkey mainly concentrates on food waste which occurs in the retail and consumption processes. However, there are no sufficient practices for food loss which is in the process up to the retail process. Thus, it is necessary to redesign FLW approaches to gain circular features in AFSC in terms of SDGs.

# 6 Conclusion

The redesign from linear to circular in AFSC will ensure successful sustainability from farm to fork. The emerging points in AFSC have also been highlighted by distinguishing the definitions of food loss and food waste. According to these definitions, this chapter proposed a conceptual framework for circularity (6Rs) in AFSC considering FLW to accomplish SDGs. To make AFSCs more robust and competitive, CE applications are evolving this sector into modern and more advanced infrastructures, such as clean energies, green water, and waste/resource management, reverse logistics, research support. Besides, a SWOT analysis is provided for Turkey related to circular AFSCs in terms of SDGs. Key results showed that there is misunderstanding in Turkey about food loss and waste problems. In addition, FLW problem solutions and innovations are usually based on food waste. Although Turkey is very rich in terms of biodiversity, also it has a dense population. This dense population increases the FLW level. Furthermore, the lockdown and travel restrictions caused by the COVID-19 outbreak affect greatly agri-culture and the food sector in many ways such as reducing the food sales to dining restaurants and hotels in the tourism in Turkey. However, circular steps have been taken and should be expected to provide results on FLW. The limitations of the study are the number of CE principles (6Rs) used during this framework. In this context, CE principles can be extended to offer a more precise solution to the FLW issue and eliminate the environmental effects. Furthermore, packaging processes, which have various processes depending on the characteristics of the food, are not included in this study because it requires a more specific focus on the reasons underneath FLW resulting from the packaging process. For further studies, the packaging process can be examined in more detail as an indicator of FLW.

# References

1. AA (2017a) Türkiye'nin karasal alanının yüzde 30,8'inde tarım yapılıyor. Anadolu Ajansı (AA), 17 Feb 2021 retrieved from https://www.aa.com.tr/tr/ekonomi/turkiyenin-karasal-ala ninin-yuzde-30-8inde-tarim-yapiliyor/812502

2. AA (2017b) Migros, 'Fazla Gıda'yı ihtiyaç sahiplerine ulaştıracak. Anadolu Ajansı (AA). 20 Feb 2021 retrieved from https://www.aa.com.tr/tr/sirkethaberleri/perakende/migros-fazla-gid ayi-ihtiyac-sahiplerine-ulastiracak/637730
3. Aivazidou E, Tsolakis N, Iakovou E, Vlachos D (2016) The emerging role of water footprint in supply chain management: a critical literature synthesis and a hierarchical decision-making framework. J Clean Prod 137:1018–1037. https://doi.org/10.1016/j.jclepro.2016.07.210
4. Aldaco R, Hoehn D, Laso J, Margallo M, Ruiz-Salmón J, Cristobal J, Kahhat R, Villanueva-Rey P, Bala A, Batlle-Bayer L, Fullana-I-Palmer P (2020) Food waste management during the COVID-19 outbreak: a holistic climate, economic and nutritional approach. Sci Total Environ 742:140524
5. BBC (2020) Coronavirus crisis forces farmers to throw milk away, 15 Feb 2020 retrieved from https://www.bbc.com/news/av/uk-52205163
6. Balanay R, Halog A (2016) Charting policy directions for mining's sustainability with circular economy. Recycling 1(2):219–231. https://doi.org/10.3390/recycling1020219
7. Bennett EM, Carpenter SR, Caraco NF (2001) Human impact on erodable phosphorus and eutrophication: a global perspective: increasing accumulation of phosphorus in soil threatens rivers, lakes, and coastal oceans with eutrophication. Bioscience 51(3):227–234. https://doi. org/10.1641/0006-3568(2001)051[0227:HIOEPA]2.0.CO;2
8. Borrello M, Lombardi A, Pascucci S, Cembalo L (2016) The seven challenges for transitioning into a bio-based circular economy in the agri-food sector. Recent Pat Food Nutr Agric 8(1):39–47. https://doi.org/10.2174/2212798408011160304143939
9. Cakar B, Aydin S, Varank G, Ozcan HK (2020) Assessment of environmental impact of FOOD waste in Turkey. J Clean Prod 244:118846. https://doi.org/10.1016/j.jclepro.2019.118846
10. DHA (2019) Türkiye tarım alanlarının yüzde 8,3'ünü kaybetti. Demirören Haber Ajansı (DHA), 17 Feb 2021 retrieved from https://www.ntv.com.tr/ekonomi/turkiye-tarim-alanlarinin-yuzde-8-3unu-kaybetti,uIiFwiRUhEu6frqPQmfxwQ
11. EC (2020) Communication from the Commission to the European Parliament, the Council, the Economic and Social Committee and the Committee of the Regions. Our life insurance, our natural capital: an EU biodiversity strategy to 2020, 1–14. European Commission (EC)
12. EEB (2014) Advancing resource efficiency in Europe: indicators and waste policy scenarios to deliver a resource efficient and sustainable Europe. European Environmental Bureau (EEB), Brussels
13. Ecofys & Circle Economy (2016) Implementing circular economy globally makes Paris targets achievable, 17 Feb 2021 retrieved from https://circulareconomy.europa.eu/platform/en/knowle dge/implementing-circular-economy-globally-makes-paris-targets-achievable
14. Ellen MacArthur Foundation (2013) Towards the circular economy. Ellen MacArthur Foundation, Cowes
15. Esposito B, Sessa MR, Sica D, Malandrino O (2020) Towards circular economy in the agri-food sector: a systematic literature review. Sustainability 12(18):7401
16. FAO (2013) Food wastage footprint: impacts on natural resources. Food and Agriculture Organization (FAO), 17 Feb 2021 retrieved from http://www.fao.org/docrep/018/i3347e/i3347e. pdf
17. FAO (2015) FAO statistical pocketbook. Food and Agriculture Organization of the United Nations, Rome, 2015. ISBN 978-92-5-108802-9. Food and Agriculture Organization (FAO)
18. FAO (2018) Transforming Food and Agriculture to Achieve the SDGs, Rome, 2018. ISBN 978-92-5-130626-0. Food and Agriculture Organization (FAO)
19. FAO (2019) The state of food and agriculture, moving forward on food loss and waste reduction; Food and Agriculture Organization of the United Nations. Rome, Italy, 2019. Food and Agriculture Organization (FAO)
20. FAO (2020) Türkiye'nin gıda kayıpları ve israfının önlenmesi azaltılması ve yönetimine ilişkin ulusal strateji belgesi ve eylem planı. Food and Agriculture Organization (FAO). Ankara. 20 February 2021 retrieved from http://www.fao.org/3/cb1074tr/CB1074TR.pdf
21. Farooque M, Zhang A, Liu Y (2019) Barriers to circular food supply chains in China. Supply Chain Manage: Int J 24(5):677–696. https://doi.org/10.1108/scm-10-2018-0345

22. GTKD (2020a) Gıdayı boşa harcama, 20 Feb 2021 retrieved from https://gktd.org/
23. GTKD (2020b) Atığı önleyen teknoloji çözümleri, 20 Feb 2021 retrieved from https://sif
    iratik.gov.tr/content/files/uploads/9/EK-11,%20Say%C4%B1n,%20Olcay%20S%C4%B0L
    AHLI,%20G%C4%B1da%20Kurtarma%20Dern.%20Kuru.%20%C3%9Cye.Fazla%20G%
    C4%B1da%20A.pdf
24. Garske B, Heyl K, Ekardt F, Weber LM, Gradzka W (2020) Challenges of food waste
    governance an assessment of European legislation on food waste and recommendations for
    improvement by economic instruments. Land 9(7):231. https://doi.org/10.3390/land9070231
25. GlobalGoals (2021) The 17 goals, 14 Feb 2021 retrieved from https://www.globalgoals.org/
26. Gustavsson J, Cederberg C, Sonesson U, van Otterdijk R, Meybeck A (2011) Global food
    losses and food waste—extent, causes and prevention. FAO, Rome
27. Güneş E (2001) Türkiye'de tarıma dayalı sanayinin durumu ve sorunları. Türktarım Dergisi
    140:16–19
28. HAYTAP (2021) Migros-Haytap İşbirliği. Animal Rights Federation in Turkey (HAYTAP),
    20 Feb 2021 retrieved from https://www.haytap.org/tr/sosyal-sorumluluk-projesini-geniletti
    galeri-2
29. Hussain MI, Muscolo A, Farooq M, Ahmad W (2019) Sustainable use and management of
    non-conventional water resources for rehabilitation of marginal lands in arid and semiarid
    environments. Agric Water Manag 221:462–476
30. IPCC (2014) Climate change. Intergovernmental Panel on Climate Change (IPCC). Geneva,
    Switzerland.
31. Jurgilevich A, Birge T, Kentala-Lehtonen J, Korhonen-Kurki K, Pietikäinen J, Saikku L,
    Schösler H (2016) Transition towards circular economy in the food system. Sustainability
    8(1):69. https://doi.org/10.3390/su8010069
32. Kalkınma Bakanlığı (2018) Onbirinci kalkınma planı (2019–2023) tarım ve gıdada rekabetçi
    üretim özel ihtisas komisyonu raporu taslağı, 18 Feb 2021 retrieved from https://sbb.gov.
    tr/wp-content/uploads/2020/04/Tarim_ve_GidadaRekabetciUretimOzelIhtisasKomisyonuR
    aporu.pdf
33. Kayikci Y, Ozbiltekin M, Kazancoglu Y (2020) Minimizing losses at red meat supply chain
    with circular and central slaughterhouse model. J Enterprise Inf Manage 33(4):791–816. https://
    doi.org/10.1108/JEIM-01-2019-0025
34. Kayikci Y, Kazancoglu Y, Lafci, C, Gozacan N (2021) Exploring barriers to smart and sustain-
    able circular economy: the case of an automotive eco-cluster. J Clean Prod 314:127920. https://
    doi.org/10.1016/j.jclepro.2021.127920
35. Keskin B, Demirbas N, Gunes E (2018) Sustainable development of agri-food sector in Turkey.
    UARD Collective Monograph Series, 1
36. Kim MH, Song HB, Song Y, Jeong IT, Kim JW (2013) Evaluation of food waste disposal
    options in terms of global warming and energy recovery: Korea. Int J Energy Environ Eng
    4(1):1–12. https://doi.org/10.1186/2251-6832-4-1
37. Lipinski B, Hanson C, Lomax J, Kitinoja L, Waite R, Searchinger T (2013) Reducing food
    loss and waste, installment 2 of creating a sustainable food future. World Research Institute
    Working Paper, 40
38. Morgan J, Mitchell P (2015) Employment and the circular economy: job creation in a more
    resource efficient Britain. WRAP and Green Alliance, London
39. OECD (2016) Agricultural policy monitoring and evaluation. OECD Publishing
40. Östergren K, Gustavsson J, Bas-Brouwers H, Timmermans T, Hansen O-J, Møller H, Anderson
    G, O'Connor C, Soethoudt H, Quested T et al (2014) FUSIONS definitional framework for
    food waste. Full Report. The Swedish Institute for Food and Biotechnology. Göteborg, Sweden
41. Perakende (2020) Metro Türkiye 2025'e kadar gıda kaybını azaltmayı hedefliyor, 18 Feb 2021
    retrieved from https://perakende.org/public/metro-turkiye-2025-e-kadar-gida-kaybini-azaltm
    ayi-hedefliyor
42. Qingbin WANG, Liu CQ, Zhao YF, Kitsos A, Cannella M, Wang SK, Lei HAN (2020) Impacts
    of the COVID-19 pandemic on the dairy industry: lessons from China and the United States
    and policy implications. J Integr Agric 19(12):2903–2915. https://doi.org/10.1016/S2095-311
    9(20)63443-8

43. Rodias E, Aivazidou E, Achillas C, Aidonis D, Bochtis D (2021) Water-energy-nutrients syner-gies in the agrifood sector: a circular economy framework. Energies 14(1):159. https://doi.org/10.3390/en14010159
44. Salihoglu G, Salihoglu NK, Ucaroglu S, Banar M (2018) Food loss and waste management in Turkey. Biores Technol 248:88–99. https://doi.org/10.1016/j.biortech.2017.06.083
45. Schroeder P, Anggraeni K, Weber U (2018) The relevance of circular economy practices to the sustainable development goals. J Ind Ecol 23(1):77–95. https://doi.org/10.1111/jiec.12732
46. Secondi L, Principato L, Ruini L, Guidi M (2019) Reusing food waste in food manufacturing companies: the case of the tomato-sauce supply chain. Sustainability 11(7):2154. https://doi.org/10.3390/su11072154
47. Stuart T (2009) Waste: uncovering the global food scandal. WW Norton & Company
48. TGDF (2018) Sıfır gıda atığı liderler ağı zirvesi. Food and Drink Industry Associations of Turkey Federation (TGDF), 18 Feb 2021 retrieved from https://www.tgdf.org.tr/tgdf-sifir-gida-atigi-liderler-agi-zirvesi-ankarada-yapildi/
49. TarımOrman (2020) Cari fiyatlarla tarımsal GSYH ve tarımın payı. Türkiye Cumhuriyeti Tarım ve Orman Bakanlığı (TarımOrman). https://www.tarimorman.gov.tr/SGB/Belgeler/Ver iler/GSYH.pdf
50. TurkStat (2019) Turkish greenhouse gas inventory report. Turkish Statistical Institute (Turk-Stat), 14 Feb 2021 retrieved from https://unfccc.int/sites/default/files/resource/tur-2019-nir-13apr19.zip
51. TurkStat (2021) Adrese dayalı nüfus kayıt sistemi sonuçları, 2020. Turkish Statistical Institute (TurkStat), 12 Feb 2021 retrieved from https://data.tuik.gov.tr/Bulten/Index?p=Adrese-Dayali-Nufus-Kayit-Sistemi-Sonuclari-2020-37210
52. UN (2015) A/RES/70/1, Resolution adopted by the General Assembly on 25.09.2015. Trans-forming our World: The 2030 Agenda for Sustainable Development; United Nations. United Nations (UN). New York, NY, USA
53. UNEP (2012) The global outlook on SCP policies. Taking action together. Nairobi Kenya. United Nations Environment Program (UNEP)
54. Wang Y, Serventi L (2019) Sustainability of dairy and soy processing: a review on wastewater recycling. J Clean Prod 237:117821. https://doi.org/10.1016/j.jclepro.2019.117821

# Circular Economy in Agri-Food Sector: Food Waste Management Perspective

Umair Tanveer, Shamaila Ishaq, and Andrew Gough

**Abstract** Reducing the food waste is the greatest challenge in the present times for sustainable food management systems that have significant economic, environmental and social impact on the food supply chain. The Circular Economy (CE) paradigm advocates the concept of the closed-loop economy endorsing more responsible utilization and appropriate exploitation of resources in contrast to the open-ended linear economic system of take-make-use and dispose. This chapter has explored Agri-Food waste in the context of CE, triple bottom line (TBL), and sustainability. An alignment of circular strategies with the food waste hierarchy is proposed that indicates practical application of the gradations of circularity in the food waste management that could lead to the development of sustainable food management system targeting the sustainable development goals of Zero Hunger and Responsible consumption and production. This chapter also highlights some opportunities and challenges of Agri-Food waste in the application of circular bio-economy.

**Keywords** Food waste (FW) · Circular economy (CE) · Triple bottom line (TBL) · Sustainability · Sustainable development goals (SDGs) · Circular strategies · Food supply chain (FSC)

U. Tanveer (✉)
University of Bristol, Bristol, England
e-mail: Umair.tanveer@bristol.ac.uk

S. Ishaq
University of Derby, Derby, England
e-mail: s.ishaq@derby.ac.uk

A. Gough
University of Northampton, Northampton, England
e-mail: Andrew.gough@northampton.ac.uk

# 1  Introduction

Food waste reduction is one of the biggest sustainability challenges faced in the present times. According to an estimate of The UN Food and Agriculture Organisation (FAO) approximately 1.3 billion tonnes of food is wasted each year, amounting to one-third of all food produced globally for human consumption [1]. Two common terms "food loss" and "food waste" are commonly used to represent the waste generated at different stages in the food supply chain [2]. World Resources Institute defined Food Loss (FL) and Food Waste (FW) as "the unintended result of an agricultural process or technical limitation in storage, infrastructure, packaging, or marketing" and as "food that is of good quality and fit for human consumption but that does not get consumed because it is discarded" respectively [3]. According to European Union (EU) definition, FW is a "fractions of food and inedible parts of food removed from the food supply chain to be recovered or disposed" [4].

According to FAO (2021), food loss refers to the decrease in the quantity and quality of food lost at different stages of growing (pre-harvest), post-harvest and processing stages but not included the retail level whereas food waste is associated with the decrease in the quantity or quality of food that is fit for human consumption but is discarded due to decisions and actions of retailers, food service providers and consumers.

In addition to food waste and food loss a third term Food surplus (FS) is also described in the literature that represents the leftover edible food fit for human consumption. FS is generated at the retail and consumption stages of the food supply chain (FSC) [5], but also refers to overproduction at the agricultural/primary production stage [6]. Some of the surplus food changes into food waste due to ineffective management of food surplus [6] that need to be managed at different stages of food supply chain either by recovering for human consumption or preventing at source to limit the unnecessary use of natural resources [5].

A common expression of Food Loss and Waste (FLW) is also introduced in literature that combines the concept of food loss and food waste and represents the total share of food produced, retailed or served for human consumption but is not consumed and redirected to feed people, animals or used for new edible products [5].

All these definitions are considering the decrease in the quantity and quality of food at different stages of food production that starts from the pre-harvesting till the food is available for consumption.

From the last few years, the rate of food production has grown faster than the human population growth rate which resulted in food surplus that gets lost or wasted. According to FAO [1] one- third of the food is lost or wasted while flowing through the food supply chain (FSC). This has a significant impact on the triple bottom line (economic, environmental and social) of the supply chain for many institutions (public and private sector). Food Waste (FW) also contributes to supply chain risks and food insecurity as well as greenhouse gases arising from their decomposition if landfilled. Now, there is an urgency to stop this significant depletion of critical assets in food losses and waste at all tiers of supply chain (from manufacturer to end consumer).

The Circular Economy (CE) paradigm advocates the concept of the closed-loop economy endorsing more responsible utilization and appropriate exploitation of resources in contrast to the open-ended linear economic system of take-make-use and dispose. CE aims to recover the enviromental damage and improve the well beings of human that is highlighted by the practitioners [7] and academics. The most dominant aspect of CE is effective waste management practices where some of the practices could be more effective in certain sectors/conditions but fail in other sectors/situations.

One such system of circular strategies is expressed in the form of gradations of circularity (shown in figure below). Gradations of circularity represent circular strategies (10Rs) in hierarchical order consist of refusing, rethinking, reducing, reusing, repairing, refurbishing, remanufacturing, repurposing, recycling, and recovering [8]. The current study has proposed an alignment of these circular startegies with the food waste management hierarchy framework identified in literature (these are discussed in detail in Sects. 4 and 5).

The chapter has explored the concept of CE, TBL, Sustainability, Food waste Hierarchies and proposed the alignment of CE practices with the food waste management hierarchy. The last section describes some challenges and opportunities for Agri-Food based CE.

## 2   The Origin/Emergence of Circular Economy (CE)

Circular Economy (CE) advocates the concept of the closed-loop economy in contrast to the open-ended linear economic system of take-make-use and dispose [7]. The term Circular Economy was coined first time by Pearce and Turner in relation to the inter-linkages between the environmental and economic activities. Pearce and Turner derived inspiration for Circular Economy from the work of Boulding which described the earth as the closed and circular system having limited natural resources for the human activities emphasizing the need for the existence of an equilibrium between the economy and environment. Circular Economy has emerged from a variety of concepts proposed by different authors in the past and recent times and developed the broad spectrum of postulates and principles of Circular Economy. The concept of Circular Economy was initially introduced during the 1970s by the Swiss architect and economist Walter Stahel who proposed that materials can be processed in a 'closed loop' that transforms 'waste' into a resource [9]. Stahel [9] focused on the industrial ecology and proposed the industrial strategies of waste prevention, resource efficiency, regional job creation and dematerialisation of the industrial economy to conceptualize the loop economy. Stahel [9], proposed the sustainable business model for the loop economy (termed as spiral loop system) by; defining this loop system as a 'Cradle-to-Cradle' system and the linear model as Cradle-to-Grave, proposing the need for product life extension through reuse, repair, reconditioning and recycling and introducing the idea of selling utilization instead of the ownership of goods allowing industries to earn profit without externalizing the risk and cost of waste [10]. All these concepts are now considered integral to the Circular Economy. Some other most relevant theoretical developments in this dimension are regenerative design [11], industrial ecology [12, 13], Cradle to Cradle [14] and looped and performance economy [9].

The work of Ellen MacArthur Foundation is very important in the context of Circular Economy as it has a range of publications on the topic including a series of reports and a book by Webster. The Ellen MacArthur Foundation is also serving as a hub for business, academia and policymakers. Now the consultancies such as McKinsey & Co. are working in collaboration with the Ellen MacArthur to tap into opportunities of CE.

The concept of CE evolved gradually and can be divided into three distinct stages. The first stage is the linear economy stage that initiated with the industrial revolution, technological development, overexploitation of resources and economic growth. However, in 60s the concerns were raised at this developmental stage as a result of increasing interest on the environment by the ecologists such as Carson and the economists such as Boulding who suggested that earth is a closed and circular system with limited natural resources for human activities. Both ecologists and economists emphasized the need to recirculate the natural resources for developing an equilibrium between the economy and environment.

The second stage started with the increasing awareness of researchers and policymakers towards environmental protection. It led to the emergence of the concepts of loop economy [9] and the industrial metabolism[1] that led to the development of the strategies for the resource efficiency and waste prevention by stabilizing the control of the economic system through human component. The environmental protection awareness at this stage stimulated the development of environmental strategies by the government and institutions that played a key role in the emergence of the concept of green economy[2] and sustainability.

The third stage started in 90s when Pearce and Turner coined the term "circular economy" by stating that Earth is a closed economic system in which economy and environment are characterized by a circular relationship instead of linear interlinkage where everything is an input of everything else. They also mentioned entropy as the physical obstacle in the way of redesigning the economy as a closed and sustainable system. Since its inception CE has been enriched through multiple concepts such as regenerative design, industrial ecology [12, 13], Cradle to Cradle [14] looped and performance economy [9] etc.

The circular economy has gained tremendous attention of the academic researchers in last few years which is evident from a large number of reviews

---

[1] It can be described schematically as a sequence of processing stages between extraction and ultimate disposal, with a number of actual or hypothetical intermediate loops that would permit the system to be closed with respect to mass flows (Ayres 1989).

[2] United Nations Environment Programme (UNEP) has defined it as the one that results in improved human wellbeing and the social equity while reducing the environmental risks and ecological scarcities [15].

published on the topic in last few years [16–19]. The major topics discussed in relation to CE include; the circular business models [20], closed-loop and supply chains [21–23] product design [24].

The concept has also received the attention of policy makers, governments and intergovernmental agencies at local, regional, national and international level [25]. Germany was the pioneer in incorporating the concept of CE into national laws with the enactment of "Closed Substance Cycle and Waste Management Act" in 1996. This trend was followed by Japan in 2002 with the formulation of "Basic Law for Establishing a Recycling-based Society" and by China in 2009 through "Circular Economy Promotion Law of the People's Republic of China".

- *Conceptualisation of Circular Economy on the basis of Definitions*

The definitions of Circular Economy developed over time. The prominent trends in the definitions are demonstrated in Table 1.

There are a few trends that are obvious in these definitions. In the beginning, definitions focused on the replacement of the linear economic system with closed loop of material, energy and waste flow through reduced input, reuse, recycle (the 3R's concept) to achieve resource efficiency as can be seen from definitions by [26–29].

After that definitions added the concept of the sustainable economic development by focusing mostly on the environmental protection [30–33, 41, 42] which is mentioned as the concept of closing the economic and ecological loops of resource flow. The linkage of CE to sustainability is not new rather it acted as a stimulus for the initiation of the CE concept as expressed by Boulding and Pearce and Turner who suggested that sustainability requires circularity in the economic system. This dimension has become most important with the growing importance of Sustainable development Goals (SDGs) set by United Nations in 2015. In the context of Food waste management, reducing food loss and food waste is critical to creating a Zero Hunger world and reaching the world's Sustainable Development Goals (SDGs), especially SDG 2 (Z Hunger) and SDG 12 (Responsible Consumption and Production) FAO [1].

The major contribution in this dimension is made by the Ellen MacArthur Foundation that described CE as the "industrial system that is restorative and regenerative by intention and design" [43]. This definition indicated the need for a systematic shift and innovation in the economic and industrial system ranging from the design of the product to the restoration and value creation after the end of life of the product.

Since 2012, the CE definitions are derived from the idea of CE given by the Ellen MacArthur Foundation and this trend can be observed in the definitions given by [17, 35, 37].

The Circular Economy definitions have shown rapid development and diversity during the last few years as presented in Table 1. Some of the recent definitions have addressed the 3R's concept in the production, distribution and consumption processes [39, 40] whereas [38] extended the definition by describing CE as the

**Table 1** Trends in CE definitions

| Author year | Definitions |
|---|---|
| *Theme 1—replacement of the linear economic system with closed loop* | |
| Yang and Feng [26] | Circular economy is an abbreviation of "Closed Materials Cycle Economy or Resources Circulated Economy" (…) "The fundamental goal of circular economy is to avoid and reduce wastes from sources of an economic process, so reusing and recycling are based on reducing." |
| Geng and Doberstein [27] | "mean the realization of a closed loop of materials flow in the whole economic system." (…) "implying a closed-loop of materials, energy and waste flows" |
| Peters et al. [28] | "The central idea is to close material loops, reduce inputs, and reuse or recycle products and waste to achieve a higher quality of life through increased resource efficiency |
| Yuan et al. [29] | "Although there is no commonly accepted definition of CE so far, the core of CE is the circular (closed) flow of materials and the use of raw materials and energy through multiple phases." |
| *Theme 2—sustainable economic development* | |
| Geng and Doberstein [27] | "The concept of CE has the same essence as industrial ecology, implying a closed-loop of materials, energy and waste flows... It presents a new concept of more sustainable urban economic and industrial development." |
| Park et al. [30] | "The Chinese CE policy originated with the IE policy and is built upon the concept of industrial supply chain loop closing" |
| Hass et al. [31] | The circular economy is a simple, but convincing, strategy, which aims at reducing both inputs of virgin materials and output of wastes by closing economic and ecological loops of resource flows |
| Ma et al. [32] | A circular economy is a mode of economic development that aims to protect the environment and prevent pollution, thereby facilitating sustainable economic development |
| Ma et al. [33] | CE is specifically based on both resource efficiency and eco-efficiency, and its purpose is to acquire a set of key measures to move towards a more circular, green, and sustainable economy |
| *Theme 3—derived from the CE concept of Ellen McArthur* | |
| Ghisellini et al. [17] | "Circular economy is defined by Charonis [34], in line with The Ellen Macarthur Foundation vision (2012), as a system that is designed to be restorative and regenerative." |
| Hobson [35] | The CE has been defined as an industrial system that is restorative or regenerative by intention and design, and aims for the elimination of waste through the superior design of materials, products, systems and business models |
| Moreau et al. [36] | A circular economy is restorative and regenerative by design that preserves and enhances natural capital, optimizes resource yields, and minimizes system risks by managing finite stocks and renewable flow |

(continued)

**Table 1** (continued)

| Author year | Definitions |
|---|---|
| Niero et al. [37] | The circular economy, defined as a restorative or regenerative industrial system by intention and design |
| *Theme 4—CE over the supply chain* | |
| Murray et al. [38] | "The Circular Economy is an economic model wherein planning, resourcing, procurement, production and reprocessing are designed and managed, as both process and output, to maximize ecosystem functioning and human well-being." |
| Naustdalslid [39] | 'The term "circular economy" as mentioned in these measures is a generic term for the reducing, reusing and recycling activities conducted in the process of production, circulation and consumption' |
| Blomsma and Brennan [40] | Circular economy is a general term covering all activities that reduce, reuse, and recycle materials in production, distribution, and consumption processes |

economic model with planning, resourcing, procurement, production and reprocessing of output/processes to maximize the ecosystem functioning and the human wellbeing.

Other trends observed are; the extension of the 3R's (reduced input, reuse, recycle) concept of CE to 6R's by enriching with repair, refurbishment and remanufacturing [44], the development of long-lasting design of the product to extend the product life [9], operation at multiple levels such as micro, meso and macro [45] to achieve the long term sustainability referred as the balanced integration of economic, environmental and social aspects [25]. This concept is further extended in the form of nine circular strategies [8] presented as (R1–R9) in Fig. 1.

In waste management systems CE is going further ahead towards the circular bioeconomy. In circular bioeconomy, the unavoidable fraction of food waste creates the huge opportunity for the bioconversion in useful materials (such as chemicals, fertilizers) and energy (biofuels and electricity).

## 3  Sustainability, Triple Bottom Line (TBL) and CE

The concept of sustainability is derived from the French verb soutenir, which means "to hold up or support" [46] and its modern conception has its origins in forestry based on the "silvicultural principle" and was already written in early eighteenth century in "Sylvicultura oeconomica". There seem to be some older sources following the same principle used for the shortage of wood supply and husbandry of cooperative systems.

Despite its existence since the eighteenth century, the concept of Sustainability has gained prominence since the global-scale environmental risks (including climate

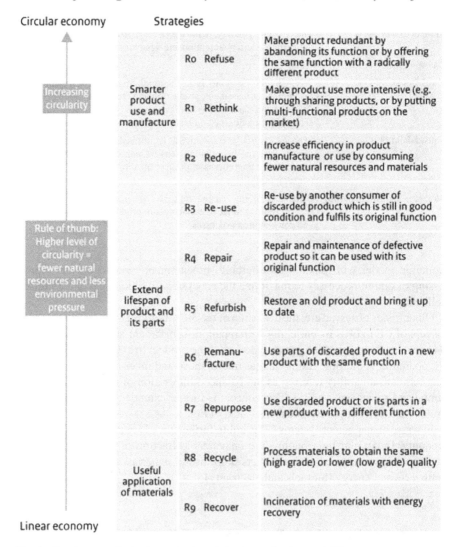

Circularity strategies within the production chain, in order of priority

| Circular economy | Strategies | | |
|---|---|---|---|
| **Increasing circularity** | Smarter product use and manufacture | Ro Refuse | Make product redundant by abandoning its function or by offering the same function with a radically different product |
| | | R1 Rethink | Make product use more intensive (e.g. through sharing products, or by putting multi-functional products on the market) |
| | | R2 Reduce | Increase efficiency in product manufacture or use by consuming fewer natural resources and materials |
| **Rule of thumb: Higher level of circularity = fewer natural resources and less environmental pressure** | Extend lifespan of product and its parts | R3 Re-use | Re-use by another consumer of discarded product which is still in good condition and fulfils its original function |
| | | R4 Repair | Repair and maintenance of defective product so it can be used with its original function |
| | | R5 Refurbish | Restore an old product and bring it up to date |
| | | R6 Remanu-facture | Use parts of discarded product in a new product with the same function |
| | | R7 Repurpose | Use discarded product or its parts in a new product with a different function |
| | Useful application of materials | R8 Recycle | Process materials to obtain the same (high grade) or lower (low grade) quality |
| Linear economy | | R9 Recover | Incineration of materials with energy recovery |

**Fig. 1** Circular strategies in the production chain. *Source* Potting et al. [8]

change, the ozone depletion, biodiversity loss and the change in the biogeochemical cycles) have been identified in the twentieth century. These risks have been analyzed since the 1960s, raising questions as to whether the current sociological trends can be upheld in the future [47, 48] subsequently causing many issues for concern, such as the limitation of resources and the way in which they are unevenly distributed geographically as well as occupationally [49].

The Brundtland Commission defined sustainability as "development that meets the needs of the present without compromising the ability of future generations to meet their own needs".

The negative impacts of increasing population, consumption, technology and economic growth on the environment and human well-being gained considerable attention of the scholars who highlighted the role of science, technology and innovation for initiating social inclusion and environmental resilience in the economic development. This literary inclination led to the development of the present concept of sustainability.

Although sustainability was initially driven by environmental concerns, it has combined a variety of concepts and accommodated a range of expectations for desirable progress [25]. The most relevant fusion in the term is the triple bottom line principle of Elkington [50] consisting on 3P's: Profit, People and Planet [50] which he particularly coined for the corporate world. Elkington [50] proposed TBL as the concrete tool for the companies who committed themselves to sustainable values. Although three aspects of sustainability—environmental, social and economic—existed already but this approach played a crucial role for shaping initiative towards corporate social responsibility, climate change and fair trade [51].

After the World Summit 2002 [52], it was considered as the balanced combination of social, environmental and economic performance. These three dimensions are systematically interlinked and continuously and mutually influence each other. In other words, these three dimensions are "interdependent and mutually reinforcing pillars [53], that are adaptive to a broad range of context and time horizons [54]. According to [55], sustainability is "the designing and employing human systems as well as industrial systems in order to use natural resources and to make sure that the normal cycles do not have a negative impact on social conditions, human health and ecosystem".

Based on these views it can be deduced that triple bottom line is embedded in the sustainability as TBL's three pillars: people, profit and planet are analogous to environmental, social and economic aspects of sustainability.

However, the limitation of the TBL is that it does not protect the human and natural capital [56] and also lacking the fourth pillar of sustainable economic approach (that accounts for the future-oriented dimension of sustainability) [57]. Based on this concept and keeping in view the flexible, adaptive and holistic nature of sustainability it can be stated as "the balanced and systematic integration of intra and intergenerational economic, social and environmental performance" [25].

Sustainability is a broader concept than TBL and is not just confined to setting common operational goals in three dimensions (economic, environmental and social) rather it opens up the scope for multiple aspects such as what should be developed, what should be sustained, for how long and whom should it benefit.

Circular strategies, proposed for transition to circular economy, contribute towards reducing the consumption of natural resources and virgin materials and consequently fewer environmental effects by introducing innovation across entire network system of supply chain (economical, technological, political, socio-institutional and environmental among others) [8] that lead to transformation of the societies and economies toward sustainability. Therefore, in terms of relationship between CE and sustainability we can say that CE is means and sustainability is an end.

- *TBL and Sustainability in Food Waste Rethinking*

Food loss and waste (FLW) is an issue in economic, environmental and social terms (triple bottom line) that pose a challenge for food sustainability. FLW is a significant economic loss in terms of time and resources (agriculture inputs and associated cost) invested on the production and food supply chain. Reducing food loss and waste have a significant impact on savings (time and resources) throughout up streams and down streams on the food supply chain [58]. Initiatives taken to reduce the food loss at its roots are beneficial for producers who aim to have a high volume of sales as well as for consumers who could have access to economical food [59]. However, there is a lot of criticism and debate in the literature about extra operational cost linked with managing food surplus.

Regarding the food loss and waste impact on environment, it is evident that a high level of agricultural inputs such as fertilizers and water are used to produce, process, transport and deliver food to make it available for end user. Food loss and waste is also a waste of resources such as water, energy, land and other inputs. All those actions which help to manage the surplus food as well as preventing its generation are directly or indirectly contribute towards reducing the burden on natural resources. Moreover, the positive attitude towards prevention of food loss and waste can have a considerable impact on society.

Reduction in food loss and waste pledge towards a better and enhanced food security system which extends its benefits for end users. For example, the redistribution of surplus food by food banks or food aid organisations provide food to the people who otherwise have no access to nutritional food and avoid the food waste [5] that create a positive social impact.

The total food wasted every year could feed one in nine people all over the world, who have low income or suffer from food hunger, especially in developing and third world countries. In terms of sustainability there is a need to create awareness about the "food paradox", i.e. the waste of food in a world that is still food deficient (FAO, 2021) that will serve as social driver to reduce the food waste. Moreover, food production and consumption exploit the environment through resource utilization and waste generation therefore sustainable production and consumption are vital for sustainable development (SDG-12). Sustainable production leads to sustainable consumption that is viewed as satisfying customers need by reducing the negative impacts on environment. CE is a link between Sustainable production and sustainable consumption [60] that aims to avoid and minimize product and resource consumption through multiple material loops [45].

In addition to food surplus, reuse and recovery of material (such as agriculture material residue) wasted along the supply chain is crucial for fostering the CE practices in the food sector. Waste of residues in many agricultural productions causes serious sustainability problems because of their production in large quantities in limited time period and being of a particular organic matter.

Based on the concept of reuse and recovery of nutritional benefits of residues, a new frontier of Agri-Food research is emerging that is related to the reuse of waste and by-products to increase the nutritional power of food or its shelf-life extension [61, 62]. A new term of waste-to-value (WTV) products is used for these novel food products [61] to highlight the circular bioeconomy approach that transforms wastes or surplus ingredients, obtained during the manufacturing of other foods, into new value-added food with higher nutritional properties [63]. However, after the development of WTV food products, the final market uptake depends on consumers' purchase choices. Particularly, in the Agri-Food sector, consumer acceptance is decisive in the development of successful novel foods. Coderoni and Perito [64] described that products origin, nutritional value, consumers inclination towards the sustainable production and consumption (such as organic food) are the drivers behind buying WTV food. However, consumer income, socio-demographic characteristics and trust on the environmental and health benefits of WTV food are the potential barriers in this regard.

## 4  Food Waste Management Hierarchies and Circular Economy

Food waste hierarchies are developed by different authors after the Waste Framework Directive 2008/98/EC [65]. The Food Waste Hierarchy (FWH) framework introduced by Papargyropoulou [6], proposed different options and the prioritization of those options based on the environmental and social aspect of food surplus and waste by giving least importance to financial aspect. According to this framework strategies for avoiding surplus food generation and strategy of reusing surplus food for human consumption possess high priority because these strategies contribute to reduce the depletion of natural resources as well as limit the negative social and ethical implications of food waste. They were of the view that food supply chain has the greatest potentials to prevent the generation of food surplus in the upstream through new infrastructures, skills, storage and transportation technologies. Garrone [5], was of the view that prevention of food waste through different redistribution and reuse strategies still targets human consumption, therefore minimizing the waste from a social perspective. Garrone [5], extended the idea of reuse and redistribution by describing different options. Reuse options include, sales with promotions and discounts, remanufacturing and repacking, sales in secondary channels as ad-hoc distributor for surplus food and redistribution covers both internal (to the employees of a company) and external channels (through the collaboration with food aid organizations). In terms of avoid or prevent strategy, Garrone [5], and Vandermeersch [66] refer to the prevention of food waste and loss but not to the reduction or avoidance of surplus food that is a distinct feature in Papargyropoulou [6], framework.

Similar to Papargyropoulou [6], framework, the Food Recovery Hierarchy by United States Environmental Protection Agency (EPA, 2012) priorities the strategy of reduction of the volume of surplus food generated at source. Rood [67], classified different redistribution and reuse options and added distinct layers in the Moerman's ladder such as converting into human food (in which food products are transformed into new edible products) in addition to human food (usually considered as redistribution in original form), raw material for industry and turned into fertilizer through fermentation.

Teigiserova [68], has provided six distinct categories in food waste comprising one edible and five inedible food categories. Based on these categories they proposed a waste hierarchy and expanded it by material recycling and nutrient recovery to reflect the future food waste biorefineries in the circular bioeconomy.

An interesting thing to note in these frameworks is that most of the frameworks have similar options in the lower parts of the waste hierarchy (such as recycling of food products into non-edible alternatives such as food for animals and fertilizers, recovery by energy generation, incineration and disposal) but some differences can be seen moving up in the waste hierarchy.

Another interesting aspect of these hierarchies is that these focus primarily on managing food surplus, harvesting losses considering those unavoidable and food loss and waste along the food supply chain but didn't account for low-yields or low productivity in the farming operations.

All these frameworks provide the application of reduce, recycle and reuse (3Rs) that support the aims of the circular economy. Table 2 provides a summary of different frameworks and their priorities of actions and relevant policies to follow for food waste management.

# 5   Aligning the Food Waste Hierarchy with Circular Strategies

The food waste hierarchies used in the literature have mostly described "Reduce", "Reuse", "Recycle", and "Recover" strategies in the context of CE. However, the food waste management system can be extended to other circular strategies described in literature. In this study, we have proposed the alignment of Moerman's Ladder developed by Rood [67], for food waste management with the circular strategies suggested by Potting et al. [8] for production chain (presented in Table 3).

In the waste management hierarchy, prevention is the first priority for surplus food that is aligned with the circular strategy of "Refuse" and "Reduce" that emphasize abandoning the production if not required and reducing the consumption of natural resources. Redistribution of surplus food stated as *Human Consumption* is aligned with the circular strategies of "Rethink" and "Reuse" because food use is intensified by sharing it in its original form with the people in need (alternative channels of consumptions) that otherwise would not have access to nutritional food. The third level in food hierarchy, *Converted into Human Food*, stresses the food processing of surplus food to extend its shelf life that is in line with the circular strategies of

**Table 2** Summary of food waste hierarchies presented in literature ( Adapted from Ciccullo et al. 2021)

| Name of the framework | Prioritization of actions in the hierarchy (from most preferable to least preferable) | References |
|---|---|---|
| Waste hierarchy food waste hierarchy | 1. Prevention<br>2. Prepare for reuse<br>3. Recycling<br>4. Other recovery (e.g. Energy recovery)<br>5. Disposal | [65]<br>[6] |
| Food recovery hierarchy | 1. Source reduction<br>2. Feed hungry people<br>3. Feed animals<br>4. Industrial use<br>5. Composting | United State Environmental Protection agency-EPA (2012) |
| Availability-surplus-recoverability-waste model (ASRW) | 1. Recover surplus food to feed humans<br>2. Recover surplus food to feed animals<br>3. Waste recovery<br>4. Waste disposal | [5] |
| Food waste management hierarchy | 1. Prevention<br>2. Conversation for human nutrition<br>3. Use of animal feed<br>4. Use as raw material in industry<br>5. Process into fertilizer<br>6. Use as a renewable energy<br>7. Incineration<br>8. Landfill | [66] |
| Moerman's Ladder | 1. Preventing food losses<br>2. Human food<br>3. Converted into human food (food processing)<br>4. Use in animal feed<br>5. Use as raw material in industry<br>6. Process into fertilizers through fermentation<br>7. Process into fertilizers through composting<br>8. Applied for sustainable energy<br>9. Incineration | [67] |

(continued)

**Table 2** (continued)

| Name of the framework | Prioritization of actions in the hierarchy (from most preferable to least preferable) | References |
|---|---|---|
| Hierarchy for food surplus and waste | 1. Prevention<br>2. Reuse-H<br>3. Reuse-A<br>4. Material recycling<br>5. Nutrient recovery<br>6. Energy recovery<br>7. Disposal | [68] |

**Table 3** Alignment of circular strategies with food waste management hierarchy

| Circularity strategies within production chain | Moerman's Ladder | | |
|---|---|---|---|
| Refuse (Ro), Reduce (R2) | 1. Prevention (preventing food losses) | Edible food | Surplus food |
| Rethink (R1), Reuse (R3) | 2. Human food | | |
| Repair (R4), Refurbish (R5) Remanufacture (R6), and Repurpose (R7) | 3. Converted into human food (food processing) | | |
| Reuse (R3), and Repurpose (R7) | 4. Used in animal feed | Inedible | Food waste |
| Repurpose (R7) | 5. Raw materials for industry (bio-based economy) | | |
| Recycle (R8) | 6. Turned into fertilizer through fermentation (and for energy generation) | | |
| Recycle (R8) | 7. Turned into fertilizer through composting | | |
| Recover (R9) | 8. Applied for sustainable energy (purpose is energy generation) | | |
| Recover (R9) | 9. Incineration as waste (incineration/discharge with energy and mineral recovery) | | |
| Disposal | 10. Landfill (avoid if possible) | | |

Repair, Refurbish, Remanufacture and Repurpose that work towards extending the life of product/part with its original functionality by carrying out varying degrees of processing. All the strategies having high degree of circularity are embedded in the Food surplus management options detailed in the Moerman's ladder.

The next level in food management hierarchy is food waste that is inedible for human therefore alternative options for use are suggested to minimize it. The first inedible option is to use it as the animal food that is in line with the circular strategies of Reuse and Repurpose because the discarded inedible food is diverted to non-human consumption. The next level in waste management is the use of by products or food residues as the raw material for industries that coverts it into products with different functionality and is in line with the circular strategy of Repurpose. The next two levels (level 7 and 8 in food waste) describe the retrieval of organic nutrients from FW and reintroducing them to the ecosystem to restore the depletion of natural resources (agricultural land) that is in line with the recycling where the product remains in circulation with value addition (may be of high or low-grade quality). The last two grades of food waste have value addition through recovering energy in line with the Recovery strategy of CE that is considered the least circular strategy.

The explanation above demonstrates the alignment of circular strategies with the waste management hierarchy indicating that circular strategies have the universal application for the waste management systems. In the context of food waste management systems, food surplus management by preventing excess food production and maximizing surplus consumption among human by redistribution and increasing food life through processing leads to sustainable consumption and production systems that can contribute to achieve the goals of "Zero Hunger" and "Responsible Consumption and Production".

# 6   Challenges and Opportunities

The above section described the circular ways of managing food waste to minimize the negative impacts on the environment alongside leveraging the societal benefits. However, this circular food waste management system presents both opportunities and challenges when put into practice in real life and need careful considerations and collaboration of academia, practitioners and policy makers to overcome the challenges and tackling the FW-related financial and environmental constraints. The challenges and opportunities are categorized as technological, economical, and cultural dimensions.

- *Technological opportunities and challenges*

Food waste is one of the biggest sustainability challenges and could be reduced by better waste management throughout the supply chain in addition to redistribution of surplus food and sustainable food management. Technology is an important resource that can play a positive role in preserving surplus food through innovative packaging and storage. In addition, the FW is an important source of energy and chemicals where appropriate technology can enable the recovery of materials and energy by disposing the FW in environmental-friendly way. FW and FL occur at all stages in the supply chain due to technical and infrastructural reasons that need technology application at large scale by having minimum environmental impact. However, the challenge is technology solutions are developed and assessed at the laboratory where large-scale application require huge amount of resources. Moreover, for the recovery technology the biggest challenge is determining the product quality and hygiene that make it hard to assess the actual yield (chemical, bio gas, etc.) of biomass. Therefore, government support and incentivization is necessary for the application of technology at large scale in the long run.

- *Economic opportunities and challenges*

Food redistribution, reuse and recovery strategies feedback to the economy with additional by- products and economic benefits. Moreover, the use of food waste as feedstock reduces the cost of disposal for the industry. For biorefinery, the high initial cost of set up is balanced by the availability of cheap FW feedstock. In terms of challenge, the cost of generating energy from FW (considering the availability and transportation cost) cannot be estimated precisely due to absence of real biorefinery implementations. Moreover, FW feed stock usage for generating commodities and energy can only be achieved through the development of proper incentive system by the Government and collaboration among different actors across the entire food supply chain.

- *Cultural opportunities and challenges*

One of the benefit for the effective food management system is food surplus redistribution that provides nutritional food to underprivileged citizens by relevant associations and charities and create awareness for sustainable consumption. The food waste use for the energy generation can reduce power scarcity and the decline of wood burning in low-income countries or in the countryside [69]. FW can also provide a low-cost alternative for generating energy instead of using new and costly raw material. However, reducing the food surplus and food waste require substantial change in production and consumption patterns at industry and market levels that is hard to change due to complex relationship between suppliers and distributers, contractual agreements, food standards, and inaccurate food demand forecast Eriksson [70]. Moreover, variable definitions of food waste hinder the development of standard regulatory metrics for the qualification. Another challenge is that the recovery technology is mostly available in the developed economies whereas the feedstock market is in the developing economies. In terms of WTV products, consumer acceptance is a crucial factor and serve as the decisive strategy for the development of such products. Substantial FW is caused at the household level that require educating individuals for zero food waste and introducing responsible consumption habits to reduce this waste.

# 7    Conclusion

Food waste poses the greatest challenge for sustainability of the food management systems. Circular economy is redefining the ways for waste management including the food waste and creating new business opportunities by circulating food waste into various closed loops. Food waste can be used to produce various biomaterials, bioenergy and high-value products.

A systematic literature review is conducted in the study to describe the initiation of CE and different emerging themes in the CE. Moreover, literature on the Agri-Food waste is explored for identification of different food management frameworks, their similarities and differences. The study also derived an alignment between the food waste management hierarchy and circular strategies of production chains. This alignment highlights that the efforts for application of food waste management hierarchy are actually a functional implementation of circular strategies that lead to achievement of food sustainability with the minimum environmental effects. The study has also explored the concept of TBL and sustainability and draws similarities and differences between these two concepts. CE is discussed as a way towards the achievement of sustainability particularly in the context of food waste management that could contribute towards the achievement of sustainable development goals of "Zero Hunger" and "Responsible Production and Consumption".

However, there are several challenges for the food waste prevention as a sustainable waste management system in the emerging circular bioeconomy that needs standard regulatory infrastructure, Government incentive systems, collaboration among different actors in the food supply chain, cooperation between academics and practitioners and change in the way of production and consumption at individual, institutional and market levels.

# References

1. FAO (2014) Developing sustainable food value chains-guiding principles, 89 p
2. Vilariño MV, Franco C, Quarrington C (2017) Food loss and waste reduction as an integral part of a circular economy. Front Environ Sci 5
3. Lipinski B, Hanson C, Waite R, Searchinger T, Lomax J, Kitinoja L (2013) Reducing food loss and waste|world resources institute [internet]. In: Creating a sustainable food future, installment two (cited 2021 Feb 14). Available from https://www.wri.org/publication/reducing-food-loss-and-waste
4. Stenmarck Å, Jensen C, Quested T, Moates G, Cseh B, Juul S et al. (2016) FUSIONS—estimates of European food waste levels [internet]. Fusions, 1–80 p. Available from https://www.eu-fusions.org/phocadownload/Publications/EstimatesofEuropeanfoodwast elevels.pdf%5Cn; https://phys.org/news/2016-12-quarter-million-tonnes-food-logistics.html#nRlv
5. Garrone P, Melacini M, Perego A (2014) Opening the black box of food waste reduction. J Food Policy [Internet] 46:129–139. Available from https://doi.org/10.1016/j.foodpol.2014.03.014
6. Papargyropoulou E, Lozano R, Steinberger JK, Wright N, Bin UZ (2014) The food waste hierarchy as a framework for the management of food surplus and food waste. J Clean Prod 76:106–115
7. Ellen MacArthur Foundation (2014) Towards the circular economy, vol 3. In: Accelerating the scale-up across global supply chains. Ellen MacArthur Found [Internet], pp 1–64. 2014 (Jan). Available from https://www.ellenmacarthurfoundation.org/publications/towards-the-circular-economy-vol-3-accelerating-the-scale-up-across-global-supply-chains
8. Potting J, Hekkert M, Worrell E, Hanemaaijer A (2017) Circular economy: measuring innovation in the product chain—policy report. PBL Netherlands Environ Assess Agency 2544:42
9. Stahel WR (2010) The performance economy [Internet]. London, Palgrave Macmillan UK (cited 29 Dec 2017). Available from https://doi.org/10.1057/9780230274907

10. Stahel W, Reday G (1976) The potential for substituting manpower for energy. Report to the Commission of the European Communities
11. Lyle JT (1994) Regenerative design for sustainable development. Wiley, 338 p
12. Graedel TE, Allenby BR (1995) Industrial ecology and sustainable engineering [Internet]. Prentice Hall (cited 29 Dec 2017], 403 p. Available from http://agris.fao.org/agris-search/sea rch.do?recordID=US201300142472
13. Erkman S (1997) Industrial ecology: an historical view. J Clean Prod [Internet] 5(1–2):1–10. Available from http://linkinghub.elsevier.com/retrieve/pii/S0959652697000036
14. McDonough W, Braungart M, Anastas PT, Zimmerman JB (2003) Peer reviewed: applying the principles of green engineering to cradle-to-cradle design. Environ Sci Technol [Internet] 37(23):434A-441A. Available from https://doi.org/10.1021/es0326322
15. UNEP (2011) Towards a green economy: pathways to sustainable development and poverty eradication. Available at http://web.unep.org/greeneconomy/sites/unep.org.greeneconomy/files/field/image/green_economyreport_final_dec2011.pdf
16. Tukker A (2015) Product services for a resource-efficient and circular economy—A review. J Clean Prod [Internet] 97:76–91. Available from https://doi.org/10.1016/j.jclepro.2013.11.049
17. Ghisellini P, Cialani C, Ulgiati S (2016) A review on circular economy: the expected transition to a balanced interplay of environmental and economic systems. J Clean Prod [Internet] 114:11–32. Available from https://doi.org/10.1016/j.jclepro.2015.09.007
18. Merli R, Preziosi M, Acampora A (2017) How do scholars approach the circular economy? A systematic literature review. J Clean Prod [Internet] 2017. Available from http://linkinghub.els evier.com/retrieve/pii/S0959652617330718
19. Su B, Heshmati A, Geng Y, Yu X (2013) A review of the circular economy in China: moving from rhetoric to implementation. J Clean Prod [Internet] 42:215–227. Available from https://doi.org/10.1016/j.jclepro.2012.11.020
20. Bocken NMP, Short SW, Rana P, Evans S (2014) A literature and practice review to develop sustainable business model archetypes. J Clean Prod [Internet] 65:42–56. Available from https://doi.org/10.1016/j.jclepro.2013.11.039
21. Govindan K, Soleimani H, Kannan D (2015) Reverse logistics and closed-loop supply chain: a comprehensive review to explore the future. Eur J Oper Res [Internet] 240(3):603–626. Available from https://doi.org/10.1016/j.ejor.2014.07.012
22. Pan SY, Du MA, Huang I Te, Liu IH, Chang EE, Chiang PC (2014) Strategies on implementation of waste-to-energy (WTE) supply chain for circular economy system: a review. J Clean Prod 108:409–421. Available from https://doi.org/10.1016/j.jclepro.2015.06.124
23. Genovese A, Acquaye AA, Figueroa A, Koh SCL (2017) Sustainable supply chain management and the transition towards a circular economy: evidence and some applications. Omega [Internet] 66:344–357. Available from http://linkinghub.elsevier.com/retrieve/pii/S03050483 15001322
24. Lieder M, Rashid A (2016) Towards circular economy implementation: a comprehensive review in context of manufacturing industry. J Clean Prod [Internet] 115:36–51. Available from https://doi.org/10.1016/j.jclepro.2015.12.042
25. Geissdoerfer M, Savaget P, Bocken NMP, Hultink EJ (2017) The circular economy—a new sustainability paradigm? J Clean Prod [Internet] 143:757–768. Available from https://doi.org/10.1016/j.jclepro.2016.12.048
26. Yang S, Feng N (2008) A case study of industrial symbiosis: Nanning Sugar Co., Ltd. in China. Resour Conserv Recycl 52(5):813–820
27. Geng Y, Doberstein B (2016) Developing the circular economy in China: challenges and opportunities for achieving 'leapfrog development.' Int J Sustain Dev World Ecol 2008(15):231–239
28. Peters GP, Weber CL, Guan D, Hubacek K (2007) China's growing $CO_2$ emissions—a race between increasing consumption and efficiency gains. Environ Sci Technol 41(17):5939–5944
29. Yuan Z, Bi J, Moriguichi Y (2016) The circular economy 10(1):1–7
30. Park J, Sarkis J, Wu Z (2010) Creating integrated business and environmental value within the context of China's circular economy and ecological modernization. J Clean Prod [Internet] 18(15):1492–1499. Available from https://doi.org/10.1016/j.jclepro.2010.06.001

31. Haas W, Krausmann F, Wiedenhofer D, Heinz M (2015) How circular is the global economy? An assessment of material flows, waste production, and recycling in the European union and the world in 2005. J Ind Ecol 19(5):765–777

32. Ma SH, Wen ZG, Chen JN, Wen ZC (2014) Mode of circular economy in China's iron and steel industry: a case study in Wu'an city. J Clean Prod [Internet] 64:505–512. Available from https://doi.org/10.1016/j.jclepro.2013.10.008

33. Ma S, Hu S, Chen D, Zhu B (2015) A case study of a phosphorus chemical firm's application of resource efficiency and eco-efficiency in industrial metabolism under circular economy. J Clean Prod 87(1):839–849

34. Charonis G-K (2012) Degrowth, steady state economics and the circular economy: three distinct yet increasingly converging alternative discourses to economic growth for achieving environmental sustainability and social equity. In: World Economics Association (WEA) Conferences. Available at https://sustainabilityconference2012.weaconferences.net/papers/degrowth-steady-state-economics-and-the-circular-economy-three-distinct-yet-increasingly-converging-alternative-discourses-to-economic-growth-for-achieving-environmental-sustainability-and-social-eq/

35. Hobson K, Lynch N (2016) Diversifying and de-growing the circular economy: Radical social transformation in a resource-scarce world. Futures [Internet] 82:15–25. Available from https://doi.org/10.1016/j.futures.2016.05.012

36. Moreau V et al (2017) Coming full circle: why social and institutional dimensions matter for the circular economy. J. Ind. Ecol. 21(3):497–506. Available at http://doi.wiley.com/10.1111/jiec.12598

37. Niero M, Hauschild MZ, Hoffmeyer SB, Olsen SI (2017) Combining eco-efficiency and eco-effectiveness for continuous loop beverage packaging systems: lessons from the Carlsberg circular community. J Ind Ecol 21(3):742–753

38. Murray A, Skene K, Haynes K (2017) The circular economy: an interdisciplinary exploration of the concept and application in a global context. J Bus Ethics 140(3):369–380

39. Naustdalslid J (2014) Circular economy in China—the environmental dimension of the harmonious society. Int J Sustain Dev World Ecol [Internet] 21(4):303–313. Available from https://doi.org/10.1080/13504509.2014.914599

40. Blomsma F, Brennan G (2017) The emergence of circular economy: a new framing around prolonging resource productivity. J Ind Ecol 21(3):603–614

41. Wu HQ, Shi Y, Xia Q, Zhu WD (2014) Effectiveness of the policy of circular economy in China: a DEA-based analysis for the period of 11th five-year-plan. Resour Conserv Recycl [Internet] 83:163–175. Available from https://doi.org/10.1016/j.resconrec.2013.10.003

42. Li J, Yu K (2011) A study on legislative and policy tools for promoting the circular economic model for waste management in China. J Mater Cycles Waste Manage 13(2):103–112

43. MacArthur E (2013) Towards the circular economy: opportunities for the consumer goods sector. Ellen MacArthur Found:1–112

44. Zink T, Geyer R (2017) Circular economy rebound. J Ind Ecol 21(3):593–602

45. Kirchherr J, Reike D, Hekkert M (2017) Resources, conservation and recycling conceptualizing the circular economy: an analysis of 114 definitions 127:221–232

46. Brown BJ, Hanson ME, Liverman DM, Merideth RW (2017) Global sustainability: toward definition. Environ Manage [Internet] 11(6):713–719. 1987 Nov (cited 2017 Dec 29). Available from https://doi.org/10.1007/BF01867238

47. Clark W, Crutzen P (2005) Science for global sustainability: toward a new paradigm. KSG Work Pap No [Internet] 120:1–28. Available from http://www.hks.harvard.edu/var/ezp_site/storage/fckeditor/file/pdfs/centers-programs/centers/cid/publications/faculty/wp/120.pdf%5Cn, http://papers.ssrn.com/sol3/papers.cfm?abstract_id=702501

48. Rockström J, Steffen W, Noone K, Persson Å, Chapin FS III, Lambin EF et al (2009) Planetary boundaries: exploring the safe operating space for humanity. Ecol Soc 14(2):32

49. Georgescu-Roegen N (1977) Inequality, limits and growth from a bioeconomic viewpoint. Rev Soc Econ [Internet] 35(3):361–375. Available from https://doi.org/10.1080/00346767700000041

50. Elkington J (1997) Cannibals with forks. Cannibals with forks triple bottom line 21st century. The triple bottom line 21st century [Internet], Apr, pp 1–16. Available from http://pdf-release. net/external/242064/pdf-release-dot-net-148_en.pdf

51. Govindan K, Kannan D, Shankar KM (2014) Evaluating the drivers of corporate social responsibility in the mining industry with multi-criteria approach: a multi-stakeholder perspective. J Clean Prod [Internet] 84(1):214–232. Available from https://doi.org/10.1016/j.jclepro.2013. 12.065

52. McMichael AJ, Butler CD, Folke C (2003) New visions for addressing sustainability. Science (80), 302(5652):1919–1920

53. UN General Assembly (2005) Resolution adopted by the general assembly. 60/1. 2005 world summit outcome, New York

54. Wise N (2016) Outlining triple bottom line contexts in urban tourism regeneration. Cities [Internet] 53:30–34. Available from https://doi.org/10.1016/j.cities.2016.01.003

55. Seuring S, Müller M (2008) From a literature review to a conceptual framework for sustainable supply chain management. J Clean Prod 16(15):1699–1710

56. Rambaud A, Richard J (2015) The "triple depreciation line" instead of the "triple bottom line": towards a genuine integrated reporting. Crit Perspect Account [Internet] 33:92–116. Available https://doi.org/10.1016/j.cpa.2015.01.012

57. Rodger JA, George JA (2017) Triple bottom line accounting for optimizing natural gas sustainability: a statistical linear programming fuzzy ILOWA optimized sustainment model approach to reducing supply chain global cybersecurity vulnerability through information and communications t. J Clean Prod [Internet] 142:1931–1949. Available from https://doi.org/10.1016/j. jclepro.2016.11.089

58. Chaboud G, Daviron B (2016) Food losses and waste: navigating the inconsistencies. Glob Food Sec 2017(12):1–7

59. De Steur H, Wesana J, Dora MK, Pearce D, Gellynck X (2016) Applying value stream mapping to reduce food losses and wastes in supply chains: a systematic review. Waste Manage [Internet] 58:359–68. Available from https://doi.org/10.1016/j.wasman.2016.08.025

60. Tunn VSC, Bocken NMP, van den Hende EA, Schoormans JPL (2019) Business models for sustainable consumption in the circular economy: an expert study. J Clean Prod [Internet] 212:324–333. Available from https://doi.org/10.1016/j.jclepro.2018.11.290

61. Aschemann-Witzel J, Peschel AO ()2019 How circular will you eat? The sustainability challenge in food and consumer reaction to either waste-to-value or yet underused novel ingredients in food. Food Qual Prefer [Internet] 77:15–20. Available from https://doi.org/10.1016/j. foodqual.2019.04.012

62. Cavaliere A, Ventura V (2018) Mismatch between food sustainability and consumer acceptance toward innovation technologies among millennial students: the case of shelf life extension. J Clean Prod [Internet] 175:641–650. Available from https://doi.org/10.1016/j.jclepro.2017. 12.087

63. Bhatt S, Lee J, Deutsch J, Ayaz H, Fulton B, Suri R (2018) From food waste to value-added surplus products (VASP): consumer acceptance of a novel food product category. J Consum Behav 17(1):57–63

64. Coderoni S, Perito MA (2020) Sustainable consumption in the circular economy. An analysis of consumers' purchase intentions for waste-to-value food. J Clean Prod [Internet] 252:119870. Available from https://doi.org/10.1016/j.jclepro.2019.119870

65. European Commission (2014) Circular economy scoping study

66. Vandermeersch T, Alvarenga RAF, Ragaert P, Dewulf J (2014) Environmental sustainability assessment of food waste valorization options. Resour Conserv Recycl [Internet] 87:57–64. Available from https://doi.org/10.1016/j.resconrec.2014.03.008

67. Rood T, Gena C, Chiesa I, Cietto V (2018) IoT for the circular economy:95–102

68. Teigiserova DA, Hamelin L, Thomsen M (2020) Towards transparent valorization of food surplus, waste and loss: clarifying definitions, food waste hierarchy, and role in the circular economy. Sci Total Environ [Internet] 706:136033. Available from https://doi.org/10.1016/j. scitotenv.2019.136033

69. Breitenmoser L, Gross T, Huesch R, Rau J, Dhar H, Kumar S et al (20) Anaerobic digestion of biowastes in India: Opportunities, challenges and research needs. J Environ Manage [Internet] 236:396–412. Available from https://doi.org/10.1016/j.jenvman.2018.12.014
70. Eriksson M, Strid I, Hansson PA (2015) Carbon footprint of food waste management options in the waste hierarchy—a Swedish case study. J Clean Prod [Internet] 93:115–125. Available from https://doi.org/10.1016/j.jclepro.2015.01.026

# Sustainable Food Value Chains and Circular Economy

Simmi Ranjan Kumar, Saugat Prajapati, and Jose V. Parambil

**Abstract** Food waste and loss are major threats to our food systems' long-term sustainability. Researchers, government organizations, non-governmental organizations, and the agricultural industry are actively proposing, researching, and introducing creative and diverse strategies to solve the issue of food waste. In the conventional linear food supply chain model, billions of tons of food end up in landfills. Food waste is a global sustainability problem that plagues food systems and is an urgent concern from an economic, environmental, and social perspective. Integrating the circular economy principles into the food supply chain can make the food industry more sustainable. The circular economy in the food supply chain perspective involves an economic-industrial system that utilizes the six-R concepts, such as reducing, reusing, repurpose, repair, refurbishment, and recovery. Thus, it will open up various opportunities within the so-called food waste hierarchy. Different hierarchies follow different approaches to food waste management. While some focus on reducing waste generation early in the supply chain, others focus on later stages in the supply chain. This study focuses on the need for a sustainable food value chain to avoid loss during processing and its role in the circular economy and gain insights into how the implementation of circular economic models and tools can achieve sustainability in food loss and food waste in the food chain process.

**Keywords** Food waste · Food supply chain · Circular economy · Food waste hierarchy · Sustainable food supply chain · Food loss

S. R. Kumar
Asian Institute of Technology, Khlong Luang District, Thailand

S. Prajapati
College of Applied Food and Dairy Technology, Kathmandu, Nepal

J. V. Parambil (✉)
Chemical and Biochemical Engineering, Indian Institute of Technology, Patna, Bihta, Bihar 801106, India
e-mail: josevparambil@iitp.ac.in

# 1 Introduction

The demand for food and other agricultural necessities is increasing day by day with the world population. Food production will have to increase by at least two-thirds by 2050 to ensure sufficient nutrition for all [1]. Meeting this demand sustainably will not be possible by focusing only on organic agricultural practices, i.e., in the production section of the supply chain alone. There is a considerable need for developing a sustainable approach to the entire supply chain in agriculture. Food production, processing, and supply have to be based on the principles of the circular economy. Such a change will be able to overcome the inefficiency and ethical issues with the current agricultural system. It would also help mitigate the possible decline in crop yields because of climate change and pandemics such as COVID-19 [2].

In the existing food supply chain, wastage and loss are severe challenges for sustainability. Worldwide, nearly one-third of food supply is wasted and destroyed [3, 4]. In 2015, United Nations set a sustainable development goal to "halving the per capita food waste and loss globally at the level of consumer and retailer along with the production, supply chain, and post-harvest loss" by 2030 [5]. Governments have initiated food policies to address this concern. Researchers, policy-makers, retailers, food industries, non-government, and government organizations work together to minimize food waste through social movements that impart food appreciation to the public.

In food supply chain, the terms 'food loss' and 'food waste' may or maynot be distinguished as separate entities [6]. According to FAO [7], food loss is defined as the food that is discarded during post-harvest, production, and processing periods of the food supply chain. As per Edjabou et al. [8], food waste is termed as the food waste generated through food processing, post-harvest handling, storage, agriculture production, retail and wholesale distribution, household kitchen, and large-scale consumers. In this article, there will not be any differentiation between these terms. Hence, food loss and waste (FLW) would refer to any food material that is discarded at any stage of the food chain, either by producer, processor, transporter, or consumer.

A circular economy is a system that can utilize waste as a resource and generate value-added materials, products, or business opportunities continuously, with minimal losses. The food waste can be utilized to produce bioenergy, other biomaterials, and high-value products, both edible and in-edible [9]. The circular economy concept is vital in the food loss and waste context because it has various solutions to deal with food waste using behavioral, cultural, and technological practices and policy recommendations [10].

This chapter focuses on sustainable food supply chain and its role in the circular economy. By utilizing the circular economy and sustainable food value chain, food loss and waste can be reduced effectively.

## 2 The Sustainable Food Supply Chain

Sustainable food production can be defined as an efficient and healthier food production system that employs effective and non-polluting techniques. It favors the conservation of non-renewable natural resources without compromising future generations' needs [11]. The global population is projected to become 9.1 billion by 2050. This is 16% more than the current population. Consequently, agricultural production has to be increased by 50% from 2012 levels [12]. Unfortunately, around 1.3 billion tons of agricultural produce is lost or wasted yearly before reaching the consumers. In most developing countries, the major loss in the supply chain, i.e., collectively during harvesting, storing, processing, or distributing to final consumers. However, in developed countries, the FLW is mostly accounted for at the retailer and consumer level [13]. Massive land space, a high volume of water, and tremendous energy are spent in the food supply chain. Furthermore, the negative impact of waste disposal such as emission of Green House Gases (GHGs), reduction in soil quality, water, and air pollution, and loss of biodiversity result in environmental depletion and deteriorate human life quality [14]. Hence, the world is seeking for a sustainable food supply chian which is defined by UN Global Compact as to develop, conserve, and sustain long-term environmental and socio-economic merits to all stakeholders dedicated to fetch products and services to market. More precisely, sustainable food value chains can be expressed as the integration and coordination of all the diversified farms and firms, their agricultural contribution for production of raw materials with high quality and quantity without depleting the natural resources and their transformation into high-value products being readily available to final consumers [15]. The reduction in FLW needs to be an integral part of the answer to the future of food security and sustainability. Further, food waste minimization could contribute to reducing hunger, contributing to a sustainable and resilient food supply chain [16].

## 3 The Primary Sustainability Issues with Food Loss and Waste

The accountability of FLW is a significant issue to the sustainable food production and supply. The higher the FLW, the higher will be the pressure on the environment. Identifying points of the supply chain which is the most responsible for environmental degradation is difficult. This is important since attempts to reduce FLW at various points in the value chain may have varying impacts on the environment. According to Cattaneo et al. [17], the consequence of the reduction in FLW at the consumer level has a positive effect on conserving the environment and resources. Counter-intuitively, the reduction in FLW at the early stages of production may result in an increased movement of food in the supply chain and thereby increase the aggregate GHG emission, thereby reducing the environment's quality. However, as a general perspective, for the sustainable use of land, water, and energy sources for food production, losses and wastes have to be avoided, where not possible to avoid, should be controlled by reusing, recycling, and reducing the resources at all possible sections of the food supply chain.

GHG emissions from food waste can be accounted in two ways: first, the emission due to decomposition in landfills, and secondly, the emissions embedded in activities such as food production, processing, transportation, and distribution, also termed as the life cycle view of food waste [18]. It is assessed that the household food waste in the UK (8.3 Mt) roughly equates to $18.6 million in economic costs and accounts for 3% of the overall GHG emission of the country. Similarly, the USA's food waste is estimated to be 40% of annual production, which accounts for 25% and 4% of fresh water and petroleum usages, respectively [19]. The consumption patterns in developing countries are changing rapidly. With improved economic conditions, preferences move towards easily perishable fruits and vegetables from non-perishable staple foods. Estimates suggest that by 2050, the carbon dioxide emission from FLW will reach between 5.7 and 7.9 Gt per year, which is 2.5 times than 2011 [20, 21].

Similarly, the consumption of animal and poultry-based food products is increasing at an alarming rate, which will increase by 44% by the year 2030. Animal-based food adds a large quantity of GHGs, mainly methane, nitrous oxide, and carbon dioxide, to the environment. The aggregate annual emission of carbon dioxide from global livestock is 7.1 (GT), about 14.5% of the world's GHG emissions. Therefore, the growing demand for animal-based foods, byproducts, and associated food waste contributes significantly to global warming and environmental degradation [22].

In a linear supply chain model, raw materials extracted from the land are processed into products, consumed by people, and waste is discarded. Waste food is mostly discarded to the landfill with no or less recovery of the products. Considering the strain on natural resources, the increase in populations with varied consuming demands, insufficient agricultural land for processing, climate change, and rising food insecurity worldwide, it is important to ensure that the amount of food lost in the food supply chain is negligible. To ensure food protection for all people globally, food provisioning must be achieved sustainably [23]. The circular economy concept gains momentum due to the need to create sustainable food supply chains [24]. The circular economy is a brigher substitute to the conventional linear economy. In this, the resources are retained in operation for longer duration to extract maximum utilization. Further, the goods are recycled or repurposed at the end of their utility, giving them further life in the supply chain [25]. The circular economy aims an economic model that achieves ecosystem optimization by redesigning and controlling resources, procurement, output, and recycling processes more efficiently and effectively [26].

# 4  Barriers to Sustainable Supply Chain Processes

Adopting sustainable food supply chain and a circular economy is a multidimensional problem where numerous stakeholders need to actively participate. Governments, organizations, producers, and general public (consumers) need to work together to achieve this. Although several studies have focused on the barriers that hinder sustainability in the supply chain of corporations [27], such studies are scarce in the perspective of food supply chains where multiple organizations and individuals are involved. Hence, we will first review the challenges in an organizational level and then interpret them in the context of food supply chain.

Business or country-specific studies have established the major challenges that organizations face while adopting sustainable supply chain management [28]. Common hurdles comprise absence of interest from higher authorities, strain in aligning short- and long-term objectives, inhibitions to modifying business processes and strategies, high expenditure expectations, unavailability of environmental guidelines and legislation, lack of knowledge of consumers, and difficulties in generating such awareness, lack of capital for vendors, etc. These obstacles are classified into different categories such as internal, external, social, technical, financial, political, managerial, and behavioral [28]. Within the internal barriers, the dependency on conventional accounting methods that do not allow triple bottom line reporting, which would provide importance to environmental and social concerns along with financial calculations, is a major component [29]. Typical external challenges include the need of customers to lower costs, competitive pressure, and greenwashing. Table 1 enlists the barriers to a sustainable supply chain.

Kumar et al. [31] highlights the danger of mis-investment, insufficient plan for industry development and circular economy, and lack of waste management as barriers to a sustainable supply chain on a broader scale. For large-scale changes to the supply chain which involve multiple players at different levels, strong and focused political and economic leadership is required. Practically, the lack of political leadership that helps and nudges the retailers to adopt sustainable practices has been described as the most important impediment to circular economy. Mixed and unclear retailer transparency messages from politicians, a lack of consistent vision for retail sector sustainability, and the absence of an action plan established in consultation with companies all create uncertainties that hamper long-term investments in this direction [32].

In case of food supply management, restrictions and directives provided by the licensing and regulatory authorities can also affect the adoption of sustainable practices. Through proactive measures, regulatory authorities, in association with government agencies and private industry players could catalyze the transformation to a circular economy. For example, government policies and incentives for the production of biodiesel from waste cooking oil can make the supply chain more sustainable [33]. In India, a government initiative termed RUCO (Repurpose Used Cooking Oil), headed by the Food Safety and Standards Authority of India (FSSAI) has enrolled 26 biodiesel manufacturers across the country. The program aims to recover 220 crore

**Table 1** Barrier to adopting sustainable supply chain processes in food industry. Adapted from [30]

| Area | Barriers |
|---|---|
| Creation of sustainable supply chain | Financing new technologies<br>Regulatory frameworks and government policies<br>Lack of expertise and support structure for collaborative ventures where multiple stakeholders with different products or services work together |
| Product development | Capital cost for technology adoption<br>Reluctance to fund in collaborative sustainable supply chains due to high level of risk<br>Lack of specialized knowledge and expertise in creating supporting services (IT, accounting, branding) for a collective cause |
| Access to market | Mistrust between the retail sellers, farmers, and intermediaries, where sub-par products produced unsustainably might be mislabelled and added to the supply chain<br>Lack of constant connectivity between market places and production centers<br>Capital and running costs of stalls, space, and cooling systems at markets<br>Lack of diversity in local production<br>Lack of accountability on customer service due to the presence of a larger number of organizations in a supply chain |

liter of used cooking oil by 2020 [34]. Such large-scale initiatives need the support of apex bodies of industries, both from the public and private sectors, to be successful. On the contrary, lack of vision and support from regulatory agencies may deter the corporations and individuals from adopting novel and sustainable ideas that align with the circular economy concept. Unfortunately, despite the increasing attention, effective implementation of environmental policies in many countries is questionable and skepticized as reactive to major environmentally catastrophic events.

## 5 Circular Economy and Sustainable Food Supply Chain

Stahel and Reday [35] first introduced the term "circular economy." A circular economy encompasses a regenerative structure based on zero-waste viewpoint. The idea is that waste generated within an organization or a community can be used as a useful source by the same or another organization or community. This will reduce the overall dependency on new resources in an economy. A circular economy plays an essential role in addressing environmental sustainability as supply chains become complex. It focuses on minimizing lowering waste generation and optimizing resource consumption for economic performance [36]. It can be seen as a business model that utilizes resources via components and retain materials in a closed-loop commercial structure. Thus, it mainly focuses on producing new components by reusing waste and end-of-life materials to boost both economic and environmental benefits [37].

Various motives make it challenging to develop a theoretical consensus on the circular economy's nature and implementation. Within a single organization, circular economy practices may emphasis on production and consumption practices at the micro-level. For instance, these may include eco-design, reuse, and recycling of products. It may be implemented at a mesoscale by practicing industrial symbiosis and resource utilization and optimization in business communities and industrial parks [26]. Regionally and globally, the circular economy has a broader scope at macro levels, such as sharing economies and eco-cities. This level can focus on a performance level, such as collective municipal consumption and zero waste systems [38].

As discussed previously, FLW appears in different stages of food supply chains, including production, storage and handling, packaging, and processing [3, 4]. In the initial stages of a food supply chain, FLW can occur due to a reduction in the nutritional value or dry matter. Inadequate infrastructure, lack of technology, insufficient skills, lack of access to markets immediately after harvest, and natural disasters may cause food loss [39]. Consequently, food prices may increase and their quality decreases for human consumption [3, 4]. In the last two stages of a food supply chain, wastage of food is observed due to consumer behavior and abandonment or a deliberate choice to throw food by retailers [40]. Millions of perfectly palatable food pounds are discarded every year, all over the supply chain, while 815 million people remain hungry and malnourished in underdeveloped countries [41]. In the USA alone, 15.8 million homes are likely food insecure. Reducing wastage in these last phases of the supply chain alone by 15% can address food insecurity to a large extent. Furthermore, about 1 billion people can be fed by halving the food losses in the initial stages [40].

Although various factors contribute to FLW, the most concerning one is the conscious rejection of edible food based on aesthetics. A significant portion of wastage occurs due to appearance and high selection among consumers and farmers in European countries. At every step of the supply chain, it is expected that 1–40% of products are wasted due to various industry guidelines and are considered suboptimal. The fruits and vegetables are categorized into A or B class in Europe based on color, skin, shape, weight, and size. This specification leads cultivators and others involved in the supply chain to dispose of absolutely edible food because they are not up to specifications. Besides, in developed economies, many farmers have individualized product specifications, according to which the retailers purchase the farmer's goods. Thus, they choose only the perfect ones that fit the required specifications and leave the rest to trash [42]. Around 75% of consumers are unwilling to purchase agricultural produce that do not look good [43]. Hence, this is essential to ensure that management standard for food avoidance and food waste use is sustainable at either the inflection point among linear and circular paradigms, i.e., to remain inside planetary boundary guardrails [9].

In parallel, a change from a linear to a circular economy is demanded by the increasing need for energy and resources for a rapidly-growing and resource-hogging population. Circular economy manages waste as a secondary resource [9]. The EU has adopted a circular economic action plan, which aligns with the Global Sustainable Development Goal [44]. Overall sustainability in the food cycle can be achieved if each step in the action plan is followed correctly. It aims to satisfy consumer needs, including food quality and safety, with less impact on the environment. The envisioned circular economy fosters sustainability, environmental and socio-economic benefits, and resource-efficient policies [45]. Existing food supply chains neglect the waste generated in various stages. Present supply chains use the "take-produce-consume-discard" model, which suggests that the economic development is based on resource abundance and unrestricted disposal of waste [46]. This is in clear contradiction to sustainable development. However, integrating food waste management into existing food supply systems can enhance and optimize food system's sustainability. Five stages in a sustainable food cycle would include food processing, production, distribution, consumption, and food waste management [40]. However, food waste management is easier said than done! Adaptation of various mechanisms for waste management would depend on the nature of the wasted material and the locality's social, economic, and technological maturity. Some of these considerations and the concept of food waste hierarchy is discussed in the next section.

# 6 Food Waste Hierarchy

Within a circular economy, the focus of food waste management is not similar to any other solid or liquid waste management. In conventional waste management, the aim is to treat the waste materials before releasing them into the environment in such a way that they do not pose any threat when released. In a circular economy, waste management focuses on reusing, repair, refurbishing, and recycling the materials that are generally regarded as waste [46]. This approach redefines the structure requirements for the management of food waste and creates innovative business prospects. Wastes from foodstuff can be utilized to generate different value-added products. However, not all nutrients and components can be cycled. Energy recovery or final decomposition of the environment could be implemented in such cases. Nonetheless, there are opportunities to increase the amount of nutrients and matter circulating by utilizing food waste, thereby reducing the absorption of fresh nutrients and matter and partly closing the loops [46].

For food waste administration, a circular economy presents a convincing reference structure [9]. Apart from food valorization alternatives, the concept of reuse, recycle, and reduction of food components and food waste can be implemented in supply chains [47]. A well-defined waste hierarchy was first defined by the European legislation in the Community Strategy for Waste Management in 1989. Different policies and actions required for food waste management are rooted in various hierarchical structures established following the Waste Framework Directive 2008/98/EC

[48]. Papargyropoulou et al. [49] introduced a food waste hierarchy with strategies for avoiding wastage and reusing excess food for human consumption. Adaptation of such strategies will reduce the exploitation of resources and lessen the ethical and societal implications of FLW. Since then, this hierarchy has been embraced widely as a principal food waste management framework.

Japan and other countries across Asia promote other frameworks such as the 3R in which waste management gives priority to reuse, recycle, and reduce waste [49]. Reuse and redistribution options that target human consumption minimizes FLW from a social perspective [50]. Some of the options include repackaging, remanufacturing, promotion and discount with sales, and redistribution through alliance with food-aid organizations [51]. The Food Recovery Hierarchy of the Environmental Protection Agency (EPA) establishes a mechanism for prioritizing action to decrease food waste to help the environment. As per this hierarchy, reducing the overall amount of excess food produced is the most preferred way of reducing food waste. This intervention takes priority over incineration of landfills, composting, agricultural uses, and animal feeding since it affects the environment least [52]. When food is not lost, businesses and the environment profit the most. There are major environmental benefits and cost reductions that occur as the emphasis is given to the most important Food Recovery Hierarchy. From an environmental standpoint, food waste prevention saves ten times more GHG emissions than food reallocation. Simultaneously, eliminating food waste generates three times the net economic impact of retrieval and reusing together in society [40].

Similar to other waste hierarchies, food waste hierarchies illustrate a guideline for professionals and decision-makers to consider new measures to be implemented. Table 2 illustrates different policies and the main concern of actions to pursue waste management rooted in distinct hierarchical frameworks. Though the Food Recovery Hierarchy of EPA specifically treats reducing surplus produced food as the highest priority approach, many other hierarchies have a different approach. Some of them rely mainly on goals for action to handle excess food, seeing losses from harvesting as possible while giving little consideration to non-yields or less efficiency in agricultural operations. Garrone et al. [51] and Vandermeersch et al. [47] suggest hierarchies that focus on eliminating food waste and depletion, not minimizing or avoiding excess food generation. This strategy focuses on reducing post-harvest shortfalls and thereby considers the potential creation of food waste from the instant when food goods are ready for harvesting [53].

Rood et al. [54] emphasize various reuse and redistribution alternatives for human consumption. It suggests transforming discarded food into human food on a different level on the ladder. In such situations, the food products that are not eaten in their earliest form are turned into new palatable products. Only after all the possibilities of reusing for further human consumption is ruled out, the waste is reprocessed or used for animal feed. Interestingly, most of the frameworks are consistent with the solutions available in the lower sections of the waste hierarchy. They generally include food product recycling to non-edible alternatives and food recovery for energy production, followed by incineration and disposal. Major variations among hierarchies are observed only in the higher sections (Table 2).

**Table 2** Various food waste hierarchy (from Ciccullo et al. [53])

| Name of the framework | Hierarchy of food waste | References |
|---|---|---|
| Waste hierarchy | 1. Prevention<br>2. Preparation for reuse<br>3. Recycle | [48] |
| Food waste hierarchy | 4. Another recovery such as energy recovery<br>5. Disposal | [49] |
| Food waste management hierarchy | 1. Avoid<br>2. Reduce<br>3. Reuse<br>4. Recover<br>5. Treat | [55] |
| Food recovery hierarchy | 1. Source reduction<br>2. Feed hungry people<br>3. Feed animals<br>4. Industrial uses<br>5. Composting | [56] |
| Availability-surplus-recoverability-waste model (ASRW) | 1. Recover surplus food to feed humans<br>2. Recover surplus food to feed animals<br>3. Recovery of waste<br>4. Disposal of waste | [50] |
| Food waste management hierarchy | 1. Prevention<br>2. Conversion for human nutrition<br>3. Use of animal feed<br>4. Use as raw materials in the industry<br>5. Process into fertilizer<br>6. Use as a renewable energy<br>7. Incineration<br>8. Landfill | [57] |
| Moerman's Ladder | 1. Preventing food losses<br>2. Human food<br>3. Converted into human food (food processing)<br>4. Used in animal feed<br>5. Use as raw materials in industry<br>6. Process into fertilizer through fermentation<br>7. Process into fertilizer through composting<br>8. Applied for sustainable energy<br>9. Incineration | [54] |

# 7 Parameters of Sustainability in the Context of Circular Economics

Due to the global population explosion and limited resources, demand for sustainable supply chain management has grown significantly. As disruptions such as economic disasters, terrorist attacks, earthquakes, extreme climatic events, and pandemics have become more serious and recurrent, stability and resilience are becoming more important for the supply chain. Global companies face greater risk levels as delays greatly impair the supply chain's output [58]. Several methodologies to achieve sustainability in the food sector have been proposed. One of the preliminary steps in this initiative would be to conduct sustainability evaluations. Such an evaluation is intended to guide the decision-making processes, structure sustainability assessments, and promote food chain sustainability implementations. All of these would be accounted for with a triple bottom line consideration [59].

Various methods and tools exists for assessing sustainability of food processing technologies. The commonly recognized evaluation technique is the life cycle evaluation. This methodology performs the quantitative evaluation of utilized resources, environmental impacts, and energy flow of the systems, products, and services associated with the supply chain. ISO 14040 standard provides a framework for the life cycle assessment. It consists of four general stages: (a) describe the goal and scope of the assessment, (b) develop a lifecycle inventory, (c) conduct impact evaluation, and (d) analysis of the outcome. A complete cradle-to-grave life cycle assessment has to be considered at each stage in a food system, starting with raw material production, refining, packaging, distribution, use, and end-of-life management [5]. Subsequently, cradle-to-cradle assessments utilizing the end-of-life material to produce other products such as biodegradable packaging materials or animal feed have to be undertaken to estimate such operations' environmental benefits.

The circular economy is revived by nature and aims to preserve the greatest usefulness and value of goods, parts, and materials at all times. Performance economy, industrial agriculture, cradle-to-cradle architecture, commodity service networks, natural capitalism, industrial symbiosis, biomimicry, circular material flow, industrial ecosystems, eco-efficiency, regenerative design, zero emissions, and industrial ecology are some of the most significant components of the circular economy in the literature [31]. Some of the main sustainability measures are presented in Table 3.

# 8 Managerial Implications

According to the findings, managerial orientation has a direct impact on strategic sustainability orientation. As a result, the organization's policy measures to achieve environmental protection are dependent on top management thought and policy maker's thoughts. Customers expect goods and services that are delivered without harming the ecosystem. Supply chain managers can make decisions that promote

**Table 3** Sustainability parameters in the context of the circular economy [31]

| Sustainability parameters | Remarks |
| --- | --- |
| Resource circularity | The definition of cradle-to-cradle and biomimicry is based on resource circularity. This promotes the removal of waste |
| Cost-saving through product quality | Improved production cuts, scrap, shutdown, inspection, and testing prices, etc. Thus, by enhancing the efficiency of the systems, these costs can be reduced |
| Decreasing emission | Cutting emissions is one of the conditions for determining the success of viable operations |
| Waste reduction and pollution monitoring | Waste in different ways (waste products, packaging materials, wastewater, and gaseous emissions) should be minimized, and contamination for sustainability should be better controlled |
| Process design for resource and energy efficiency | The circular economy results in the greater utility of commodities, components, as it is restorative and regenerative by nature and capacity |
| Increasing profit from green products | The profit margin should be strong for consumer longevity for green goods. Reducing commodity prices contributes to a rise in demand for goods |
| Improving green logistics | Green logistics includes all green activities, such as green wrapping materials, low vehicle emissions and proper vehicle maintenance |
| Improved employees and community health | Global sustainability is concerned about that. Corporate social responsibility requires the proper care of workers and general welfare |
| Improving green purchasing | It means to purchase green materials, components, and subcomponents and services for producing the goods and services |

the incorporation and coordination of supply chain management processes in the supply chain [60]. As a result, sustainability principles will be effectively applied in the supply chain when management agrees that the trend towards sustainability and efficiency solutions will contribute to improved results and long-term competitive advantages [61]. Financial assistance, human resources, and relevant laws throughout the organization are all needed to implement environmental initiatives throughout the supply chain, coordinated by management and implemented across the enterprise.

# 9   Conclusion

Production of enough food and its appropriate distribution while minimizing its wastage is a major challenge of the food industry that can be correlated to the rising population [62]. This chapter aimed to look into the major challenges faced in the food industry to adopt a sustainable supply chain and, more specifically, see how these major challenges interacted with one another and the relationships between them. The main findings of this study are that FLW occurs at all stages of the supply chain and faces several stage-dependent causes. Based on the so-called take-make-dispose system, the unsustainability of the present economic model has arisen over the past decade. In particular, issues such as resource shortages and FLW production along the supply chain have seriously affected the agro-food sector. Climate change and the depletion of biodiversity helped identify an imperative shift in the circular economy's mindset. Recently, scientific research investigating the implementation of Circular Economic models and tools has expanded into publishing Sustainable Development Goals. The importance of moving toward a circular economy has become vital in this sense. A systematic literature review was undertaken in this chapter to analyze state-of-the-art research related to the implementation of circular economy models and instruments along the supply chain of agro-food. This analysis also illustrates that it is almost unrealistic, due to the Agri-Food supply chain's complexities, to describe a single model of the circular economy for the whole industry. Moreover, it emerges that future studies should focus on combining the various stages of the supply chain with models of the circular economy and tools for the development of a closed-loop Agri-Food system.

# References

1. Desa U (2015) World population projected to reach 9.7 billion by 2050. UN DESA United Nations Department of Economic and Social Affairs: New York, NY, USA. Retrieved from https://www.un.org/en/development/desa/news/population/2015-report.html, 29 July 2015
2. Mahroof K, Omar A, Rana N, Sivarajah U, Weerakkody V (2020) Drone as a Service (DaaS) in promoting cleaner agricultural production and circular economy for ethical sustainable supply chain development. J Clean Prod:125522
3. Food and Agriculture Organization of the United Nations (2011a) Food wastage footprint and climate change. Retrieved from http://www.fao.org/3/a-bb144e.pdf
4. Food and Agriculture Organization of the United Nations (2011b) Global food losses and food waste–extent, causes and prevention. SAVE FOOD: an initiative on food loss and waste reduction
5. Ojha S, Bußler S, Schlüter OK (2020) Food waste valorisation and circular economy concepts in insect production and processing. Waste Manage 118:600–609
6. Kennard NJ (2019) Food waste management. Zero Hunger. Encyclopedia of the UN sustainable development goals. Springer, Cham
7. Food and Agriculture Organization of the United Nations (2019) Food systems at risk: new trends and challenges. http://www.fao.org/3/ca5724en/CA5724EN.pdf
8. Edjabou ME, Petersen C, Scheutz C, Astrup TF (2016) Food waste from Danish households: generation and composition. Waste Manage 52:256–268
9. Teigiserova DA, Hamelin L, Thomsen M (2020) Towards transparent valorization of food surplus, waste and loss: clarifying definitions, food waste hierarchy, and role in the circular economy. Sci Total Environ 706:136033
10. Vilariño MV, Franco C, Quarrington C (2017) Food loss and waste reduction as an integral part of a circular economy. Front Environ Sci 5:21

11. Senker P (2011) Foresight: the future of food and farming, final project report. Prometheus 29(3):309–313
12. Mekonnen MM, Gerbens-Leenes W (2020) The water footprint of global food production. Water 12(10):2696
13. Blakeney M (2019) Food loss and food waste: causes and solutions. Edward Elgar Publishing, Cheltenham, UK. https://doi.org/10.4337/9781788975391
14. Mateo-Sagasta J, Zadeh SM, Turral H, Burke J (2017) Water pollution from agriculture: a global review. The Food and Agricultural Organization
15. Neven D (2014) Developing sustainable food value chains guiding principles. Food and Agriculture Organization of the United Nations. http://www.fao.org/3/a-i3953e.pdf
16. Derqui B, Fayos T, Fernandez V (2016) Towards a more sustainable food supply chain: opening up invisible waste in food service. Sustainability 8(7):693
17. Cattaneo A, Federighi G, Vaz S (2020) The environmental impact of reducing food loss and waste: a critical assessment. Food Policy:101890
18. Pradhan P, Reusser DE, Kropp JP (2013) Embodied greenhouse gas emissions in diets. PloS One 8(5):e62228
19. WWF (2017) Food loss and waste: facts and futures taking steps towards a more sustainable food future. World Wide Fund 32. Retrieved from www.d4d.co.za
20. Porter SD, Reay DS, Higgins P, Bomberg E (2016) A half-century of production-phase greenhouse gas emissions from food loss and waste in the global food supply chain. Sci Total Environ 571:721–729
21. Ramukhwatho FR, Du Plessis R, Oelofse S (2016) Household food wastage by income level : a case study of five areas in the city of Tshwane Metropolitan Municipality, Gauteng Province, South Africa. In: Proceedings of the 23rd WasteCon conference 17–21 Oct 2016, Emperors Palace, Johannesburg, South Africa, pp 57–64
22. Choudhary A, Kumar N (2017) Environmental impact of non-vegetarian diet: an overview. Int J Eng Technol Sci Res 6:251–257
23. Otles S, Despoudi S, Bucatariu C, Kartal C (2015) Food waste management, valorization, and sustainability in the food industry. In: Food waste recovery. Academic Press, pp 3–23
24. Despoudi S (2020) Challenges in reducing food losses at producers' level: the case of Greek agricultural supply chain producers. Ind Marketing Manage
25. WRAP (2018) WRAP and the circular economy. https://www.wrap.org.uk/about%E2%80%90us/about/wrap%E2%80%90and%E2%80%90circular%E2%80%90economy
26. Murray A, Skene K, Haynes K (2017) The circular economy: an interdisciplinary exploration of the concept and application in a global context. J Bus Ethics 140(3):369–380
27. Giunipero LC, Hooker RE, Denslow D (2012) Purchasing and supply management sustainability: drivers and barriers. J Purch Supply Manage 18(4):258–269
28. Baig SA, Abrar M, Batool A, Hashim M, Shabbir R (2020) Barriers to the adoption of sustainable supply chain management practices: moderating role of firm size. Cogent Bus Manage 7(1):1841525
29. Tay MY, Abd Rahman A, Aziz YA, Sidek S (2015) A review on drivers and barriers towards sustainable supply chain practices. Int J Soc Sci Human 5(10):892
30. Jarzębowski S, Bourlakis M, Bezat-Jarzębowska A (2020) Short food supply chains (SFSC) as local and sustainable systems. Sustainability 12(11):4715
31. Kumar P, Singh RK, Kumar V (2021) Managing supply chains for sustainable operations in the era of industry 4.0 and circular economy: analysis of barriers. Resour Conserv Recycl 164:105215
32. Chkanikova O, Mont O (2015) Corporate supply chain responsibility: drivers and barriers for sustainable food retailing. Corp Soc Responsib Environ Manage 22(2):65–82
33. Zheng T, Wang B, Rajaeifar MA, Heidrich O, Zheng J, Liang Y, Zhang H (2020) How government policies can make waste cooking oil-to-biodiesel supply chains more efficient and sustainable. J Clean Prod 263:121494
34. FSSAI (2021, Jan 30) Food safety and standards authority of India. RUCO. Retrieved from https://fssai.gov.in/ruco/index.php

35. Stahel WR, Reday-Mulvey G (1981) Jobs for tomorrow: the potential for substituting manpower for energy. Vantage Press
36. Ghisellini P, Cialani C, Ulgiati S (2016) A review on circular economy: the expected transition to a balanced interplay of environmental and economic systems. J Clean Prod 114:11–32
37. Alkhuzaim L, Zhu Q, Sarkis J (2021) Evaluating emergy analysis at the nexus of circular economy and sustainable supply chain management. Sustain Prod Consump 25:413–424
38. Liu J, Feng Y, Zhu Q, Sarkis J (2018) Green supply chain management and the circular economy. Int J Phys Distrib Logist Manag 48(8):794–817
39. Food and Agriculture Organization of the United Nations (2013) Food wastage footprint and its effects on natural resources: summary report. Retrieved from http://www.fao.org/docrep/018/i3347e/i3347e.pdf.
40. Wunderlich SM, Martinez NM (2018) Conserving natural resources through food loss reduction: production and consumption stages of the food supply chain. Int Soil Water Conserv Res 6(4):331–339
41. United Nations (2018) Goal 2: end hunger, achieve food security and improved nutrition and promote sustainable agriculture. Retrieved form https://www.un.org/sustainabledevelopment/hunger/
42. De Hooge IE, van Dulm E, van Trijp HC (2018) Cosmetic specifications in the food waste issue: supply chain considerations and practices concerning suboptimal food products. J Clean Prod 183:698–709
43. De Hooge IE, Oostindjer M, Aschemann-Witzel J, Normann A, Loose SM, Almli VL (2017) This apple is too ugly for me!: consumer preferences for suboptimal food products in the supermarket and at home. Food Qual Prefer 56:80–92
44. Flanagan K, Clowes A, Lipinski B, Goodwin L, Swannell R (2018) SDG target 12.3 on food loss and waste: 2018 progress report. Ann Update Behalf Champions 12
45. Milios L (2018) Advancing to a circular economy: three essential ingredients for a comprehensive policy mix. Sustain Sci 13(3):861–878
46. Jurgilevich A, Birge T, Kentala-Lehtonen J, Korhonen-Kurki K, Pietikäinen J, Saikku L, Schösler H (2016) Transition towards circular economy in the food system. Sustainability 8(1):69
47. Geueke B, Groh K, Muncke J (2018) Food packaging in the circular economy: overview of chemical safety aspects for commonly used materials. J Clean Prod 193:491–505
48. European Commission Directive 2008/98/EC of the European Parliament and of the Council (2008) Retrieved from https://eur-lex.europa.eu/legal-content/EN/TXT/?uri=celex%3A32008L0098
49. Papargyropoulou E, Lozano R, Steinberger JK, Wright N, Bin Ujang Z (2014) The food waste hierarchy as a framework for the management of food surplus and food waste. J Clean Prod 76:106–115
50. Garrone P, Melacini M, Perego A (2014) Opening the black box of food waste reduction. Food Policy 46:129–139
51. Garrone P, Melacini M, Perego A, Sert S (2016) Reducing food waste in food manufacturing companies. J Clean Prod 137:1076–1085
52. Environmental Protection Agency (EPA) (2018) Food recovery challenge (FRC). Retrieved from https://www.epa.gov/sustainable-management-food/food-recovery-challenge-frc.
53. Ciccullo F, Cagliano R, Bartezzaghi G, Perego A (2021) Implementing the circular economy paradigm in the agri-food supply chain: the role of food waste prevention technologies. Resour Conserv Recycl 164:105114
54. Rood T, Muilwijk H, Westhoek H (2017) Food for the circular economy: policy brief. PBL Netherlands Environmental Assessment Agency
55. Kosseva MR (2009) Processing of food wastes. Adv Food Nutr Res 58:57–136
56. United States Environmental Protection Agency (EPA) (2012) Food recovery hierarchy. Retrieved from https://www.epa.gov/sustainable-management-food/food-recovery-hierarchy
57. Vandermeersch T, Alvarenga RAF, Ragaert P, Dewulf J (2014) Environmental sustainability assessment of food waste valorization options. Resour Conserv Recycl 87:57–64

58. Bui TD, Tsai FM, Tseng ML, Tan RR, Yu KDS, Lim MK (2021) Sustainable supply chain management towards disruption and organizational ambidexterity: a data driven analysis. Sustain Prod Consump 26:373–410
59. Yontar E, Ersöz S (2020) Investigation of food supply chain sustainability performance for Turkey food sector. Front Sustain Food Syst 4:68
60. Emamisaleh K, Rahmani K (2017) Sustainable supply chain in food industries: drivers and strategic sustainability orientation. Cogent Bus Manage 4(1):1345296
61. Sen S (2009). Linking green supply chain management and shareholder value creation. IUP J Supply Chain Manage 6
62. Stella D (2019) Optimized food supply chains to reduce food losses. In: Saving food. Academic Press, pp 227–248

# Rethinking the Physical Losses Definition in Agri-Food Chains from Eco-Efficiency to Circular Economy

**Ricardo Alberto Cravero**⊙**, María de las Mercedes Capobianco-Uriarte**⊙**, and María del Pilar Casado-Belmonte**⊙

**Abstract** This study analyzes different expressions for Physical Losses and proposes an integration of concepts to approach the economic-accounting language of profits and losses to the Agri-Food chain context. Methodologically, a literature review identified the scientific and institutional documents that use the Agri-Food Physical Losses definitions and a content analysis was adopted. The main contributions of the concept come from institutions such as the Food and Agriculture Organization, the United Nations Environmental Programme, the European Commission, among others. In this context, the perspectives identified are Food Loss and Waste, Eco-Efficiency/Cleaner Production, Circular Economy, Lean Thinking and Accounting systems. The main contribution of this study is to offer a definition of Physical Losses in an Agri-Food chain context so as to help Top Management to make decisions regarding the reduction, elimination or transformation of Physical Losses and the achievement of a more sustainable business strategy.

**Keywords** Food waste · Cleaner production · Circular economy · Accounting system · Lean thinking

## 1 Introduction

The interest in achieving sustainable development has experienced several milestones mainly stemming from the endeavours of the United Nations (UN). From 1987

R. A. Cravero
National Technological University - Qinnova, Santa Fe, Argentina
e-mail: ricardo@qinnova.com.ar

M. de M. Capobianco-Uriarte (✉) · M. P. Casado-Belmonte
Department of Economics and Business, University of Almería, Almería, Spain
e-mail: mercedescapobianco@ual.es

M. P. Casado-Belmonte
e-mail: mbelmont@ual.es

93

onwards with the "Report of the World Commission on Environment and Development: Our Common Future" [5], there have been a number of proposals. In 1992, the UN Conference on Environment and Development in Rio de Janeiro established the "Agenda 21", which was aimed to reach the commitment for sustainable development. Later, in 2002, the "Report of the World Summit on Sustainable Development", held in Johannesburg, contributed to the adaptation of the development goals, including the ones in UN Millennium Declaration and the implementatiton of "Agenda 21 and its review". When it comes to clime change policy, the Copenhagen Climate Change Conference in 2009 was a milestone at political level. Subsequently, all UN Member States approved the 2030 Agenda for sustainable development in 2015. As a result, the 17 Sustainable Development Goals (SDG) stood for peace as well as prosperity of people and the planet. Thus, SDGs are deemed to be a way to raise awareness of all developed and developing countries in the overarching framework of sustainable development, where the economic growth must go hand to hand with global plans to end poverty, access to education, improve health and increase equality. In 2019, the UN Climate Action Summit, held at the UN New York City, focused on the struggle to preserve the environment, namely oceans and forests. The primary summit's goal was to take further climate action to decrease greenhouse gas emissions so as to avoid rising the mean global temperature by more than 2.7 °F above pre-industrial levels.

In this context, the Food and Agriculture Organization (FAO) [19] raised food waste awareness in the food chain to overcome hunger in the most-needy countries. As was stated, one-third of food produced around the world for human consumption was thrown away. By this initiative, an attempt was made to promote one of the SDGs, namely the 12.3 goal, by the "Global Initiative on Food Loss and Waste Reduction—Save Food".

Several world organizations and researchers endorsed these proposals, applying various approaches connected to different perspectives. For instance, Cleaner Production from the UN Environment Programme (UNEP) and Eco-Efficiency from the World Business Council for Sustainable Development (WBCSD 1998), optimizing resources and reducing waste at supply chains enterprises. The Triple Bottom Line (TBL) approach founded in the 1990s by Elkington, tried to emphasize not only the economic perspective of a firm, but also the social and environmental impact. Later, [13] heralded the Triple Bottom Line approach as a tool to influence reporting platforms such as the Global Reporting Initiative (GRI) and the Dow Jones Sustainability Indexes. Accordingly, Triple Bottom Line may impact the accounting information system, stakeholder engagement and the strategy of the firm. What is more, Triple Bottom Line should be understood as an influencer in the future of capitalism and not as an accounting tool [13].

Another approach is the Food Losses and Waste Accounting and Reporting Standard (FLW Standard), launched in 2016 [26], which highlighted the relevance of the measurement of food loss and waste since it enables the development of reduction strategies. Subsequently, [61] analyses business sustainability movements, from the green economy [66], the blue economy [50], industrial ecology [21], the sharing economy [39] to the Circular Economy (CE) [51]. This last approach, the Circular

Economy, is supported by many schools of thought, being regarded as "an industrial system that is restorative or regenerative by intention and design. It replaces the 'end-of-life' concept with restoration, shifts towards the use of renewable energy, eliminates the use of toxic chemicals, which impair reuse, and aims for the elimination of waste through the superior design of materials, products, systems, and, within this, business models" [41]. In this regard, the Waste and Resources Action Programme (WRAP) [75] promotes more CE economic growth potential. Recently, the new Circular Economy Action Plan for a cleaner and more competitive Europe highlights that the European Union (EU) must not provide the European Green Deal's ambition for a environment-neutral, efficiency in resources and Circular Economy. Accordingly, the plan includes the action throughout the life cycle of products. In such a way, it implies the product's design, processes based on Circular Economy, promotion of sustainable consumption and the commitment that resources used to stay in the EU economy for long time [15]. What is more, in the plan, EU is said to take over the way to a Circular Economy and the implementation of the SDGs of 2030 Agenda for sustainable development.

To sum up, sustainable development is promoted by the UN, through several call for action proposals. Many organizations are working in the battle against "hunger", through raising awareness about Food Waste, such as Food Losses and WasteStandard, in different countries. Complementary to this, other proposals such as Cleaner Production and Eco-Efficiency or the Triple Bottom Line have proposed reducing pollution and improving the efficient use of resources in the value chain and in enterprises, encouraging the inclusion in financial results of the 3P (Profit, People and Planet). Later, the Circular Economy introduces a cyclic model, like nature does, promoting a cradle-to-cradle design, where "waste can be considered as part of a new cycle".

The launch of Agenda 2030 [63] spurs firms to face sustainability issues at the management level while growing economically and taking into account stakeholder's expectations, requiring highly integrated sustainable management practices and tools [69]. In this context, sustainability and supply chain transparency concepts are more critical to the Agri-Food sector. Sustainability performance goals focus on efficiency and how to balance ecological, economic, and social aspects of the Agri-Food business. Accordingly, business managers have a new demand from the stakeholders [43], namely the way to plan and coordinate value chain innovations with significant impact on the Agri-Food system, to capture returns from initiatives. Ross et al. [52] propose a solution for this trade-off by adopting innovations that align participants' incentives in the value chain by creating shared value. In the Agri-Food value chain, [35], emphasize the current importance of food waste management, based on their impact on costs, environment and food security, proposing a value chain approach to find profitable opportunities from promoting overall efficiency and reducing wastage.

Thus, in an increasingly competitive world, there are several ways to generate sustainable value in organizations. In most of them, Top Management (TM) is the critical factor in strategic decision-making [76] that could lead to competitive sustainability [59, 68] and assure the optimal use of resources throughout the value chain

[36, 58]. In such a way, the decision-making process may be improved through the use of an Environmental Management Accounting system [38].

This study aims to analyze the different perspectives of expressing Physical Losses (PL) of those resources and proposes the integration of concepts. The importance lies in expressing these Physical Losses in a way that approaches the economic-accounting language of profit and loss. Thus, it enables the visualization of information for Top Management to promote its awareness and actions as well as those of the stakeholders, in line with the work proposed by the UNEP, through the 17 SDGs.

**Chapter Structure**

First, a brief state of the art is presented. Then, we explore the various sources of Physical Loss definitions published in scientific documents and reported by public and private inter/intra-national organizations. The five perspectives detected for the different definitions of Physical Losses identified (Food Waste, Eco-Efficiency and Cleaner Production, Lean Thinking, Circular Economy and Accounting System) are described, and finally summarised through a timeline. Subsequently, Top Management's decision-making tools to identify Physical Losses and apply actions in this regard are set out. Afterwards, a proposal for the definition of Physical Losses in the context of Agri-Food value chains is presented. In conclusion, a summary of the research's central ideas and future research lines are presented.

## 2 Physical Losses Review Literature and Sources

A brief state of the art is presented, exploring the definitions used in indexed scientific literature and their sources. Definitions used in academic research were analyzed bibliometrically using content analysis through keywords used by researchers. Most of the origins of the various PL definitions are from different international organizations, both public and private.

### 2.1 Bibliometric Search of Definitions

A bibliometric analysis is conducted in order to identify the critical elements in the PL research field, showing relevant information regarding keywords [7]. As in

previous bibliometric studies [60], this study undertakes five sections: (1) description
of the research field, (2) selection of the database, (3) research criteria adjustment, (4)
codification of retrieved material, and (5) analysis of the information. This research
identified a central focus on the relationship between different concepts involved with
definitions of PL and value chains in the Agri-Food sector. The results may differ
due to the database selected, thus in this research, two data sources are employed,
Web of Science (WoS) and Scopus since they are thought to be widely used [1].
In order to get an accurate search and enable the extraction of large data, Boolean
operators were established. The research formula includes three different blocks.
Block 1 brings together the main concepts related to PL, the sectorial context is in
Block 2. Finally, Block 3 considers the main topics referring to the different PL used
in the Agri-Food context (Fig. 1). The search was limited to 1992–2020, with the
limit at the end of December 2020.

Concerning the criteria for inclusion and exclusion, only articles, reviews, books
and book chapters were considered, the number of documents from WoS were 649
and from Scopus 525. However, there were 342 common to both databases, and the
final sample consisted of 1516 documents. The codification of recovered data was
made by Scimat and processed by Vosviewer to get keyword networks. Therefore,
to examine the research field and analyze the most important aspects, we conducted
scientific mapping based on the co-word analysis. Moreover, the keywords of an
article reflect its primary content. Their occurrence and co-occurrence frequency
represent the most relevant topics addressed by studies in a research area and how they
are connected to each other [2, 77]. The sum of all common co-occurrences between

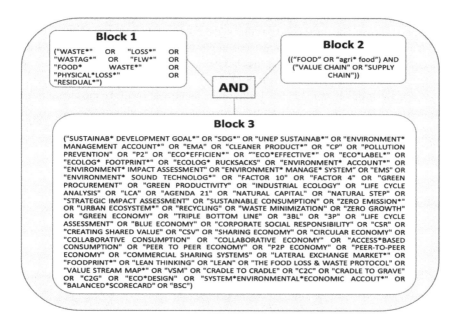

**Fig. 1** Research formula used for search in scientific literature. *Source* Author

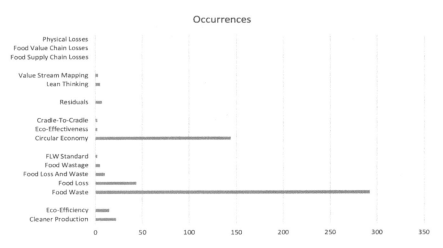

**Fig. 2** Main keywords used in the scientific literature. *Source* Author

keywords permits a network map of the co-occurrence matrix to be established, which enables the recognition of several thematic clusters. Finally, maps are produced for different periods to trace the field's conceptual space [9] and future research paths.

The keyword occurrences (Block 2 and Block 3) are analyzed together with the five perspectives detected for the different physical loss definitions identified (Food Waste, Eco-efficiency and Cleaner Production, Lean Thinking, CE and Accountant System)—developed in the third section of this study. Figure 2 shows the most recurrent concepts in the scientific literature analyzed through the search formula mentioned above. The three concepts that stand out from the rest in the last 30 years and the most used as keywords in research are *food waste*, *CE* and *food losses*. Both the keywords food waste and food loss belong to the first perspective presented in Sect. 3.1, while the keyword CE is in Sect. 3.4.

The analysis of keyword tendency allows the identification of the most employed keywords in last years as well as the demanding opportunities. The overlay of keywords tendency is exhibited in Fig. 3, being the emergent topics in yellow. As stated earlier, the network of *CE* stands as a core concept in this research. Moreover, the *SDGs* are a guide to reducing *food loss and waste*.

**Bibliometric Search**

The academic environment is generally not a source of definitions applied to PL, notwithstanding the studies by [6, 28]. Instead, most researchers use

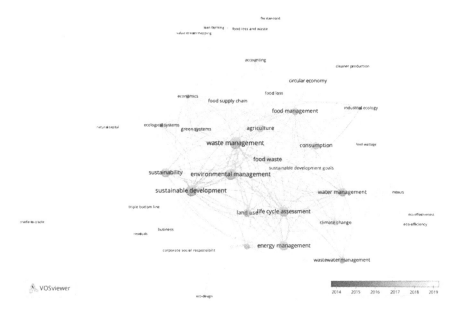

**Fig. 3** Temporal evolution of keywords network. *Source* Author

definitions generated at the governmental or supra governmental level in organizations intended to guide policy-makers in making decisions.

## 2.2 Public, Private and Mixed Organisations Involved

The scientific literature, analyzed in 2.1., was focused on concept applications more than defining food waste, food loss, food wastage or other options. Exploring the leading organizations (institutions) included in the literature search, Fig. 4 shows a matrix indicating the deployed actions. The institution matrix exhibits the organizations that contribute to the definition of physical losses in this study, differentiated at world and country level.

The world level brings together organizations working internationally, ranging from public to private, and some of them are mixed. Some organizations, mentioned at Public Country Level, are recognized in the global arena. Finally, at Country Level, Mix or Private, there are mostly locally recognized NGOs (commerce chambers, food banks, consumer organizations, among others).

**Fig. 4** Matrix of institutions and deploying level. *Source* Authors

## 3 Definition Perspectives

Regarding any process, isolated or as part of the chain value, different kinds of losses usually appear. In literature, there are many ways to express these tangible "losses", namely waste, loss, wastage, inefficiencies and residuals, unlike other intangible losses such as variation in value. For legal, economic and practical uses, countries may define a specific meaning. The WRAP [71], in accordance with the EU, highlighted the legal definition of waste as "any substance or object which the holder discards, intends to discard or is required to discard". In their analysis, they added other waste as packaging and sludge.

This study focuses on different perspectives to understand the meaning of PL in the Agri-Food chain based on sustainable development. Other references to the food supply chain and food value chain can be found in several documents. On the one hand, the food value chain is understood as a food product's movement along the supply chain, identifying the actors and their value-added activities. On the other hand, the food supply chain is the process of how food ends up on our plate [17]. In this study, both concepts are included in the Agri-Food chain.

### 3.1 Food Waste Perspective

Many organizations are involved in "food waste" perspective for Agri-Food value chains, such as FAO, EU, Food Use for Social Innovation by Optimising Waste

Prevention Strategies (FUSIONS), the WRAP, the World Resources Institute (WRI), the WBCSD, among others. They are fundamentally concerned with the primary objective "Hunger Zero" (SDG 2). Thus, they tried to provide a precise concept definition for "food waste". FAO differentiates between food losses and food waste in the value chain since they are some slight differences between the documents they produced.

The definitions of PL regarding Food Waste are analyzed with respect to their impact on poverty and hunger in the world [25] as well as their effect on water, land and fertilisers [37]. Following, the main definitions of PL in the Agri-Food chain are given.

### Food loss

- Within the Food Wastage Footprint [18], FAO explains it as a reduction in mass (dry matter) or nutritional value (quality) of food, firstly proposed for human consumption. The reasons for these losses are inefficiencies in the food suppy chain such as inapropriate infrastructure, lack of skills and knowledge on supply chain actors, difficulty in access to markets or insufficient management regarding supply chain.
- From "Think.Eat.Save" [65], when food is spoilt, spilled or lost is regarded as food. In this vein, the stages of production, post-harvest, processing and distribution may imply food losses during the food supply chain with the ensuing decrease of value and quality. The similar definition takes in a collaborative programme called "Creating a sustainable food future", from the WRI [40].
- Buzby et al. [6] define Food loss as the amount food that can be eaten as well as post-harvest available for human consumption albeit not used. It refers to cooking loss and natural reductions such as losses derived from moisture, mould, pests, inappropriate climate control, plate waste or others.

### Food waste

- According to FAO [18], food waste means food which is suitable for human consumption, but it has been discarded, regardless of its expiry date. Often this is because food has spoiled due to oversupply by the markets or individual habits of shopping or eating.
- "Think.Eat.Save" [65] uses food waste as the food that ends the food supply chain, being of good quality and suitable for human consumption albeit not consumed due to the fact that it is removed, regardless of expiry date. Moeover, food waste is thought to appear at retail and consumption stages in the food supply chain. The WRI also applies this approach, although it adds negligence or the decision to discard food as the origin of food waste.
- FUSIONS [20] tried to provide a common definition for food waste. In this regard, the food recovered from food supply chain suitable or not for human consumption is regarded as food waste. Specifically, composted, crops cultivated in/not harvested, anaerobic digestion, bioenergy production, co-generation, incineration, removal to sewer, landfill, or expulsion at sea.

- WRAP [74] introduced another point of view for Food waste "is any food that had the potential to be eaten, together with any unavoidable waste, which is lost from the human food supply chain, at any point along that chain". As their intention is to reduce waste at home [73], they differentiate between: *Avoidable* which refers to food as well as drink discarded that was previously suitable for eating (e.g. slices of bread, apples, meat). *Possibly avoidable* refers to food and drink that could be eaten by some people but not for others (e.g. bread crusts), or that can be consumed depending on the way is prepared (e.g. potato skins). *Unavoidable* refers to waste derived from food or drink elaboration that is not fit for consumption under normal circumstances (e.g. meat bones, eggshells, tea bags).

**Food wastage** means any food missed by deterioration or waste. Therefore, the word "wastage" includes both food loss as well as waste [18].

**Food quality loss or waste** refers to the reduction of a quality characteristic of food (nutrition, aspect, among others), connected to the degradation of the product, during all phases of the food chain from harvest to consumption [62].

**Food loss and waste (FLW)**

- Refers to the parts of plants and animals that can be eaten and are produced or harvested for human use although not ultimately consumed by people [40].
- Food loss and waste means the reduction in mass (quantitative) or nutritional value (qualitative) of food—parts fit to be eaten—along the supply chain that was aimed to human consumption [65]. Nevertheless, for several reasons, instead of using for human chain, food loss and waste are discarded from human consumption and employed in a non-food use (animal feed, bio-energy).

**Food Chain Inefficiencies** give a new concept to food loss and waste by the conversion efficiency used in the whole system of the Agri-Food supply chain [28].

**Food Supply Chain Losses** as "the total losses and waste within the different steps of the Food Supply Chain (production, post-harvest, processing, distribution, and consumption)". It includes natural resources in food crops [37].

**Agri-Food Value Chain Losses** can be interpreted from the perspective of CORDIS EC [11] when evaluating the food lost and wasted in the Agri-Food value chain. FAO has a similar perspective [17].

All these efforts to find the right meaning for Food waste and related were analyzed in various studies. Buzby et al. [6], mentioning that definitions of food waste and food loss differ worldwide, found some food waste definitions, including inedible portions. FUSIONS [20] support this variation, showing the tendency by research community and policy-makers to distinguish expressions. For instance, "food waste", "food loss", "avoidable food waste", "unavoidable food waste" or "potentially avoidable food waste". Besides, the definition of food waste has an influence on the elaboration of policies and on the assessment of food waste across the different sectors of the food supply chain. Östergren and Holtz [49] studied the application of different definitions. Although the authors detected several perspectives, such as FAO on food supply (edible food), EU on resource efficiency, and recognized the inedible parts of food as a resource, they concluded that it was not feasible to agree on a universal definition of food waste. The main reason is likely to be because of the conflicting interests in each institution. Garcia-Garcia [22] put the existence of multiple terminologies down to there being an increasing interest in food waste matter. He also identified that there is no global consensus of the definition of food waste.

**Food Waste Perspective**

All these organizations and researchers mentioned above, focus primarily on food waste to beat hunger, optimizing natural resources mainly in crops. Some mention packaging and sludge waste. Probably these approaches are more guided to policy-makers than to TM as decision-makers.

## 3.2 Eco-Efficiency and Cleaner Production Perspective

A complementary focus comes from UNEP and WBCSD. Both organizations state that Eco-efficiency and Cleaner Production complement each other on the road to sustainable development [70]. On the one hand, Cleaner Production is understood as the adaptation of a preventive environmental strategy to products, processes, as well as services. Accordingly, waste, pollution and risks to human health are minimized by more efficient use of natural resources, avoiding the end-of-pipe approach. On the other hand, Eco-efficiency is thought to be achieved by producing competitive goods and services that cover human needs, trying to reduce ecological impacts and intensity of resources according to the earth's estimated capacity. Thus, three issues arise from the concept: economic advancements, more cost-effective use of resources and deterrence of emissions. Van Berkel [67] summarizes these concepts from the Australian experience. In their case, initially planned for industry, the strategy is based on optimizing entire resources (FLW, natural resources, packaging, other supplies, waste disposal, emissions). In this regard, importance is given to the way to communicate, under an approach that includes: business improvement opportunity, reduction of costs, risks and liabilities, and promotion of efficiency, productivity and profitability.

In this regard, the International Organization for Standardization (ISO) launched ISO 14040/41/42/43/44 standards for Life Cycle Assessment (LCA). Later, ISO 14045 [31] explains the standards, conditions and guidelines for Eco-efficiency assessment of product systems—a quantitative management tool. Energy management systems are established by ISO 50001:2011. Complementary, ISO 14006 for Eco-Design is another tool for sustainable development. ISO has other standards from the 14000 series.

Great support comes from LCA, a systematic set of processes for collecting and analyzing the inputs and outputs of materials and energy as well as the ensuing environmental influences directly attributable to a product or service system's operating throughout its life cycle. This tool improves the environmental performance of goods and services, involving products that belong to the Agri-Food sector [3]. LCA entails cradle-to-grave explorations of production systems and offers comprehensive estimations of all upstream and downstream energy inputs as well as multimedia environmental emissions.

Similarly, the European Community established the MAESTRI project (Total Resource and Energy Efficiency Management System for Process Industries) [44]. This project was aimed to develop a platform on which the European manufacturing and process industries had a sustainability management system, the Total Efficiency Framework, to guide and simplify any innovative approach. Also, to contribute to the promotion of a culture of improvement within process industries in the decision-making process, develop improvement strategies, and identify the main concerns to enhance the company's environmental and economic performance. This framework has several perspectives: an overall efficiency performance evaluation from environmental, energy and resource efficiency, value, and cost. The integration includes several models with a life cycle perspective.

Benoit et al. [4] reviewed the Eco-efficiency concept and developed the dairy case using the life cycle perspective (a key component in the eco-efficiency concept). In the same line and as an example, Fig. 5 proposes a simplified model of inputs-

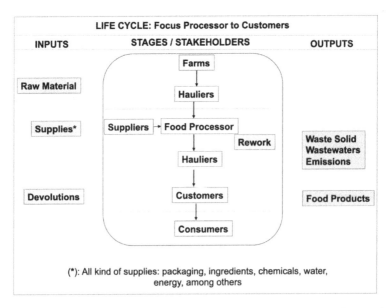

**Fig. 5** Simplified model life cycle in Agri-Food chain, Inputs-Stages/Stakeholders-Outputs. *Source* Authors

stages/stakeholders-outputs in a generic Agri-Food chain, showing the physical components from the incoming raw material until the products arrive at the customers.

**Eco-efficiency and Cleaner Production Perspective**

This approach explains other components of PL, namely supplies (of any kind, such as packaging, energy, water, ingredients, cleaning solutions), wastewaters, emissions, waste solid, rework and others. Also, at this point, strategies can be understood to apply in the supply chains and organizations to reduce inefficiencies, being useful for the decision-making process, guiding TM to be aware and proactive.

## 3.3 Lean Thinking Perspective

Taking into account the productivity and efficiency improvements, Lean production principles and tools are regarded to play a key role in reinforcing the competitive progress in organizations [16]. In particular, Value Stream Mapping (VSM) allows companies to detect value-added activities and the current value of waste in the production system, introducing a culture of continuous improvement. Therefore, VSM stands as a tool that pictures the processes and may support track waste. There are three types of processes: (1) the ones that create value, (2) those which do not create value but that cannot be eradicated in the current situation and, (3) the processes that do not create value and are potential to be reduced [72].

Mor et al. [47] reviewed scientific literature about lean production, finding the importance of lean principles to reduce waste and enhance the business value. Based on practical evidence, Garza-Reyes et al. [23] explored the application of VSM as a way to improve the environmental sustainability performance of operations and stated that the deployment of lean methods in environmental performance could encourage manufacturing companies to bear in mind the business benefits from a green lean approach in their operations. These authors found studies where these lean tools, such as VSM, contributed to addressing the sustainability challenges from requirements of customers for more environmentally friendly products/services, in accordance with governmental environmental regulations and the TBL framework. The contributions of Garza-Reyes et al. [24] are of particular relevance for manufacturing managers and their firms who endeavour to improve their operations' green performance by applying lean principles, tools, techniques and key performance indicators (KPI).

In Agri-Food, [57], explored synergies between the TBL, lean thinking and food supply chains. In this regard, it is possible to summarise Lean Thinking in food supply chains, as a Lean way of thinking that allows companies to add value. This assertion leads to the five principles of lean thinking: Value, Value Stream, Flow, Pull and Perfection. Lean Thinking is underlined as a sustainable development strategy for food Agri-Food supply chains. De Steur [10], reviewed documents about the application of VSM in the food losses and waste in supply chains. They defined "food supply chain losses" in the same way as [37]: "Food losses" and "Food wastes", or as "FLW". They found many studies about the lean tools applied in the Agri-Food context, one of the most highly rated being VSM. They were able to identify lean waste as: unnecessary inventory, defect in the product, inappropriate processing, overproduction. The lean application in the Agri-Food chain is considered under development, and the VSM is of great value since it improves the visibility of the whole value stream.

**Lean Thinking Perspective**

This point of view introduces other aspects of food waste but mainly presents a tool similar to VSM, which can be adapted to any sectorial chain or organization's needs. VSM can apply the KPIs along the lines of the TBL (3P) approach and adapted for different decision levels (operative/tactical/strategic).

## 3.4　Circular Economy Perspective

The existing food system is based on an inefficient model where opportunities are lost and negative impacts on social and environmental aspects appear. In this regard, there is a call for new principles and mechanisms derived from a system providing all food supplies in nature. The CE consciously follows these processes, where waste is used as feedstock in the next step in the cycle. The bio-cycle of the CE implies the gradually breakdown of oganic matter, free of toxic contaminants, and the fall by different value-extracting phases, prior to reverting safely to the soil [32].

From the Ellen MacArthur Foundation [41], the main CE principles are designs based on free waste and pollution, the preservation of products and materials in use and the regeneration of natural systems. A very important support comes from [46], promoting the cradle-to-cradle design and presenting a model that mirrors nature's cycle in which "waste equals food". In such a way, *eco-effectiveness* is prioritized over *eco-efficiency*.

Research has uncovered several conceptual contributions to CE. Kirchherr et al. [34] state that CE is an economic system where the end-of-life concept is replaced by reduction, reusing, recycling and recovering materials from different processes, namely production, distribution and consumption. In this regard, it can be applied at micro (firms, consumers, concepts), meso (industrial parks) and macro levels (regions). Muscio and Sisto [48] point out the importance of adopting CE for firms so as to be competitive and adapt their business strategy to the increasing awareness of shareholders. In this regard, the adoption of CE models and tools is deemed to be a shift from necessity to opportunity. Other authors link the CE processes to sustainable development [14, 55], showing that CE can encourage the achievement of SDGs targets, especially those regarding the problems of food waste and food loss. In this way, the CE model stands as a solution to overcome the Agri-Food problems [14] by the creation of a closed-loop supply chain model [27].

**Circular Economy Perspective**

The present paradigm introduces a significant change when compared with the others: "waste is food", as in nature. LCA and Eco-Design are very important tools. A great challenge comes from the constitution of closed-loop supply chains able to deal with CE business models. The European Commission presented several business opportunities arising from the CE.

## 3.5 Accounting System Perspective

Any head of an organization, lucrative or otherwise, needs to have PL expressed in their accounting, and this study found two levels of categorization.

### 3.5.1 National Level

i.  The System of Environmental-Economic Accounting 2012—Central Framework (SEEA Central Framework) is a multipurpose conceptual approach that defines the connections between the economy and the environment, and the stocks and changes in stocks of environmental assets, suggesting a uniform method of accounting at the national level in collaboration with the UN, European Community, FAO, International Monetary Fund, Organization for Economic Co-operation and Development, World Bank [64]. From the SEEA

Central Framework, some definitions of interest are shown. First, *Physical flows* are defined as the change and use of materials, water and energy. Second, *Natural inputs* are all physical inputs that is removed from their place in the environment as a component of economic production processes or directly used in production. Third, *Products* are said to be goods and services that come from a production process in the economy. Finally, *Residuals* are flows of solid, liquid as well as gaseous substances, and energy, that are removed, discharged or emitted to the environment (e.g., emission to air) by businesses and family units through different processes, namely production, consumption or accumulation.

ii.   Scialaba et al. [56], introduces a procedure that allows the *full-cost accounting* (FCA) of the food wastage footprint, mainly applied at country level. They define "food waste", "food loss" and "food wastage" in the same way as FAO does. This document expresses that the economic cost of food wastage is substantial, around USD 1 trillion per year. This is the visible wastage; there are also much more extensive hidden costs. Food produced but never consumed still has environmental effects on the atmosphere, water, land and biodiversity. These environmental costs have to be paid by the public and future generations.

### 3.5.2  Organizations/Agri-Food Chain Level

Jovanović and Janković [33] point out that the environmental accountant's primary role at the business level is to generate relevant information to make strategic and operational decisions, considering the effect of operations on the environment and providing a solid base to be sustainably competitive. Corporate sustainability has to integrate TBL principles with corporate management in company operations and decision-making systems [8]. Some research explored the Sustainable Balanced Scorecard approach in a decision support system [45]. In this regard, [53] states that the Environmentally Balanced Scorecard (EBS) can identify indicators, deemed as an appropriate instrument to communicate information. This study focuses on two approaches, the IFAC guidance [29, 30] and the TBL.

i.    The IFAC [30] guidance gives both practical introductory support for professional accountants and companies that want to delve into and implement Environmental Management Accounting. This document applied in accounting in Cleaner Production [54] provides some definitions, based on ISO 14031, such as cost categories: materials costs of products and non-products output, waste and emission control costs, prevention and other environmental management expenditures, research and development expenses or less tangible costs.

ii.   Later, in 2015, IFAC launched "Accounting for sustainability" linking sustainability to business resilience, suggesting that a CE must drive management accounting in this century, taking into account generated waste and pollution.

iii.  The TBL is the combined assessment of economic, social and environmental sustainability, which constitute the basis of sustainability and efficient use of resources [12]. A sub-section of accounting deals with processes, techniques, and systems of the firm to record, analyze, and inform the relations between social, environmental and economic aspects such as "Sustainability Accounting" [8].

**Accounting System Perspective**

Although the Accounting System primarily focuses on regulatory standards looking for 1P (Profit), as possible. In evolution, SEEA-FCA were developed as Environmental Management Accounting, guiding countries and supra organizations, promoting a joint base for policy-makers. The IFAC also proposed one for organizations. Later, other proposals talked about sustainability connected to the 17 SDGs, as IFAC in 2015 or Corporate Sustainability through the TBL approach. These systems had to define "waste" and "inefficiencies" in their language, e.g., residuals. Of course, they included the components related to Social aspects (People in TBL context). TM needs accountability to be adapted to include the TBL point of view, achieving awareness and action capacity to be a sustainable competitive organization or Agri-Food chain.

## 3.6   Integrating Perspectives

The Physical Losses are expressed in the timeline, beginning with the Brundtland Report (the '80s) until the SDGs to the 2030s. Figure 6 shows that the Cleaner Production and Eco-Efficiency (the '90s) focus on improving resource use and reducing waste. In the early 2010s, several programmes laid the guidelines to overcome hunger globally, with FAO leadership, defining food loss and waste in the food supply chain. Those definitions follow a linear path and, in the late 2010s, the CE paradigm introduced a circular path with "waste is food".

To give support to the preceding definitions, the Accounting System offered proposals to national and organizational levels. Lean Thinking is a new proposal to consider regarding waste, with a valuable tool like the VSM to give TM visualization. The TBL approach was used in accountability, although it has a broader scope promoting sustainable development for the organizations. It has given support to the GRI and other reporting formats. As an example of emerging platforms, r3.0 promotes redesign for resilience and regeneration. ISO launched environmental-related standards such as Eco-Design, LCA, guiding the organizations on standardizing when they need to apply those tools. The MAESTRI platform is the development to integrate lean, ISO standards, the internet of things, industrial symbiosis and efficiency assessment tools. In summary, Fig. 6 synthesises the timeline history of PL definitions

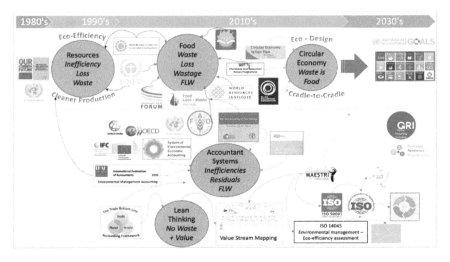

**Fig. 6** Temporal evolution of support systems and organizations that collaborated with the different definition perspectives. *Source* Authors

from several perspectives along with the tools to apply in accountability, reporting, decisions and standards.

## 4 Decision-Making at Top Management Level

While policy-makers take decisions on regulations to macro level, TM team is in charge of the decision-making process at the microlevel, that is firms. In this regard, [42] highlighted the role of professional accountants in the transformation of business processes towards sustainable development. In this vein, the existence of SDGs and indicators of sustainable development helps to restructure the current information system so as to provide information based on social, economic and environmental dimensions to the decision-making process.

The evolution of a sustainable organization also requires a clear comprehension in the TM about the strategy to apply so as to keep evolving and being competitive. In practice, the information system, based on sustainable accountability, is vital to supply a dashboard or scorecard with the main, and critical KPIs. They could assume different forms of presentation, but they need to fulfil the TM's needs. A challenging tool, coming from the Lean Thinking concept, is the VSM, and this can be adapted to express the TBL approach. In Fig. 7, an illustrative and straightforward example is shown in the dairy value chain, from the factory to the customer, excluding the farm and the consumer and expressing the wastes and inefficiencies.

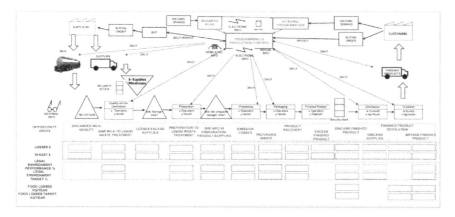

**Fig. 7** VSM adapted to dairy supply chain wastes, each in three levels as TBL proposes, including targets. *Source* Authors

## 5 Agri-Food Chain Losses Definition Proposal

In this research, a definition of PL adapted to Agri-Food Chain Losses (AFCL) is offered. In this regard, PL in the Agri-Food chain is understood as the overconsumption of resources, either through the use of more inputs-supplies/raw materials or through non-optimized product design at the consumer level. Thus, AFCL takes into account not only the raw material and transformed products but also all inputs/supplies (packaging, added ingredients, energy, water, industrial services, elements of general use, among others) that are lost in the value chain. In this sense, *AFCL can be defined as the raw material and/or its transformed products, as well as all inputs/supplies (packaging, added ingredients, energy, water, industrial services, elements of general use, among others) that are lost or wasted or overused for some reason in the value chain (overconsumption of resources, either through the use of more inputs-supplies/raw materials or through non-optimized product design at the consumer level).* In such a way, this approach provides a perspective from AFCL lenses and an Eco-Efficiency point of view (e.g. water and energy use) to CE concept (product design optimized for a responsible consumption). Figure 8 depicts the closed-loop Agri-Food chain with the main stages and the different food waste perspectives, all integrated into AFCL.

## 6 Conclusions

The perspectives related to PL reveal different points of view that can distort their identification and quantification to manage the actions aimed at their reduction, elimination or transformation. These PL occur along the entire Agri-Food chain,

**Fig. 8** Integrating perspectives and PL definitions in the Agri-Food chain. *Source* Authors

from the land to the plate, covering mostly the productive sector, where TM has a decisive role.

In the analysis of keywords in the scientific literature related to PL, food waste, CE and food losses, stand out in the last three decades. They reflect the CE trend as a core concept in the literature and the SDGs as a way to guide the reduction of food loss and waste. Taking into consideration that academic research is generally not a source of the definitions applied to PL, most researchers use definitions generated at the governmental or supra governmental level in organizations that intend to guide policy-makers.

In this study, five perspectives have been identified that cover the different definitions of PL detected. The Food Waste perspective is oriented to beat global hunger and focused on optimizing natural resources, mainly in crops. The UN leads this approach and is more guided to policy-makers than to TM as decision-makers. The Eco-Efficiency and Cleaner Production perspective explain other components of PL: supplies, wastewaters, emissions, waste solid, rework, and many others. This perspective is oriented to reducing inefficiencies in Agri-Food chains, and it is useful for a decision-making approach, guiding TM to be aware and proactive. The Lean Thinking perspective introduces other aspects of PL. However, it mainly presents a tool similar to VSM viable for adaption to any value chain or any organization's needs, for different decision levels, operative, tactical and strategic, especially for the TM. The CE perspective introduces a significant change when compared with the others, "waste is food". LCA and Eco-Design are essential tools. A significant challenge comes from the constitution of closed-loop supply chains able to deal with CE business models, and the European Commission has presented several business opportunities coming from the CE. Finally, the Accounting System perspective used to focus on regulatory standards looking for 1P (Profit), although initiatives such as

Environmental Management Accounting by SEEA-FCA contribute to guiding countries and supra organizations and promoting a common base for policy-makers. Later, other proposals deal with sustainability connected to the 17 SDGs, such as IFAC or corporate sustainability through the TBL approach. TM needs accountability adapted to the TBL perspective, raising awareness and action to be a sustainable competitive organization.

By and large, the conclusions regarding this study are the following. Through the SDGs proposed in Agenda 2030, the sustainable development requires to engage TM to get precise results in the line of the TBL approach. Therefore, organizations need clear definitions about the AFCL. This study defines AFLC as the raw material and/or its transformed products, as well as all inputs/supplies (packaging, added ingredients, energy, water, industrial services, elements of general use, among others) that are lost or wasted or overused for some reason in the value chain (overconsumption of resources, either through the use of more inputs-supplies/raw materials or through non-optimized product design at the consumer level). This definition should guide the accountant system to express the bottom-line results, from Profit, People and the Planet's perspective. The accountants should express the figures in both formats, regulatory and management (such as KPIs). The organizations must also adapt the information system at three levels: strategic, tactical and operative. VSM is a tool to give visibility to the AFCL. Targets should express the strategic goals proposed by organizations as being sustainable competitive. Worth mentioning in particular is the need to adapt VSM for TM because it is TM who need to understand how to be aware and act to continuously evolve and be a resilient organization like the CE concept promotes, even adapting their supply chain to a closed-loop one.

Concerning the managerial implications derived from this study, holders, directors and managers may gain a complete overview of the definitions of PL in the agri food industry. Accordingly, the identification of the main perspectives could benefit managers, making them aware of the need to develop an accounting system aimed at the detection and assessment of PL. Additionally, policymakers could take advantage of the identification of perspectives to aim the Agri-Food policies towards the reduction of AFCL. In this regard, the Agri-Food industry may gain profitability since the PL would easily be identified and reduced, aligning with sustainable development.

This research can be improved through the support of leading world institutions and the adaptation of a standard general definition as well as ones adapted to particular perspectives. An important issue is related to the accounting systems where there is a need to connect operational definitions to the accounting field.

# 7    Future Research Directions

Future research lines could be aimed to explore how to improve TM awareness through primary data obtained from a specific Agri-Food chain (dairy) in order to detect valuable ideas applicable to a specific sector.

# References

1. Agramunt LF, Berbel-Pineda JM, Capobianco-Uriarte MM, Casado-Belmonte MP (2020) Review on the relationship of absorptive capacity with interorganizational networks and the internationalization process. Complexity 2020:7604579
2. Alayo M, Iturralde T, Maseda A, Aparicio G (2020) Mapping family firm internationalization research: bibliometric and literature review. Rev Manage Sci:1–44
3. Arzoumanidis I, Salomone R, Petti L, Mondello G, Raggi A (2017) Is there a simplified LCA tool suitable for the agri-food industry? An assessment of selected tools. J Clean Prod 149:406–425
4. Benoit S, Margni M, Bouchard C, Pouliot Y (2019) Eco-efficiency applied to dairy processing: from concept to assessment. Environ Manage Sustain Dev 1(8):1–26
5. Brundtland G, Khalid M, Agnelli S, Al-Athel S, Chidzero B, Fadika L et al (1987) Our common future. Brundtland report. https://sustainabledevelopment.un.org/content/documents/5987our-common-future.pdf. Accessed on 09 Dec 2020
6. Buzby JC, Farah-Wells H, Hyman J (2014) The estimated amount, value, and calories of postharvest food losses at the retail and consumer levels in the United States. USDA-ERS Econ Inf Bullet 121
7. Cobo MJ, López-Herrera AG, Herrera-Viedma E, Herrera F (2011) Science mapping software tools: review, analysis, and cooperative study among tools. J Am Soc Inform Sci Technol 62(7):1382–1402
8. Coşkun Arslan M, Kisacik H (2017) The corporate sustainability solution: triple bottom line. J Account Fin:18–34
9. Coulter N, Monarch I, Konda S (1998) Software engineering as seen through its research literature: a study in co-word analysis. J Am Soc Inf Sci 49(13):1206–1223
10. De Steur H, Wesana J, Dora MK, Pearce D, Gellynck X (2016) Applying value stream mapping to reduce food losses and wastes in supply chains: a systematic review. Waste Manage 58:359–368
11. European Commission—EC (2019) Reducing food losses and waste along the agri-food value chain. https://cordis.europa.eu/programme/id/H2020_RUR-07-2020. Accessed 04 Dec 2020
12. Elkington J (1997) Cannibals with forks: the triple bottom line of 21th century business. Capstone Publishing, Oxford
13. Elkington J (2018) 25 years ago I coined the phrase "triple bottom line". Here's why it's time to rethink it. Harv Bus Rev 25:2–5
14. Esposito B, Sessa MR, Sica D, Malandrino O (2020) Towards circular economy in the agri-food sector. A systematic literature review. Sustainability 12(18):7401
15. European Union—EU (2015) EU circular economy action plan. https://ec.europa.eu/environment/circular-economy/. Accessed 04 Dec 2020
16. Ferrera E, Rossini R, Baptista AJ, Evans S, Hovest GG, Holgado M et al (2017) Toward Industry 4.0: efficient and sustainable manufacturing leveraging MAESTRI total efficiency framework. In: International conference on sustainable design and manufacturing. Springer, Cham, pp 624–633
17. Food and Agriculture Organization—FAO (2019) Food loss and waste and value chains. http://www.fao.org/3/ca5312en/CA5312EN.pdf. Accessed 03 Dec 2020
18. Food and Agriculture Organization—FAO (2013) Food wastage footprint: Impacts on natural resources. http://www.fao.org/3/i3347e/i3347e.pdf. Accessed 03 Dec 2020
19. Food and Agriculture Organization—FAO (2011) Global initiative on food loss and waste reduction. http://www.fao.org/3/a-i7657e.pdf. Accessed 03 Dec 2020
20. Food Use for Social Innovation by Optimising Waste Prevention Strategies—FUSIONS (2016) Food waste definition. https://www.eu-fusions.org/index.php/about-food-waste/280-food-waste-definition. Accessed 04 Dec 2020
21. Frosch RA, Gallopoulos NE (1989) Strategies for manufacturing. Sci Am 261(3):144–153
22. Garcia-Garcia G (2017) Development of a framework for sustainable management of industrial food waste. Doctoral dissertation, Loughborough University

23. Garza-Reyes JA, Romero JT, Govindan K, Cherrafi A, Ramanathan U (2018a) A PDCA-based approach to environmental value stream mapping (E-VSM). J Clean Prod 180:335–348
24. Garza-Reyes JA, Kumar V, Chaikittisilp S, Tan KH (2018b) The effect of lean methods and tools on the environmental performance of manufacturing organisations. Int J Prod Econ 200:170–180
25. Gustavsson J, Cederberg C, Sonesson U, van Otterdijk R, Meybeck A (2011) Global food losses and food waste. http://www.fao.org/docrep/014/mb060e/mb060e00.pdf. Accessed 05 Dec 2020
26. Hanson C, Lipinski B, Robertson K, Dias D, Gavilan I, Gréverath P et al (2016) Food loss and waste accounting and reporting standard. https://www.wri.org/publication/food-loss-and-waste-accounting-and-reporting-standard. Accessed 07 Dec 2020
27. Holgado M, Aminoff A (2019) Closed-loop supply chains in circular economy business models. In: international conference on sustainable design and manufacturing. Springer, Singapore, pp. 203–213. https://doi.org/10.1007/978-981-13-9271-9_19
28. Horton P, Reynolds CJ, Milligan G (2019) Food chain inefficiency (FCI): accounting conversion efficiencies across entire food supply chains to re-define food loss and waste. Front Sustain Food Syst 3:79
29. International Federation of Accountants—IFAC (2015) Accounting for sustainability, from sustainability to business resilience. https://www.ifac.org/knowledge-gateway/preparing-fut ure-ready-professionals/publications/accounting-sustainability-sustainability-business-resili ence. Accessed 08 Dec 2020
30. International Federation of Accountants—IFAC (2005) International guidance document: environmental management accounting. https://www.ifac.org/about-ifac/professional-accoun tants-business/publications/international-guidance-document-environmental-management-accounting-2. Accessed 07 Dec 2020
31. International Organization for Standardization—ISO (2012) ISO 14045:2012. https://www.iso.org/standard/43262.html. Accessed 08 Dec 2020
32. Jeffries N (2018) A circular economy for food: 5 case studies. https://medium.com/circul atenews/a-circular-economy-for-food-5-case-studies-5722728c9f1e. (Ellen McArthur Foundation team member). Accessed 31 Dec 2020
33. Jovanović D, Janković M (2012). Management accounting aspect of environmental costs. Contempor Issues Econ Bus Manage 525
34. Kirchherr J, Reike D, Hekkert M (2017) Conceptualizing the circular economy: an analysis of 114 definitions. Resour Conserv Recycl 127:221–232
35. Kouwenhoven G, Reddy Nalla V, Lossonczy von Losoncz T (2012) Creating sustainable busi-nesses by reducing food waste: a value chain framework for eliminating inefficiencies. Int Food Agribus Manage Rev 15:119–138
36. Krishnan R, Agarwal R (2020) Redesigning a food supply chain for environmental sustain-ability—an analysis of resource use and recovery. J Clean Prod 242:118374
37. Kummu M, De Moel H, Porkka M, Siebert S, Varis O, Ward PJ (2012) Lost food, wasted resources: global food supply chain losses and their impacts on freshwater, cropland, and fertiliser use. Sci Total Environ 438:477–489
38. Latan H, Jabbour CJC, de Sousa Jabbour ABL, Wamba SF, Shahbaz M (2018) Effects of envi-ronmental strategy, environmental uncertainty and top management's commitment on corporate environmental performance: the role of environmental management accounting. J Clean Prod 180:297–306
39. Lessig L (2008) Remix: making art and commerce thrive in the hybrid economy. Penguin press, New York
40. Lipinski B, Hanson C, Waite R, Searchinger T, Lomax J, Kitinoja L (2013) Reducing food loss and waste. World Resour Inst Work Pap 1:1–40
41. MacArthur E (2013) Towards the circular economy. J Ind Ecol 2:23–44
42. Makarenko I, Plastun A (2017) The role of accounting in sustainable development. Account Fin Control 1(2):4–12

43. Mangla SK, Luthra S, Rich N, Kumar D, Rana NP, Dwivedi YK (2018) Enablers to implement sustainable initiatives in agri-food supply chains. Int J Prod Econ 203:379–393
44. MAESTRI (2016). MAESTRI—a H2020-Project under the SPIRE-PPP initiative. https://mae stri-spire.eu/. Accessed 20 Dec 2020
45. Marimin M, Wibisono A, Darmawan MA (2017) Decision support system for natural rubber supply chain management performance measurement: a sustainable balanced scorecard approach. Int J Supply Chain Manage 6(2):60–74
46. McDonough W, Braungart M (2010) Cradle to cradle: remaking the way we make, things. North Point Press, New York
47. Mor R, Singh S, Bhardwaj A (2015) Learning on lean production: a review of opinion and research within environmental constraints. Oper Supply Chain Manage Int J 11, 9(1):61–72
48. Muscio A, Sisto R (2020) Are agri-food systems really switching to a circular economy model? Implications for European research and innovation policy. Sustainability 12(14):5554
49. Östergren K, Holtz E (2016) Food waste prevention strategies in global food chains: Conclusions and recommendations from the SIANI Expert group on food waste 2016. (RISE Rapport). Retrieved from http://urn.kb.se/resolve?urn=urn:nbn:se:ri:diva-44553. Accessed 07 Dec 2020
50. Pauli G (2010) The blue economy—10 years, 100 innovations, 100 million, Jobs. Paradigm Publications, Brookline, MA, USA
51. Pearce DW, Turner RK (1990) Economics of natural resources and the, environment. The John Hopkins University Press, Baltimore
52. Ross RB, Pandey V, Ross KL (2015) Sustainability and strategy in US agri-food firms: an assessment of current practices. Int Food Agribus Manage Rev 18:17–47
53. Scavone GM (2006) Challenges in internal environmental management reporting in Argentina. J Clean Prod 14(14):1276–1285
54. Schaltegger S, Bennett M, Burritt R, Jasch C (eds) (2008) Environmental management accounting for cleaner, production. Springer, Dordrecht
55. Schroeder P, Anggraeni K (2019) The relevance of circular economy practices to the sustainable development goals. J Ind Ecol 23(1):77–95
56. Scialaba et al (2014) Food wastage footprint—full-cost accounting. FAO Report. www.fao.org/nr/sustainability/food-loss-and-waste, http://www.fao.org/3/a-i3991e.pdf. Accessed 09 Dec 2020
57. Sjögren P (2014) Usefulness of lean as a sustainable strategy in food supply chains. MSc Thesis, Cranfield University. http://stud.epsilon.slu.se. Accessed 12 Oct 2020
58. Straková J, Rajiani I, Pártlová P, Váchal J, Dobrovič J (2020) Use of the value chain in the process of generating a sustainable business strategy on the example of manufacturing and industrial enterprises in the Czech Republic. Sustainability 12(4):1520
59. Straková J, Talíř M (2020) Strategic management and decision making of small and medium-sized enterprises in the Czech Republic. SHS Web Conf 73:02005. https://doi.org/10.1051/shs conf/2020730
60. Terán-Yépez E, Marín-Carrillo GM, del Pilar Casado-Belmonte M, de las Mercedes Capobianco-Uriarte M (2020) Sustainable entrepreneurship: review of its evolution and new trends. J Clean Prod 252:119742
61. Tóth G (2019) Circular Economy and its comparison with 14 other business sustainability movements. Resources 8(4):159
62. Timmermans AJM, Ambuko J et al (2014) Food losses and waste in the context of sustainable food systems (No. 8). CFS Committee on World Food Security HLPE
63. United Nations—UN (2015). The 17 Goals. https://sdgs.un.org/goals. Accessed 09 Dec 2020
64. United Nations (UN), European Union, Food and Agriculture Organization of the United Nations, International Monetary Fund, Organisation for Economic Co-operation and Development and The World Bank (2014) System of environmental-economic accounting central framework 2012. United Nations, New York. https://seea.un.org/sites/seea.un.org/files/seea_cf_f inal_en.pdf. Accessed 13 Oct 2020
65. United Nations Environmental Programme-Food and Agriculture Organization UNEP-FAO (2012) Definition of food loss and waste. https://www.unenvironment.org/thinkeatsave/about/ definition-food-loss-and-waste. Accessed 09 Dec 2020

66. United Nations Environmental Programme—UNEP (2011) Towards a green economy: pathways to sustainable development and poverty eradication. https://www.unep.org/explore-top ics/green-economy. Accessed 09 Dec 2020
67. Van Berkel R (2007) Cleaner production and eco-efficiency initiatives in Western Australia 1996–2004. J Clean Prod 15(8–9):741–755
68. Vargas JRC, Mantilla CEM, de Sousa Jabbour ABL (2018) Enablers of sustainable supply chain management and its effect on competitive advantage in the Colombian context. Resour Conserv Recycl 139:237–250
69. Vitale G, Cupertino S, Rinaldi L, Riccaboni A (2019) Integrated management approach towards sustainability: an Egyptian business case study. Sustainability 11(5):1244
70. WBCSD-UNEP (1998) Cleaner production and eco-efficiency. Complementary approaches to sustainable development. World Business Council for Sustainable Development, United Nations Environment Programme: Geneva, Paris. https://bit.ly/2NQEyjq. Accessed 08 Dec 2020
71. Whitehead P, Parfitt J, Bojczuk K, James K (2013) Estimates of waste in the food and drink supply chain. Banbury: Waste and Resources Action Programme. https://bit.ly/3qIHVHM. Accessed 10 Dec 2020
72. Womack JP, Jones DT (1997) Lean thinking—banish waste and create wealth in your corporation. J Oper Res Soc 48(11):1148–1148
73. Waste and Resources Action Programme—WRAP (2020) Categories of food waste. https://wrap.org.uk/resources/guide/waste-prevention-activities/food. Accessed 09 Dec 2020
74. Waste and Resources Action Programme—WRAP (2015) Strategies to achieve economic and environmental gains by reducing food waste. https://wrap.org.uk/sites/default/files/2020-12/Strategies-to-achieve-economic-and-environmental-gains-by-reducing-food-waste.pdf. Accessed 09 Dec 2020
75. Waste and Resources Action Programme—WRAP (2013) https://wrap.org.uk/about-us/our-vis ion/wrap-and-circular-economy/wraps-vision-uk-circular-economy. Accessed 09 Dec 2020
76. Wu T, Wu YJ, Tsai H, Li Y (2017) Top management teams' characteristics and strategic decision-making: a mediation of risk perceptions and mental models. Sustainaible 9(12):2265
77. Zong QJ, Shen HZ, Yuan QJ, Hu XW, Hou ZP, Deng SG (2013) Doctoral dissertations of library and information science in China: a co-word analysis. Scientometrics 94(2):781–799

# Can CE Reduce Food Wastage?
# A Proposed Framework

**Rohini Sharma, Anjali Shishodia, Tavishi Tewary, and Rohit Sharma**

**Abstract** There is a constant tussle between sustainability and effective resource utilization. With the current consumption pattern of make, use, and dispose, we cannot think of a sustainable future. Circular economy (CE) aims at waste minimization through optimal resource utilization, environmental protection, and other socio-economic benefits. The present study presents a CE perspective for reducing food wastage. Further, the study highlights how behavior reasoning theory (BRT) plays a crucial role in shaping consumers' attitude toward wastage of food. The present study also presents a theoretical framework wherein CE is tested as a moderator leading to pro-environmental behavior. The study is expected to provide valuable contributions on the CE front and its possible implications for theory and practice. CE adoption is also critical for sustaining the current food production to meet the ever-increasing demands of an ever-growing population.

**Keywords** Circular economy · Food supply chain · Food wastage · Emerging economy · Circularity

R. Sharma
Azim Premji University, Survey No 66, Burugunte Village, Bikkanahalli Main Rd., Sarjapura, Bengaluru 562125, India

A. Shishodia
LM Thapar School of Management Dera Bassi Campus, Off, Derabassi - Barwala Rd, Chandigarh, Punjab 140507, India
e-mail: anjali.shishodia@thapar.edu

T. Tewary · R. Sharma (✉)
Jaipuria Institute of Management, Noida, A-32, Opposite IBM, Sector 62, Noida, Uttar Pradesh 201309, India
e-mail: rohit.sharma@jaipuria.ac.in

T. Tewary
e-mail: tavishi.tewary@jaipuria.ac.in

# 1 Introduction

The circular economy (CE) is a systematic method that aims to "reduce, reuse, recycle" the production and consequent consumption processes to reduce energy consumption and waste generation [1]. In recent times, many case studies, and reviews have been conducted on this topic [2]. While the linear economic system (see Fig. 1) has led to huge economic growth up to the twentieth century, it is seeming to be a weakness now and is forecasted to have an ultimate breakdown. Existing business models are not effective as business management tools seem to be heavily dependent on non-renewable sources of energy [3]. Rapid consumption of non-renewable resources has resulted in large quantities of waste production. The capacity of ecosystems has declined and it cannot deal with drastic changes [4].

The framework of a circular economic model must be built on innovative designs [5].

Academicians have laid emphasis on closed loops, while studying CE [6, 7]. The Ellen MacArthur Foundation [8] has defined CE as an alternative model that takes a holistic approach whereby the waste generated by industrial processes is not seen as the result, instead, it is seen as an opportunity to contribute toward sustainable development. The Ellen MacArthur Foundation has developed a "value circle" which illustrates the flow of materials within the system. The model for a CE has been shown in Fig. 2.

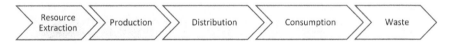

**Fig. 1** Linear economy

**Fig. 2** Circular economy

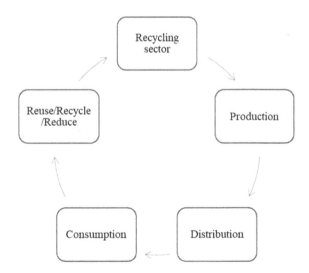

The concept of CE has been key in grouping different range of waste resource management approaches by very efficiently drawing the attention toward the capacity of prolonging resource use [9]. It has also linked an economy that attempts to work in synchronization with the socioeconomic and environmental systems in which it is embedded, for a better and sustainable future [10].

Researchers have linked certain attributes connecting the CE model and the concept of sustainability [4]. Both concepts have a global reach; both include innovations and focus on new product designs, cooperation among the various stakeholders across different sectors.

CE attempts to provide a global solution for the issue at hand [11]. The food consumption habits of humankind are currently unsustainable [12]. The inefficiency in the food economy indicates that we lose efforts, productivity, natural resources energy and incur the cost of pollution, greenhouse emissions, and throwing the food away [13]. According to recent reports, the cost of these inefficiencies is as high as USD two trillion when we account for social and environmental costs to the society [14]. The problem of food wastage creates huge problems across the triple bottom line [15] and attract attention on the academic and intuitional front [13]. Food waste is defined as the wholesome edible part of the food intended for human consumption that is still disposed off at various levels in the food supply chain [16]. A similar definition is given by [17] where they define food waste as "the healthy and edible substance that is wasted, lost and degraded in supply chain transition and include organic waste generations from households, commercial kitchens, and food processing industries". Therefore, food wastage is the biggest hindrance in realizing CE in the food and beverages industry and is an impediment in achieving the United Nation's Sustainable Development Goals (UNSDGs) of reducing the global per capita food waste to half by 2050, i.e., UNSDG2.

Food wastage is becoming prominent in developing countries as a result of rapid urbanization and global food convergence [18]. There is a dearth of studies examining these food waste habits in developing countries and refrains researchers from proposing impactful waste reduction strategies [19]. Thus, studying food wastage at the consumer level and understanding their behavior plays a vital role in food waste since it can influence sustainable consumption behavior in the long run. Therefore, increasing interest among scholars as well as practitioners toward understanding the consumer behavior issues related to food waste is observed [20].

The current study builds on this existing literature and utilizes behavioral reasoning theory (BRT) to examine food wastage behavior. The study highlights the influential role of circular economy as an important factor determining consumer's food wastage behavior, i.e., attitude against food wastage. The study proposes a framework that enables scholars and practitioners to better understand the consumer behavior issues related to food wastage phenomena and help in taking adequate measures in reducing food wastage.

## 2 Literature Review

This section presents the review studies on food wastage behavior and emphasizes on food wastage behavior in an emerging economy context.

The food discarded in retail, restaurants and food joints, and consumer households makes up the largest share of the wasted food [21]. Different stages of food waste generation can be defined as: stage 1 at the time of purchase, stage 2 at the time of storage, stage 3 at the time of preparation, stage 4 at the time of consumption, and finally, stage 5 at the time of disposal of leftovers [22]. Majority of the prior academic literature has highlighted the consumer behavior is highly influential in understanding the reasons for wasting food in different parts of the world [23].

Various studies utilized different theoretical perspectives for highlighting the consumers' food waste behavior. The theory of planned behavior is the most prominent theory applied in the literature [23]. The study by Richter and Bokelmann [24] utilized the means-end approach for identifying and reporting consumers' food waste behavior. The other theories utilized were the theory of interpersonal behavior [25], the comprehensive model of environment behavior [25], the theory of reasoned action [23].

In terms of geographical settings, the majority of the studies highlighted food waste behavior in developed countries. For example, the study by [23] was carried out in Italy, the study by was carried out in Germany, the study by Russell et al. [25] was carried out in the United Kingdom. In comparison to this, the empirical studies examining consumer behavior with respect to consumers dwelling in developing countries is relatively less [22]. This suggests the need for more studies on consumer-related food wastage in developing countries.

In terms of the study variables, a review of prior literature suggests that the focus mainly relied on attitude and intention for highlighting consumers' food waste behavior. Corsini et al. [23] used attitude, perceived behavioral control, and waste prevention measures as dependent variables and used personal norms, social norms, and awareness as independent variables. Russell et al. [25] utilized "intention to reduce food waste and food waste behavior" as dependent variables, and social norms, perceived behavioral control, emotions, and attitude as independent variables. The study by [26] found that wastage of food was directly proportional with materialism values, purchasing discipline, and food waste prevention behavior and was inversely proportional with environmental values. Furthermore, eco-friendly behavior, collaborative consumption, economic literacy, and domestic skills are associated with reduction in food wastage. The most recent study by [22] highlighted that leftovers and fresh produce are the main contributors to food waste due to prolonged storage and excess stock.

## 2.1  Food Waste and India

The studies examining food waste in context to emerging economies are scarce at present [21]. In emerging economies, the impact of perceived social norms (value orientations and cultural differences) is relatively higher as compared to developed economies which is also a major contributor to increased food wastage. Nevertheless, the rising income levels in emerging economies is leading to an increase in generation of food waste and has become an issue of major concern (Bahadur et al. 2016). Wang et al. reported that food waste has already reached the same level as in developed. Therefore, it is a necessity to explore and counter food waste at the consumer level as it is tackled in developed economies.

Some of the prominent factors that contribute to rising food wastage in low-income households include over-purchased commodities [27] and providing food abundancy for the whole family [28]. Research studies in the past have also proved that the awareness levels pertaining to food wastage and its harmful ecological effects are found to be low among emerging economies [29].

Food waste in India typically consists of the waste generated at various steps of the food supply chain (post-harvest and storage) [30] and wasted foodstuffs from households and restaurants [31]. Most of the existing literature focuses on food loss or wastage across the food supply chain (esee [32]) or investigate the quantity, type of food wastage, or concerned costs generated at various stages of the food supply chain (e.g., [33]) or for particular segments of foods such as non-perishables [34].

## 3  Theoretical Framework and Hypothesis Development

This section highlights the theoretical underpinnings, hypothesis development, and the proposed research framework.

## 3.1  BRT and Consumer Behavior

Consumer behavior (CB), owing to its complexity, often fails to follow the traditional rational economic theories. One example of that is the rational choice model [35]. There is often an attitude-behavior discrepancy in the actual intended behavior of people as opposed to their actual behavior. It varies depending on the irrationality and rationality of consumers' decision-making process (CDP). The theory is applicable on the choices of the individuals pertaining to pro-environmental behavior. According to the BRT, beliefs and motives including social norms and attitude are linked together with reasons. Individuals justify and defend their actions in order to promote and protect their self-worth with the help of behavioral reasoning [36]. Global motives in the form of attitude and social norms influence various human behavioral domains.

As per Ajzen [37], attitude, norms, and perceived control are global motives are subjective and significantly forecast CDP. BRT is superior to other existing theories such as the theory of reasoned action [38] and the theory of planned behavior [39]. The new theory provides insights into motivational mechanisms as reason concepts have displayed predictive validity in various decision-making scenarios. Reasons can function as automatic stabilizers and can influence CDP. This can be widely covered by BRT but neglected by the previous studies. One more advantage of BRT is that it can provide a wider conceptual platform for comprehending reasons. Numerous past assessments of reasons lack underlying robust theoretical justifications [40].

Reason plays the most important role in shaping CDP [41]. As mentioned above, individuals justify and support their decision-making choices with the help of reasons. This is in incongruence with theory of explanation based decision-making. They also highlight that reasons help individuals to select a particular alternative in CDP with confidence. Reasons also act as motivators for behavior by upholding consumer beliefs and provide explanation of CB and their causal relationship with the environment [40].

The application of BRT in consumer behavior studies varies from transportation [42] to branding of organic products [43], from telecommunication and banking [44] to technology acceptance [45] and charity [46]. Based on these prior studies, it can be inferred that BRT acts as a comprehensive blueprint for the prediction of CDP. According to BRT, reasons act as an important antecedent of the attitude toward performing a behavior, subjective norms toward the behavior, and perceived control toward the behavior.

Based on the BRT and the theory of explanation-based decision-making, it can be said that people from favorable evaluations toward a given alternative when they have strong reasons supporting and justifying the alternative. The justification mechanism plays a critical role in the formation of judgments. An individual will have strong reasons for performing a certain behavior and is likely to activate more behaviorally related cognitions (such as global positive attitude) toward performing that behavior. Levi and Pryor [47] report that manipulation of reasons can directly influence judgments and attitudes, highlighting the fact that reasons play an important role in the judgment process.

In the BRT framework, reasons are closely related to psychological concepts such as sense-making, functional theorizing, and psychological coherence. Reasons either "in support of" or "counter to" is used for supporting an individual's decision-making. These also help in justifying a subsequent action. Consumer attitudes toward any subject are formed based on their reasoning ability [48].

Past research has discovered that excess purchasing of food [49] is the major reason for their food waste behavior, whereas concern for the environment [22, 50] has been cited as the main motivation against food waste behavior. As these findings are consistent with the BRT framework, the current study argues that consumer reasoning has significant effects on attitude. Specifically, this research postulates that the consumer who have strong reasons for (against) food waste are likely to have a positive (negative) attitude toward it.

## 3.2 Attitude and Food Wastage

Attitude has been defined in literature as either a positive or negative feeling of an individual, while displaying a certain behavior [51]. In addition to this, attitude has also been defined as the extent an individual values favorably or unfavorably, the performance of certain behavior [52]. Prior experience of the consumer influences consumer's judgment toward a specific behavior [53]. Some studies use consumer attitude to predict consumer intention, which in turn affects consumer behavior [54]. Earlier research studies in consumer behavior have also proved that attitudes are strong indicators of consumer behavior and intention [55].

Attitudes comprise of three components viz. affective, behavioral, and cognitive [56]. The affective component is related to an individual's feelings or emotions toward a particular product or service. The behavioral component is linked to an individual's tendency to respond in a certain manner toward an activity [56]. The cognitive component is related to an individual's belief about a particular product or a service. Consumer's shopping behaviors are routinized and the purchase of too much food during shopping trips is common and hence, contributes to increased food waste. One major reason for food waste can be associated with excessive purchasing and stockpiling of food [57]. Graham-Rowe et al. [52] report that the favorable unambiguous attitudes toward food waste reduction are often related to greater intentions to reduce household food waste. As [58] report, consumers are against food waste and engage themselves in environment-friendly behaviors, most of the food waste is hence, caused by the intentional behavior of the consumers. Therefore, the reasons for food waste are consciously derived and are driven by explicit attitudes toward food waste [59].

In the context of food waste, previous research studies have found a positive relationship between attitude and food waste [60]. Similarly, [61] found attitude as a significant predictor of intention to waste food at the consumer level. Therefore, based on the evidence gathered from previous studies, and in the context of the present study, attitude is anticipated to be positively associated with the behavioral intention of not wasting food and therefore, we hypothesize that:

**H1.** *The reason will have a positive association on the Attitude toward the reduction in food waste.*

## 3.3 Moderating Variables

Shin [62] reports that the use of moderators intensifies the predictive validity in the research study and help in highlighting unpredictable findings with respect to different disciplines. As food waste is an issue of concern, many research studies in the past have used moderators such as age [20], gender [49], and income level [49] in justifying their findings for household food waste generation. The circular economy approach provides an insight into how the environment invokes economic thinking

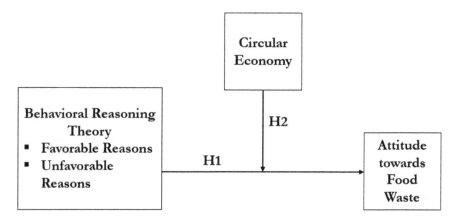

**Fig. 3** Proposed framework

[63]. The aspects of reduction, recycle and reuse have been focused upon in various studies [64]. These 3R principles indicate a connection between the environment and economics [65, 66]. The circular economy concept bridges the gap between economics and the environment in a sustainable manner [7]. Therefore, based on the evidence gathered from previous studies, and in the context of the present study, circular economy is anticipated to be a moderator in defining the relationship between the role of reasons in shaping attitude toward the reduction in food wastage.

**H2**: *The relationship between attitude and intention toward food waste will be moderated by the circular economy* (Fig. 3).

## 4    Results and Analysis

Many studies have highlighted that majority of the household food waste generated depends upon a multitude of factors such as consumer behavior and attitude, socio-economic factors, and consumers' eating preferences [29]. This is a novel approach which explores the moderating influence of CE in shaping the consumer attitude and toward food waste in a developing country scenario using a BRT framework.

The first hypothesis investigates the influence of BRT on attitude toward food waste. The study intends to establish a positive association between the two constructs, which supports the previous findings [25]. This may suggest that customers perceive food wastage as a matter of serious concern and harmful to the environment [67]. Consumers also perceive that there is also a feeling of guilt associated with wasting food [60] which is positively associated with attitudes toward food waste.

The second hypothesis examines the moderating role of the circular economy between BRT and attitude toward food waste. The hypothesis suggests that there is an increasing role of circular economy in shaping the attitude of the consumer households toward food wastage. The more the consumer is aware of the circularity, the more he is concerned about reducing food wastage.

The present study proposes a framework for understanding the role CE plays in influencing the food wastage behavior in Indian households. The model proposed in the study helps in understanding the food wastage behavior at the consumer level in an emerging economy in South Asia. Emerging economies are focusing their efforts on minimizing all sorts of waste (including food waste) therefore, it is expected that the consumer awareness campaigns targeting waste reduction will meet their goals steadily. This study is expected in devising and exploring innovative means for reducing food wastage at household levels.

## 5    Implications and Discussion

In order to survive on current world resources, effective waste reduction and management are required along with responsible consumer behavior [68]. The framework addresses the under conceptualization of the role of circular economy acting as a moderator in reducing consumer's food waste behavior. This study reveals that circular economy can be conceptualized as a major moderator and presents a snapshot of the future quantitative studies that can be designed for testing the given framework.

The implications of the study are three-fold viz. for consumers, practitioners, and policy-makers.

Firstly, it provides various suggestions to consumers to plan their shopping routines and avoid buying extra food so that does not go waste [69, 20]. Consumers should make use of the leftover food creatively [59] and should adopt food sharing initiatives for reducing food wastage. Food sharing among communities is one of the effective and upcoming ways of tackling food waste at the consumer level. Consumers should set goals for themselves for reducing food wastage and should commit to their goals [70]. Consumers should check their food stock and fridge before shopping, rather than before cooking or at the time of cleaning the fridge [22].

Secondly, for the policy-makers, in encouraging food wastage prevention behavior at the consumer level. This can boost food security and strengthen the sustainability of the local food system. Local authorities should devise measures for reducing household food waste and come out with waste reduction and minimization strategies. The local authorities can conduct campaigns, workshops, and seminars highlighting the efficacy of food waste reduction among the consumers [71]. It can help in reducing economic costs incurred along the food supply chain. It can also help in the socio-economic assessment of food waste generation.

Thirdly, food waste reduction campaigns and initiatives can work effectively at the community level therefore, choosing the right target audience can effectively help in reducing and minimizing food waste. Food sharing forums and community networks should be initiated to effectively reduce food wastage as food wastage is the problem of society as a whole. For encouraging more participation from the people, social media sites should be used for creating awareness regarding food wastage.

## 6 Conclusion and Future Scope

This paper aims to integrate the concept of circularity in the food consumption pattern and attempts to achieve a circular food system. Consumers play an integral role in the transition toward the circular economy by opting for sustainable consumption. The contribution of the household sector in reducing food wastage is huge and should be emphasized. In this research, we have highlighted that attitudes play an important role in defining the consumers; food waste behavior and circularity can improve this relationship. A framework is provided in this regard which can serve as a reference for future research. Potential solutions to achieve a reduction in food waste generation are presented and analyzed through the lens of BRT theory. This can help in the transition toward incorporating circularity in mainstream practice. In the case of food waste, it is also important to differentiate between avoidable and unavoidable food waste, as it helps in prioritizing and framing prevention strategies. The framework is vital for practitioners and future researchers because they propose to change the consumer perspective toward food waste in developing countries perspective. Assessment of the power of significance of the relationship between BRT and attitude of the households is an opportunity for quantitative studies to be developed in future. This information can be used to inform practitioners and policy-makers and may contribute to future efforts for reducing food waste at household levels.

## References

1. Morseletto P (2020) Targets for a circular economy. Resour Conserv Recycl 153:104553
2. Dodick J, Kauffman D (2017) A review of the European Union's circular economy policy. Report from project the route to circular economy. Project funded by European Union's Horizon, 2020, No. 730378. Retrieved from http://www.r2piproject.eu/wp-content/uploads/2017/04/A-Rview-of-the-European-Unions-Circular-Economy-Policy.pdf
3. Chang Y, Li Y (2014) Non-renewable resources in Asian economies: perspectives of availability, applicability acceptability, and affordability
4. Geissdoerfer M, Savaget P, Bocken NMP, Hultink EJ (2017) The circular economy—a new sustainability paradigm? J Clean Prod 143:757–768. https://doi.org/10.1016/j.jclepro.2016.12.048
5. Lieder M, Asif FMA, Rashid A, Mihelič A, Kotnik S (2017) Towards circular economy implementation in manufacturing systems using a multi-method simulation approach to link design and business strategy. Int J Adv Manuf Technol 93(5–8):1953–1970. https://doi.org/10.1007/s00170-017-0610-9
6. Camilleri MA (2019) The circular economy's closed loop and product service systems for sustainable development: a review and appraisal. Sustain Dev 27(3):530–536. https://doi.org/10.1002/sd.1909

7. Murray A, Skene K, Haynes K (2017) The circular economy: an interdisciplinary exploration of the concept and application in a global context. J Bus Ethics 140(3):369–380. https://doi.org/10.1007/s10551-015-2693-2
8. Stahel WR (2016) The circular economy. Nat News 531(7595):435
9. Allam Z, Jones D (2018) Towards a circular economy: a case study of waste conversion into housing units in Cotonou, Benin. Urban Science 2(4):118. https://doi.org/10.3390/urbansci2040118
10. Berg A, Antikainen R, Hartikainen E (2018) Reports of the Finnish environment: circular economy for sustainable development
11. Wijkman A, Skånberg K (2016) The circular economy and benefits for society. The Club of Rome, pp 1–59
12. Kummu M, De Moel H, Porkka M, Siebert S, Varis O, Ward PJ (2012) Lost food, wasted resources: global food supply chain losses and their impacts on freshwater, cropland, and fertiliser use. Sci Total Environ 438:477–489
13. Jurgilevich A, Birge T, Kentala-Lehtonen J, Korhonen-Kurki K, Pietikäinen J, Saikku L, Schösler H (2016) Transition towards circular economy in the food system. Sustainability 8(1):69
14. Gustavsson J, Cederberg C, Sonesson U, van Otterdijk R, Meybeck A (2011) Global food losses and food waste: extent, causes and prevention. FAO, Rome
15. Elkington J (1998) Partnerships from cannibals with forks: the triple bottom line of 21st-century business. Environ Qual Manage 8(1):37–51
16. FAO (2015) Global initiative on food loss and waste reduction. FAO. http://www.fao.org/3/a-i4068e.pdf
17. Yang Y, Bao W, Xie GH (2019) Estimate of restaurant food waste and its biogas production potential in China. J Clean Prod 211:309–320
18. Thi NBD, Kumar G, Lin CY (2015) An overview of food waste management in developing countries: current status and future perspective. J Environ Manage 157:220–229
19. Schanes K, Stagl S (2019) Food waste fighters: What motivates people to engage in food sharing? J Clean Prod 211:1491–1501
20. Stancu V, Haugaard P, Lähteenmäki L (2016) Determinants of consumer food waste behaviour: two routes to food waste. Appetite 96:7–17
21. Xue L, Liu G, Parfitt J, Liu X, Van Herpen E, Stenmarck Å, O'Connor C, Östergren K, Cheng S (2017) Missing food, missing data? A critical review of global food losses and food waste data. Environ Sci Technol 51(12):6618–6633
22. Aschemann-Witzel J, Ares G, Thøgersen J, Monteleone E (2019) A sense of sustainability? How sensory consumer science can contribute to sustainable development of the food sector. Trends Food Sci Technol
23. Corsini F, Gusmerotti NM, Testa F, Iraldo F (2018) Exploring waste prevention behaviour through empirical research. Waste Manage 79:132–141
24. Richter B, Bokelmann W (2018) The significance of avoiding household food waste–a means-end-chain approach. Waste Manage 74:34–42
25. Russell SV, William Young C, Unsworth KL, Robinson C (2017) Bringing habits and emotions into food waste behaviour. Resour Conserv Recycl 125:107–114
26. Diaz-Ruiz R, Costa-Font M, Gil JM (2018) Moving ahead from food-related behaviours: an alternative approach to understand household food waste generation. J Clean Prod 172:1140–1151
27. Porpino G, Parente J, Wansink B (2015) Food waste paradox: antecedents of food disposal in low income households. Int J Consum Stud 39(6):619–629
28. Porpino G (2016) Household food waste behavior: avenues for future research. J Assoc Consumer Res 1(1):41–51
29. Schanes K, Dobernig K, Gözet B (2018) Food waste matters—a systematic review of household food waste practices and their policy implications. J Clean Prod 182:978–991
30. Jha SN, Vishwakarma RK, Ahmad T, Rai A, Dixit AK (2015) Report on assessment of quantitative harvest and post-harvest losses of major crops and commodities in India. All India Coordinated Research Project on Post-Harvest Technology, ICAR-CIPHET.

31. Ong KL, Kaur G, Pensupa N, Uisan K, Lin CSK (2018) Trends in food waste valorization for the production of chemicals, materials and fuels: case study South and Southeast Asia. Biores Technol 248:100–112
32. Balaji M, Arshinder K (2016) Modeling the causes of food wastage in Indian perishable food supply chain. Resour Conserv Recycl 114:153–167
33. Song G, Li M, Semakula HM, Zhang S (2015) Food consumption and waste and the embedded carbon, water and ecological footprints of households in China. Sci Total Environ 529:191–197
34. Henz GP (2017) Postharvest losses of perishables in Brazil: what do we know so far? Hortic Bras 35(1):6–13
35. Frederiks ER, Stenner K, Hobman EV (2015) Household energy use: applying behavioural economics to understand consumer decision-making and behaviour. Renew Sustain Energ Rev 41:1385–1394
36. Kunda Z (1990) The case for motivated reasoning. Psychol Bull 108(3):480
37. Ajzen I (2001) Nature and operation of attitudes. Ann Rev Psychol 52(1):27–58
38. Fishbein M, Ajzen I (1975) Belief. In: Attitude, intention and behavior: an introduction to theory and research, vol 578
39. Ajzen I (1991) The theory of planned behavior. Organ Behav Hum Decis Process 50(2):179–211
40. Westaby JD (2005) Behavioral reasoning theory: Identifying new linkages underlying intentions and behavior. Organ Behav Hum Decis Process 98(2):97–120
41. Greve W (2001) Traps and gaps in action explanation: Theoretical problems of a psychology of human action. Psychol Rev 108(2):435
42. Claudy MC, Peterson M (2014) Understanding the underutilization of urban bicycle commuting: a behavioral reasoning perspective. J Public Policy Mark 33(2):173–187
43. Ryan J, Casidy R (2018) The role of brand reputation in organic food consumption: a behavioral reasoning perspective. J Retail Consum Serv 41:239–247
44. Gupta N, Yadav KK, Kumar V (2015) A review on current status of municipal solid waste management in India. J Environ Sci 37:206–217
45. Claudy MC, Garcia R, O'Driscoll A (2015) Consumer resistance to innovation—a behavioral reasoning perspective. J Acad Mark Sci 43(4):528–544
46. Chatzidakis A, Lee MS (2013) Anti-consumption as the study of reasons against. J Macromark 33(3):190–203
47. Levi AS, Pryor JB (1987) Use of the availability heuristic in probability estimates of future events: the effects of imagining outcomes versus imagining reasons Organ Behav Hum Decis Process 40(2):219–234
48. Myyry L, Siponen M, Pahnila S, Vartiainen T, Vance A (2009) What levels of moral reasoning and values explain adherence to information security rules? An empirical study. Eur J Inf Syst 18(2):126–139
49. Di Talia E, Simeone M, Scarpato D (2019) Consumer behaviour types in household food waste. J Clean Prod 214:166–172
50. Graham-Rowe E, Jessop DC, Sparks P (2019) Self-affirmation theory and pro-environmental behaviour: promoting a reduction in household food waste. J Environ Psychol 62:124–132
51. Dixit S, Badgaiyan AJ, Khare A (2019) An integrated model for predicting consumer's intention to write online reviews. J Retail Consum Serv 46:112–120
52. Graham-Rowe E, Jessop DC, Sparks P (2015) Predicting household food waste reduction using an extended theory of planned behaviour. Resour Conserv Recycl 101:194–202
53. Yarimoglu E, Kazancoglu I, Bulut ZA (2019) Factors influencing Turkish parents' intentions towards anti-consumption of junk food. British Food J
54. Dreezens E, Martijn C, Tenbült P, Kok G, De Vries NK (2005) Food and values: an examination of values underlying attitudes toward genetically modified-and organically grown food products. Appetite 44(1):115–122
55. Casidy R, Lwin M, Phau I (2017) Investigating the role of religiosity as a deterrent against digital piracy. Market Intell Plan
56. Hawkins JD, Smith BH, Catalano RF (2004) Social development and social and emotional learning. *Building academic success on social and emotional learning:* In: What does the research say, pp 135–150

57. Cicatiello C, Franco S, Pancino B, Blasi E (2016) The value of food waste: an exploratory study on retailing. J Retail Consum Serv 30:96–104
58. Sirieix L, Lála J, Kocmanová K (2017) Understanding the antecedents of consumers' attitudes towards doggy bags in restaurants: concern about food waste, culture, norms and emotions. J Retail Consum Serv 34:153–158
59. Ilyuk V (2018) Like throwing a piece of me away: how online and in-store grocery purchase channels affect consumers' food waste. J Retail Consum Serv 41:20–30
60. Goh E, Jie F (2019) To waste or not to waste: exploring motivational factors of generation Z hospitality employees towards food wastage in the hospitality industry. Int J Hosp Manage 80:126–135
61. Fami HS, Aramyan LH, Sijtsema SJ, Alambaigi A (2019) Determinants of household food waste behavior in Tehran city: a structural model. Resour Conserv Recycl 143:154–166
62. Shin DH (2009) Towards an understanding of the consumer acceptance of mobile wallet. Comput Hum Behav 25(6):1343–1354
63. Beckerman W (1992) Economic development and the environment: conflict or complementarity? In: world development report. Retrieved from http://eprints.ucl.ac.uk/17888/
64. Barr S, Gilg AW, Ford NJ (2001) Differences between household waste reduction, reuse and recycling behaviour: a study of reported behaviours, intentions and explanatory variables. Environ Waste Manage 4(2):69–82
65. Gertsakis J, Lewis H (2003) Sustainability and the waste management hierarchy. In A discussion paper on the waste management hierarchy and its relationship to sustainability. Retrieved from http://www.ecorecycle.vic.gov.au/resources/documents/TZW_-_Sustai nability_and_the_Waste_Hierarchy_(2003).pdf
66. Liu H (2009) Recycling economy and sustainable development. J Sustain Dev 2(1):209–212. https://doi.org/10.5539/jsd.v2n1p209
67. Mak TM, Iris KM, Tsang DC, Hsu SC, Poon CS (2018) Promoting food waste recycling in the commercial and industrial sector by extending the theory of planned behaviour: a Hong Kong case study. J Clean Prod 204:1034–1043
68. Röös E, Bajželj B, Smith P, Patel M, Little D, Garnett T (2017) Greedy or needy? Land use and climate impacts of food in 2050 under different livestock futures. Glob Environ Chang 47:1–12
69. Aydin AE, Yildirim P (2021) Understanding food waste behavior: the role of morals, habits and knowledge. J Clean Prod 280:124250
70. Klöckner CA (2015) The psychology of pro-environmental communication: beyond standard information strategies. Springer
71. Dyen M, Sirieix L, Costa S (2021) Fostering food waste reduction through food practice temporalities. Appetite 161:105131
72. Kashif M, Zarkada A, Ramayah T (2018) The impact of attitude, subjective norms, and perceived behavioural control on managers' intentions to behave ethically. Total Qual Manag Bus Excell 29(5–6):481–501
73. Khalid S, Naseer A, Shahid M, Shah GM, Ullah MI, Waqar A, Abbas T, Imran M, Rehman F (2019) Assessment of nutritional loss with food waste and factors governing this waste at household level in Pakistan. J Clean Prod 206:1015–1024
74. Andrews D (2015) The circular economy, design thinking and education for sustainability
75. Gilli M, Nicolli F, Farinelli P (2018) Behavioural attitudes towards waste prevention and recycling. Ecol Econ 154:294–305
76. Ingrao C, Faccilongo N, Di Gioia L, Messineo A (2018) Food waste recovery into energy in a circular economy perspective: a comprehensive review of aspects related to plant operation and environmental assessment. J Clean Prod 184:869–892
77. Mattar L, Abiad MG, Chalak A, Diab M, Hassan H (2018) Attitudes and behaviors shaping household food waste generation: lessons from Lebanon. J Clean Prod 198:1219–1223
78. Mondéjar-Jiménez JA, Ferrari G, Secondi L, Principato L (2016) From the table to waste: an exploratory study on behaviour towards food waste of spanish and italian youths. J Clean Prod 138:8–18

79. Parker JR, Umashankar N, Schleicher MG (2019) How and why the collaborative consumption of food leads to overpurchasing, overconsumption, and waste. J Public Policy Mark 38(2):154–171

80. Principato L, Ruini L, Guidi M, Secondi L (2019) Adopting the circular economy approach on food loss and waste: the case of Italian pasta production. Resour Conserv Recycl 144:82–89

81. van Herpen E, de Hooge IE (2019) When product attitudes go to waste: wasting products with remaining utility decreases consumers' product attitudes. J Clean Prod 210:410–418

82. Vilariño MV, Franco C, Quarrington C (2017) Food loss and waste reduction as an integral part of a circular economy. Front Environ Sci 5:21

83. Visvanathan C, Norbu T (2006) Reduce, Reuse, and Recycle : The 3Rs in South Asia 2 . Current Practices of 3Rs in South Asia. 3 R South Asia Expert Workshop, (April)

84. Wang RJH, Malthouse EC, Krishnamurthi L (2015) On the go: How mobile shopping affects customer purchase behavior. J Retail 91(2):217–234

85. Watson M, Meah A (2012) Food, waste and safety: negotiating conflicting social anxieties into the

86. Yadav D, Barbora L, Rangan L, Mahanta P (2016) Tea waste and food waste as a potential feedstock for biogas production. Environ Prog Sustainable Energy 35(5):1247–1253

87. Zhang H, Duan H, Andric JM, Song M, Yang B (2018) Characterization of household food waste and strategies for its reduction: a Shenzhen City case study. Waste Manage 78:426–433

# Modeling the Causes of Post-harvest Loss in the Agri-Food Supply Chain to Achieve Sustainable Development Goals: An ISM Approach

Mukesh Kumar and Vikas Kumar Choubey

**Abstract** In emerging economies like India, wastage along the Agri-Food chain is a huge challenge. Thus, it becomes imperative to analyze the causes of post-harvest losses. This study aims to identify the causes of wastage in Agri-Food, as well as the interaction among the causes. The driving and dependence power of every cause of wastage, as well as their level of importance, has been analyzed. Interpretive structural modeling (ISM) has been utilized for obtaining driving and dependence power and establishing the interrelationship among causes. Matrix cross multiplication and classification (MICMAC) has been applied for clustering the causes into four groups. Thirteen causes of Agri-Food wastage in the Indian context have been analyzed by applying ISM and MICMAC. The study shows that poor logistics infrastructure, large numbers of intermediates, and lack of innovative technology are the most influential causes for PHL in Indian Agri-Food supply chain. Reducing wastage in the Agri-Food supply chain in emerging economies countries like India may get big boosts in their economy as well as to achieve sustainable development goals (like zero hunger, sustainable cities, etc.). Food wastage is linked with a huge amount of environmental emission so reducing wastage will help in achieving SDG like livable land, sustainable cities, zero hunger, etc. This work may be utilized by supply chain practitioners, decision-makers, and policymakers in reduction of Agri-Food wastage.

**Keywords** Agri-Food · Post-harvest loss · Sustainable development goals · Interpretive structural modeling · MICMAC analysis

M. Kumar (✉) · V. K. Choubey
National Institute of Technology Patna, Patna 800005, India
e-mail: mukesh.me18@nitp.ac.in

V. K. Choubey
e-mail: vikas.choubey@nitp.ac.in

© The Author(s), under exclusive license to Springer Nature Singapore Pte Ltd. 2021
R. S. Mor et al. (eds.), *Challenges and Opportunities of Circular Economy in Agri-Food Sector*, Environmental Footprints and Eco-design of Products and Processes,
https://doi.org/10.1007/978-981-16-3791-9_8

133

# 1  Introduction

Fulfilling the food demand of the rising population in India (1.7 billion by 2050) while decreasing the food loss and wastage poses a huge challenge. The wastage of one kilogram of rice and wheat is interconnected to the wastage of 1500 and 3500 L of fresh groundwater respectively that were used in production. Nearly one-third of food, which equals around 1.7 billion tones, is lost every year globally [1]. These wastages account for environmental, social, and economic costs of nearly $700 billion, $900 billion, and $1 trillion per year respectively [2].

In India, food worth ₹ 92,000 crores get wasted annually. Major losses in food are associated with harvest and post-harvest losses. In the production and handling stage nearly 48% of the wastage occurs, while the consumption stage accounts for nearly 35% of food wastage. These are the main three stages of food wastage and account for 80% of total global food wastages. Previous research shows that lack of logistics infrastructure is the main issue in such food wastage [3-5]. Logistics infrastructure includes poor storage capacity, lack of refrigerated carriers for fruits and vegetables, poor road infrastructure, etc.

Post-harvest losses in the Indian Agri-Food sector account for 60% of total losses. The wastage of food surplus is also associated with greenhouse gas emissions and land degradation [6, 7]. The sustainable development goals set by the United Nation (UNICEF) (such as 'no hunger', 'sustainable cities', 'life on the planet', etc.) can't be achieved along with food wastage. To fulfill the food demand of a huge population needs to achieve SDGs ('livable planet', 'Zero Hunger', etc.) through minimizing valuable food wastages [8, 9]. Food wastage may be minimized by identifying causes of food losses and then interpreting the interrelationship among causes of food wastage to obtain the most critical cause which contributes most. The post-harvest in the Agri-Food supply chain is responsible for nearly 60% of total food wastage, so in this paper, we study the causes for the post-harvest loss. Interpretive structural modeling ISM has been applied to analyze the hierarchical model for the causes of PHL. A matrix cross multiplication and classification MICMAC analysis has been performed to classify the various post-harvest losses into four different groups.

## 1.1  Research Objectives

This paper has the following objectives which are:

RO1—To identify the causes for the post-harvest losses in the Agri-Food supply chain.

RO2—To Understand the contextual interrelationship among the causes of PHL in Agri-Food supply chain (AFSC) using ISM methodology.

RO3—To Develop the hierarchical structural model and importance level for all the causes of PHL in AFSC.

## 2 Literature Review

This section of the paper focuses on previous research on food wastage, and application of ISM approach in the Agri-Food supply chain. 30 research articles were studied to identify causes of food wastage in the PHL system. Peer-reviewed articles were included from Google scholar and Scopus database.

Post-harvest loss is quantifiable quantity (quantitatively as well as qualitatively) in the PHL system. The post-harvest loss denotes the loss of Agri-Food between harvesting stage and the final consumption [10, 11]. There are a large number of intermediate stages after post-harvest such as Harvesting, transportation, storage, packaging, processing, distribution, and consumption, etc., all these stages are accountable for some valuable food loss [12]. Food losses are accountable for financial loss, food value loss, increased hunger, environmental loss, etc.

Food wastage is accountable for direct or indirect effect on the environment, social, and economy of the country [13]. In harvesting stage, several causes of food wastages are associated such as, (a) lack of knowledge about harvesting [12], this leads to premature harvesting, (b) losses due to unsuitable weather conditions [14] accountable for the moisture contents in food product, (c) lack of harvesting technique [15]. Main causes of wastage at storage stage are insects, deterioration, shrinkage, spoilage, losses due to handling, etc. [12, 16, 17]. There are several causes for food wastage at processing stage such as packaging failure, transportation issue, shrinkage, etc. [18]. At distribution stage, various losses occur such as transportation losses, poor logistics infrastructure losses, improper handling and storage losses, demand management losses, etc. [19, 20]. Several losses associated at consumption stages such as poor storage of food grains, bulk purchasing, etc. [21].

Balaji and Arshinder [3], Garrone et al. [22], Papargyropoulou et al. [23] developed a food waste hierarchy to minimize food wastage and found that prevention of food wastage is the best way to reduce food waste, after that reuse of food, recycle the food commodity, recovery, and disposal. Balaji and Arshinder [3] observed that if food is surplus then donation of food to the needy people is the best way to avoid wastages. Eriksson and Spångberg [24] applied LCA, to reduce carbon emission and energy utilization from the fruit and vegetable waste generated in supermarkets, and observe that there was a huge potential of reducing GHG emissions and primary energy through changing food waste management techniques. Gardas et al. [25], Kumar et al. [26], Kumar And Choubey [9] applied ISM in modeling the cause of PHL in the food sector, while [12] applied DEMATEL to find critical causal factors of PHL. Balaji and Arshinder [3] applied ISM to model the cause of PHL in Indian perishable food supply chain.

After an exhaustive literature study, it is observed that food wastage is the prime concern. Although several literatures are available in identifying causes of Agri-Food wastage. Still there are some reasonable gaps in the area of food wastage. This paper owing to identifying the causes of post-harvest loss in the Agri-Food supply chain.

## 3   Research Methodology

A proposed research framework has been depicted in Fig. 1. To fulfill the research objective, several causes of food wastage were identified through literature review. Various keywords were searched in google scholar search engine and Scopus

**Fig. 1** Research framework. *Source* Author

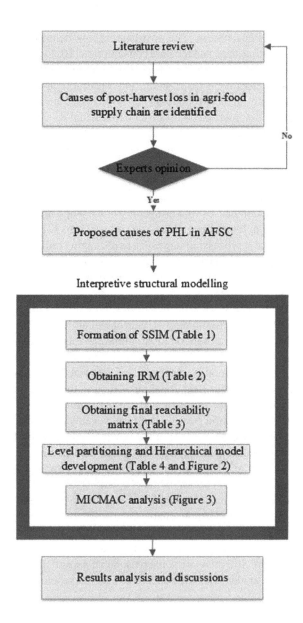

database, such as 'food wastage', 'Agri-Food wastage', 'post-harvest loss', etc. Identified causes of PHL in AFSC are sent to experts for opinion. In this study 4 experts have participated, 2 from academics and 2 from the food industry. After analysis thirteen causes were selected for analysis. An ISM approach has been applied in this paper to obtain hierarchical model along with the importance level of each cause. Following steps were followed that are given below.

*Step 1. Identification of cause for the post-harvesting losses in Agri-Food supply chain.*

Various causes for PHL in Agri-Food were identified through the available literature, articles published in newspapers and expert's opinion.

*Step 2. Verifying causes for PHL through field as well as academic experts.*

Identified causes for PHL are sent to various experts (academics as well as field) through mail and asked to select relevant causes for PHL in Indian Agri-Food sector. Four responses were collected from the expert. Thirteen causes recommended by the experts were selected for analysis in this study.

*Step 3. Proposed causes for post-harvest in Agri-Food supply chain.*

The following thirteen causes for PHL in the Agri-Food supply chain have been selected after experts' recommendations and using previous literature.

i.   *Traceable issue in transport, and storagePHL-F1*
     Traceability enables stakeholders to keep adequate information about the store inventory within the food supply chain [27]. Lack of traceability issue in transportation and storage is one of the key causes for PHL in AFSC. The traceable systems increase viability in the food supply chain so that product will be tracked and traced anywhere in the supply chain, it also increases transparency within the system [12, 28].

ii.  *Improper packaging and handling PHL-F2*
     Improper packaging and handling system leads to food wastage. Pandey [29] observed that packaging in the Agri-Food supply chain has a significant role in food wastage. Proper packaging and handling may increase the security and quality of food systems [23, 30].

iii. *Lack of cold storage infrastructure PHL-F3*
     Cold storage facility is the backbone of the perishable food supply chain. Currently, India requires a total of 61.13 million metric tons capacity of cold storage but only 32 million metric tons capacity is available, i.e., approximately 50% of required capacity [3, 31]. Zanoni and Zavanella [32] stated that cold chain infrastructure is most important in the perishable food supply chain to avoid food wastage. Balaji and Arshinder [3], Raut et al. [21] identified cold chain infrastructure as a main cause of food wastage after PHL.

iv.  *Lack of government support PHL-F4*
     Government support leads to establishing the cold storage facility, buying facilities of food (mandi), the decision on import and export, and storage

facility which are the prime causes of food wastage. Government support to farmers and intermediates is the prime factor.

v.   *Poor logistics infrastructure PHL-F5*
     Balaji and Arshinder [3] had identified that logistics infrastructure is the main cause of PHL in Indian perishable food supply chain. Nyamah et al. [33], Singh et al. [34] said that logistics infrastructure associated with refrigerated carrier's, and good road infrastructure are needed for a sustainable food supply chain.

vi.  *Lack of innovative technology (for handling, harvesting, etc.) PHL-F6*
     Application of Innovative technology in the food supply chain includes traceable systems, information communication tools, modern threshing machines, modern harvesting methods, and other modern technology in storage, harvesting, transporting, etc. helps to minimize food wastage [3, 35, 36].

vii. *Lack of communication PHL-F7*
     Communication among supply chain players helps to reduce the food wastage [37]. Gold et al. [38], Ivanov and Dolgui [39] said that communication systems within the supply chain increases the viability and transparency that helps to reduce food wastage.

viii. *Severe heat of Indian summer PHL-F8*
     The panel of experts suggests that severe heat of Indian summer affects the food supply chain badly and the primary cause of wastages of fruits and vegetables.

ix.  *Premature harvesting PHL-F9*
     Premature harvesting is the primary cause of food wastage. It directly affects the quality of food and food gets wasted. Gardas et al. [12] used premature harvesting as a causal factor of food waste after PHL.

x.   *Lack of scientific/modern harvesting method PHL-F10*
     Singh et al. [40] stated that modern harvesting methods help to decrease the PHL in the Agri-Food supply chain. In the new era, the agriculture sector needs for agri-4.0 to tackle food wastage within the supply chain [41-43].

xi.  *Large number of intermediates PHL-F11*
     Gardas et al. [12] stated that large numbers of intermediates are responsible for wastage in the Agri-Food sector because it requires lack of traceability, large number of storages, transportation channel, packaging and handling at each stage.

xii. *Poor demand management PHL-F12*
     Demand management plays an important role for the food supply chain. Failure of demand management means improper distribution of food commodity and hence wastage is there. If forecasting of any system gets wrong it leads to shortage or accumulation of food commodities.

xiii. *Lack of linkage between institution, industry, and government PHL-F13*
     Balaji and Arshinder [3], Gardas et al. [12] observed that the Linkage between institutions, industry, and Govt. helps to get new and updated technology rapidly. The problem associated with the Agri-Food supply chain may be

shared with institutions and government and subsequently they try to resolve the same.

*Step 4. Developing ISM model.*

ISM was first utilized by Warfield [44] to analyze the contextual interrelationship for a complex system, and generate hierarchical multilevel structural model. In this study, the following steps have been adopted from [3], Kamble et al. [36]; Mangla et al. [45]; Prakash et al. [46] to implement ISM methodology.

i. Development of structural self-interaction matrix—SSIM
    SSIM was formed to fill the relationship among the causal factors in a pairwise comparison matrix. The pairwise comparison matrix was filled through performing brainstorming sessions, expert's opinion, and group discussion. The SSIM matrix for causal factors for PHL is shown in Table 1. Relationship among causes in SSIM was filled by using various symbols 'V', 'A', 'X', and 'O'. The significance of the symbols are given below

    $V$—$V$ is used when cause '$i$' leads to cause '$j$'.
    $A$—$A$ is used when cause '$j$' leads to cause '$i$'.
    $X$—$X$ is used when '$i$' and '$j$' both help each other.
    $O$—$O$ is used when both the causes are unrelated.

    Note—here '$i$' denoted to row value and '$j$' denoted by column value.

ii. Obtaining Initial reachability matrix—IRM
    Initial reachability matrix was obtained by transforming the symbols used in SSIM into a binary digit. The transformation of SSIM to IRM is followed by some rules. The rules for converting SSIM symbols into binary digits are given below.

    - When the symbol $(i, j)$ is $V$ then the entry value of $(i, j)$ is 1 and $(j, i)$ is 0. For example, if $(1, 2)$ value is $V$ then the entry at $(1, 2)$ is '1' and the entry at $(2, 1)$ is 0. Here $(1, 2)$ denotes entry value at first row and second column.
    - When the symbol of cell $(i, j)$ is $A$ then the entry value of $(i, j)$ is 0 and $(j, i)$ is 1.
    - When the symbol of cell $(i, j)$ is $X$ then the entry value of $(i, j)$ is 1 and $(j, i)$ is 1.
    - When the symbol of cell $(i, j)$ is $O$ then the entry value of $(i, j)$ is 0 and $(j, i)$ is 0.

    By using the above rules, the SSIM is converted into binary IRM and shown in Table 2.

iii. Obtaining final reachability matrix
    Final reachability matrix was obtained by performing a transitivity check for each entry whose value is zero in IRM. Transitivity check is performed by checking the relationship of cause, if cause '$p$' is related to cause '$q$' and cause '$q$' is related to cause '$r$' then cause '$p$' and cause '$r$' are also related. If '$p$' and '$r$' are not related then it is replaced by value 1. The transitivity check is

**Table 1** Structural self-interaction matrix-SSIM

| | PHL-F1 | PHL-F2 | PHL-F3 | PHL-F4 | PHL-F5 | PHL-F6 | PHL-F7 | PHL-F8 | PHL-F9 | PHL-F10 | PHL-F11 | PHL-F12 | PHL-F13 |
|---|---|---|---|---|---|---|---|---|---|---|---|---|---|
| PHL-F1 | X | O | A | V | A | A | V | O | O | A | A | O | O |
| PHL-F2 | | X | A | O | O | O | O | O | O | O | O | O | O |
| PHL-F3 | | | X | A | A | A | O | O | O | A | A | O | A |
| PHL-F4 | | | | X | V | V | O | O | V | V | O | O | V |
| PHL-F5 | | | | | X | O | O | O | O | O | O | O | O |
| PHL-F6 | | | | | | X | O | V | O | V | O | V | O |
| PHL-F7 | | | | | | | X | O | V | V | A | V | V |
| PHL-F8 | | | | | | | | X | O | O | O | O | O |
| PHL-F9 | | | | | | | | | X | A | O | O | A |
| PHL-F10 | | | | | | | | | | X | O | O | A |
| PHL-F11 | | | | | | | | | | | X | V | V |
| PHL-F12 | | | | | | | | | | | | X | O |
| PHL-F13 | | | | | | | | | | | | | X |

*Source* Author

**Table 2** Initial reachability matrix IRM

| | PHL-F1 | PHL-F2 | PHL-F3 | PHL-F4 | PHL-F5 | PHL-F6 | PHL-F7 | PHL-F8 | PHL-F9 | PHL-F10 | PHL-F11 | PHL-F12 | PHL-F13 |
|---|---|---|---|---|---|---|---|---|---|---|---|---|---|
| PHL-F1 | 1 | 0 | 0 | 1 | 0 | 0 | 1 | 0 | 0 | 0 | 0 | 0 | 0 |
| PHL-F2 | 0 | 1 | 0 | 0 | 0 | 0 | 0 | 0 | 0 | 0 | 0 | 0 | 0 |
| PHL-F3 | 1 | 1 | 1 | 0 | 0 | 0 | 0 | 0 | 0 | 0 | 0 | 0 | 0 |
| PHL-F4 | 0 | 0 | 1 | 1 | 1 | 1 | 0 | 0 | 1 | 1 | 0 | 0 | 1 |
| PHL-F5 | 1 | 0 | 1 | 0 | 1 | 0 | 0 | 0 | 0 | 0 | 0 | 0 | 0 |
| PHL-F6 | 1 | 0 | 1 | 0 | 0 | 1 | 0 | 1 | 0 | 1 | 0 | 1 | 0 |
| PHL-F7 | 0 | 0 | 0 | 0 | 0 | 0 | 1 | 1 | 1 | 1 | 0 | 1 | 1 |
| PHL-F8 | 0 | 0 | 0 | 0 | 0 | 0 | 0 | 1 | 0 | 0 | 0 | 0 | 0 |
| PHL-F9 | 0 | 0 | 0 | 0 | 0 | 0 | 0 | 0 | 1 | 0 | 0 | 0 | 0 |
| PHL-F10 | 1 | 0 | 1 | 0 | 0 | 0 | 1 | 0 | 1 | 1 | 1 | 0 | 0 |
| PHL-F11 | 1 | 0 | 1 | 0 | 0 | 0 | 0 | 0 | 0 | 0 | 1 | 1 | 0 |
| PHL-F12 | 0 | 0 | 0 | 0 | 0 | 0 | 0 | 0 | 0 | 0 | 0 | 1 | 0 |
| PHL-F13 | 0 | 0 | 1 | 0 | 0 | 0 | 0 | 0 | 1 | 1 | 0 | 0 | 1 |

*Source* Author

performed for all the entry values whose score is zero. The final reachability matrix is shown in Table 3.

iv.  Level partitioning
Level partitioning was performed to get importance level and relationship hierarchy. Level partitioning is performed by matching reachability set and interaction set for the particular factor. The interaction set was obtained by matching reachability set and antecedent set, if reachability and intersection set entry value are same then the corresponding causal factors are removed and we get 1st level and the performing the iteration by following the same rule.

v.  Formation of interrelationship hierarchy
Interrelationship hierarchy or interrelationship model was generated by using level partitioning. The causal factor selected at first level was placed at top of the hierarchy and causal factors selected at least were placed at bottom of the hierarchy. The relationship vector is directed towards upward direction. The relationship hierarchy is shown in Fig. 2.

*Step 5. MICMAC analysis.*

MICMAC analysis matrix for cross multiplication and classification is performed by using the driving and dependence power scored by all the causal factors. The dependence and driving power of the entire causal factors are calculated by summing the row value and column value of corresponding causal factors.

# 4   Results and Discussion

## 4.1   *Interpretive Structural Modeling*

Thirteen causes for PHL in the Agri-Food supply chain have been analyzed by utilizing ISM and MICMAC to establish the interrelationships between the causal factors. Structural self-interaction matrix SSIM is shown in Table 1. In SSIM table, the relationship among two set of causal factors was filled in a pairwise comparison matrix by using four alphabetical numbers *V, A, X* and *O*. SSIM is transformed into initial reachability matrix IRM by converting *V, A, X* and *O* into binary digit 0 and 1 and shown in Table 2. The final reachability matrix is generated from IRM by performing transitivity check, the result after transitivity check is shown in Table 3.

Levels of importance for each cause were obtained by applying level partitioning which is shown in Table 4 and the hierarchy plot of causal factors are shown in Fig. 2. In this study, severe heat in Indian summer PHL-F2, Improper packaging and handling PHL-F8, premature harvesting PHL-F9 and poor demand management PHL-F12 are placed at the top of the hierarchy belonging to the 1st level of causal factor. 1st levels of causal factors have very less influence on the PHL causal factors and are driven by all the causal factors placed at lower levels. Lack of cold storage facility and traceability issues are situated at 2nd level of hierarchy while

**Table 3** Final reachability matrix for causal factors in PHL

|  | PHL-F1 | PHL-F2 | PHL-F3 | PHL-F4 | PHL-F5 | PHL-F6 | PHL-F7 | PHL-F8 | PHL-9 | PHL-F10 | PHL-F11 | PHL-F12 | PHL-F13 |
|---|---|---|---|---|---|---|---|---|---|---|---|---|---|
| PHL-F1 | 1 | 0 | 1* | 1 | 1* | 1* | 1 | 0 | 1* | 1* | 0 | 1* | 1* |
| PHL-F2 | 0 | 1 | 0 | 0 | 0 | 0 | 0 | 0 | 0 | 0 | 0 | 0 | 0 |
| PHL-F3 | 1 | 1 | 1 | 1* | 0 | 0 | 1* | 0 | 0 | 0 | 0 | 0 | 0 |
| PHL-F4 | 1* | 1* | 1 | 1 | 1 | 1 | 0 | 1* | 1 | 1 | 0 | 1* | 1 |
| PHL-F5 | 1 | 1* | 1 | 1* | 1 | 0 | 1* | 0 | 0 | 0 | 0 | 0 | 0 |
| PHL-F6 | 1 | 1* | 1 | 1* | 0 | 1 | 1* | 1 | 1* | 1 | 0 | 1 | 0 |
| PHL-F7 | 1* | 0 | 1* | 0 | 0 | 0 | 1 | 0 | 1 | 1 | 0 | 1 | 1 |
| PHL-F8 | 0 | 0 | 0 | 0 | 0 | 0 | 1 | 1 | 0 | 0 | 0 | 0 | 0 |
| PHL-F9 | 0 | 0 | 0 | 0 | 0 | 0 | 0 | 0 | 1 | 0 | 0 | 0 | 0 |
| PHL-F10 | 1 | 1* | 1 | 1 | 0 | 0 | 1* | 0 | 1* | 1 | 1 | 1 | 0 |
| PHL-F11 | 1 | 1* | 1 | 1* | 0 | 0 | 1 | 0 | 0 | 1* | 1 | 1 | 1 |
| PHL-F12 | 0 | 0 | 0 | 0 | 0 | 0 | 0 | 0 | 0 | 0 | 0 | 1 | 0 |
| PHL-F13 | 1* | 1* | 1 | 0 | 0 | 0 | 0 | 0 | 1 | 1 | 0 | 0 | 1 |

*Source* Author

1* denote transitive relationship

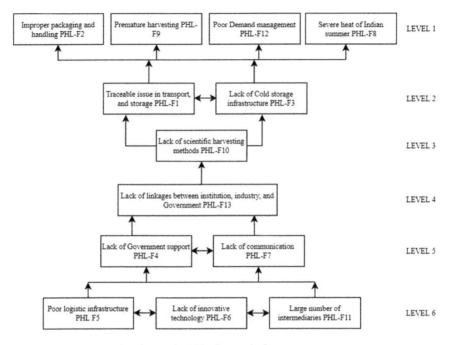

**Fig. 2** Structural hierarchy of cause for PHL. *Source* Author

**Table 4** Level partitioning

| Factors | Driving power | Dependence power | Reachability set | Antecedents set | Level |
|---------|---------------|------------------|------------------|-----------------|-------|
| PHL-F1 | 10 | 9 | 1, 3, 4, 5, 6, 7, 10, 13 | 1, 3, 4, 5, 6, 7, 10, 11, 13 | 2 |
| PHL-F2 | 1 | 8 | 2 | 2, 3, 4, 5, 6, 10, 11, 13 | 1 |
| PHL-F3 | 5 | 9 | 1, 3, 4, 7 | 1, 3, 4, 7, 10, 11, 13 | 2 |
| PHL-F4 | 11 | 7 | 4, 5, 6 | 4, 5, 6, 11 | 5 |
| PHL-F5 | 6 | 3 | 5 | 5 | 6 |
| PHL-F6 | 10 | 3 | 6 | 6 | 6 |
| PHL-F7 | 7 | 7 | 7 | 5, 6, 7, 11 | 5 |
| PHL-F8 | 1 | 3 | 8 | 6, 8 | 1 |
| PHL-F9 | 1 | 8 | 9 | 1, 4, 6, 7, 9, 10, 11, 13 | 1 |
| PHL-F10 | 7 | 7 |  |  | 3 |
| PHL-F11 | 10 | 1 | 11 | 11 | 6 |
| PHL-F12 | 1 | 6 | 12 | 1, 4, 6, 7, 11, 12 | 1 |
| PHL-F13 | 6 | 5 | 13 | 4, 7, 11, 13 | 4 |

*Source* Author

poor logistics infrastructure (PHL-F5), lack of innovative technology (PHL-F6) and large number of intermediates (PHL-F11) are placed at bottom of the hierarchy so far these are the top causal factors for PHL in Agri-Food supply chain. These causal factors are at the top priority of the organization for minimizing food wastage. Lack of government support and communication are placed at the fifth level of hierarchy which are driven by sixth level of causal factors and drive all factors placed above it. Lack of linkage between institutions, industry, and governments are placed at fourth level on the hierarchy while lack of scientific harvesting method is placed at third level of hierarchy.

Result of this study is analogues to the result derived by Balaji and Arshinder [3], Gardas et al. [12], Raut et al. [47], such that poor logistics infrastructure and lack of innovative technology are the most influenced causal PHL. Poor logistics infrastructure and innovative technology and large number of mediators are kept at top priority for practitioners as well as researchers in order to minimize the cause of PHL in the Agri-Food supply chain.

## 4.2   MICMAC Analysis

MICMAC analysis is used to classify the causes of food wastage into four clusters, namely (a) Autonomous cluster (b) dependent cluster (c) Linkage cluster (d) Independent cluster. MICMAC analysis is performed by plotting graphs between dependence power and driving power.

The factors whose dependence and driving power are low placed in autonomous clusters and it should be neglected from the study. Severe heat in Indian summer PHL-F8 is placed in autonomous clusters so this causal factor may be deleted from study and has no or very less influence on the system.

Factors whose dependence power is high but driving power are low are known as dependents cluster placed at bottom right corner. Factors placed in dependent cluster are reliant on other causes, generally placed at the top level. In this result PHL-F2, PHL-F3, PHL-F9, and PHL-F12 are placed in dependence cluster. PHL-F2, PHL-F9, and PHL-F12 are placed at the top of the hierarchy in Fig. 2. PHL-F3 is placed at 2nd level (Fig. 3).

Factors having high dependence and driving power are known as linkage variables and placed at top of the right corner in the MICMAC graph. PHL-F1, PHL-F4, PHL-F7, and PHL-F10 are placed in linkage cluster and are very sensitive in nature.

Factors whose driving power is high while dependence power is low are placed at top of the left corner and known as independent cluster. The causal factor placed within this cluster is independent. PHL-F5, PHL-F6 PHL-F11 and PHL-F13 are placed in this cluster and they are independent causal factor. Generally, indicators placed within this cluster should belong to higher level of hierarchy and PHL factors 5, 6, and 11 are placed at bottom (6th) level while PHL-F13 are placed at 4th level.

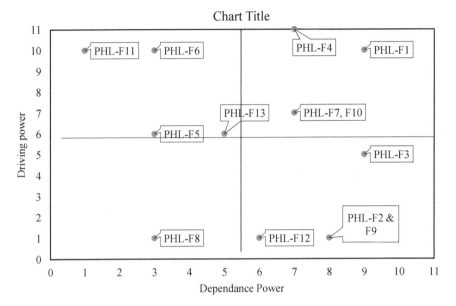

**Fig. 3** Group classification of cause of PHL in Agri-Food. *Source* Author

## 5 Implications

In covid-19 pandemic, distribution of food commodities got affected due to lock-down, boundary restrictions, fear about security of food, and many more. Food wastage is one of the key challenges in the Agri-Food supply chain. In order to minimize the food wastage several researchers have studied the causes of wastage but still there is reasonable space to identify post-harvest losses and to understand the relationship among them. By minimizing food wastage Sustainable development goal SDG set by UNICEF such as SDG 2 zero hunger, SDG 3 Good health and well-being, SDG15 life on Land, and SDG 12 sustainable city and community has to be achieved within the post-harvesting system. In this study poor logistic infrastructure PHL-F5, lack of innovative technology PHL-F6, and large number of mediators PHL-F11 are found to be the most effective cause of PHL in the Agri-Food supply chain. In MICMAC analysis PHL-F5, PHL-F6 and PHL-F11 are found at independent cluster which suggests that these are the most influential causes. The findings of this study may be utilized by the food organization to minimize food wastage along with sustainable development goals.

# 6  Conclusions

Food wastage is one of the critical challenges in India. This study identifies thirteen causes of post-harvest losses in the Agri-Food supply chain. Interaction among causes of wastages of Agri-Food is developed using ISM and a hierarchical structural model is developed. This study will help decision-makers, managers, government, and other stakeholders in understanding the interrelationships among the factors of food wastages. The study shows that poor logistics infrastructure, large numbers of intermediaries, and lack of utilization of innovative technology are the most influential cause for PHL in the Indian Agri-Food supply chain. Decision-makers and managers should first focus on these factors to minimize wastage under the Agri-Food supply chain. Reduction of food waste in the supply chain will help to achieve sustainable development goals such as zero hunger, sustainable cities, and livable land.

However, the results of ISM may be deflecting, if the set of experts are changed. Some different causes may be included for modeling. Different methods such as structural equation modeling, DEMATEL, etc. may be applied to modeling the causes of food wastages.

# References

1. Food and Agriculture Organization of the United Nations (FAO) (2012) The state of food and agriculture; investing in agriculture for a better future [online] http://www.fao.org/3/i3028e/i3028e.pdf
2. Bordoloi B (2016) Curbing food wastage in a hungry world. The Hindu Business Line 'Problem of food wastage in India—The Hindu BusinessLine'
3. Balaji M, Arshinder K (2016) Modeling the causes of food wastage in Indian perishable food supply chain. Resour Conserv Recycl 114:153–167
4. Negi S, Anand N (2017) Post-harvest losses and wastage in Indian fresh agro supply chain industry: a challenge. IUP J Supply Chain Manage 14(2)
5. SME Times (2021) Suitable infrastructure will help reduce post-harvest losses. Govt. SME Times News Bureau
6. Food and Agriculture Organization of the United Nations (FAO) (2015) Food wastage footprint & climate change
7. Noya LI, Vasilaki V, Stojceska V, Gonzalez-García S, Kleynhans C, Tassou S, Moreira MT, Katsou E (2018) An environmental evaluation of food supply chain using life cycle assessment: a case study on gluten free biscuit products. J Cleaner Prod 170:451–461
8. Sachs J, Schmidt-Traub G, Kroll C, Lafortune G, Fuller G, Woelm F (2020) The sustainable development goals and COVID-19. In: Sustainable development report, 2020
9. Kumar M, Choubey VK (2021) Modelling the interaction among the key performance indicators of sustainable supply chain in perspective of perishable food. Int J Logist Syst Manag 1(1). https://doi.org/10.1504/IJLSM.2021.10039305
10. Bendinelli WE, Su CT, Péra TG, Caixeta Filho JV (2020) What are the main factors that determine post-harvest losses of grains?. Sustain Prod Consumption 21:228–238
11. Minten B, Tamru S, Reardon T (2021) Post-harvest losses in rural-urban value chains: evidence from Ethiopia. Food Policy 98:101860

12. Gardas BB, Raut RD, Narkhede B (2018) Evaluating critical causal factors for post-harvest losses (PHL) in the fruit and vegetables supply chain in India using the DEMATEL approach. J Clean Prod 199:47–61

13. Filimonau V, Sulyok J (2021) 'Bin it and forget it!': the challenges of food waste management in restaurants of a mid-sized Hungarian city. Tour Manag Perspect 37: 100759

14. Kumar A (2014) Left out in the cold: the case of potato cold stores in West Bengal. IUP J Supply Chain Manage 11(2)

15. Parfitt J, Barthel M, Macnaughton S (2010) Food waste within food supply chains: quantification and potential for change to 2050. Philos Trans R Soc B Biol Sci 365(1554):3065–3081

16. Betz A, Buchli J, Göbel C, Müller C (2015) Food waste in the Swiss food service industry—magnitude and potential for reduction. Waste Manage 35:218–226

17. Gustavsson J, Cederberg C, Sonesson U, van Otterdijk R, Meybeck A (2017) Global food losses and food waste: extent, causes and prevention. FAO

18. Mangla SK, Sharma YK, Patil PP, Yadav G, Xu J (2019) Logistics and distribution challenges to managing operations for corporate sustainability: study on leading Indian diary organizations. J Cleaner Prod 238:117620

19. Dumont AM, Vanloqueren G, Stassart PM, Baret PV (2016) Clarifying the socioeconomic dimensions of agroecology: between principles and practices. Agroecol Sustain Food Syst 40(1):24–47

20. Hamprecht J, Corsten D, Noll M, Meier E (2005) Controlling the sustainability of food supply chains. Supply Chain Manage Int J

21. Raut RD, Gardas BB, Kharat M, Narkhede B (2018) Modeling the drivers of post-harvest losses—MCDM approach. Comput Electron Agric 154:426–433

22. Garrone P, Melacini M, Perego A, Sert S (2016) Reducing food waste in food manufacturing companies. J Clean Prod 137:1076–1085

23. Papargyropoulou E, Lozano R, Steinberger JK, Wright N, Bin Ujang Z (2014) The food waste hierarchy as a framework for the management of food surplus and food waste. J Cleaner Prod 76:106–115

24. Eriksson M, Spångberg J (2017) Carbon footprint and energy use of food waste management options for fresh fruit and vegetables from supermarkets. Waste Manage 60:786–799

25. Gardas BB, Raut RD, Narkhede B (2017) Modeling causal factors of post-harvesting losses in vegetable and fruit supply chain: an Indian perspective. Renew Sustain Energy Rev 80:1355–1371

26. Kumar A, Mangla SK, Kumar P, Karamperidis S (2020) Challenges in perishable food supply chains for sustainability management: a developing economy perspective. Bus Strategy Environ 29(5):1809–1831

27. Bosona T, Gebresenbet G (2013) Food traceability as an integral part of logistics management in food and agricultural supply chain. Food Control 33(1):32–48

28. Epelbaum FMB, Martinez MG (2014) The technological evolution of food traceability systems and their impact on firm sustainable performance: a RBV approach. Int J Prod Econ 150:215–224

29. Pandey G (2018) Challenges and future prospects of agri-nanotechnology for sustainable agriculture in India. Environ Technol Innov 11:299–307

30. Mourad M (2016) Recycling, recovering and preventing "food waste": competing solutions for food systems sustainability in the United States and France. J Clean Prod 126:461–477

31. Kumar M, Mor RS, Singh S, Choubey VK (2020) Sustainability and OEE gains in manufacturing operations through TPM. In: Circular economy for the management of operations (pp 173–185). CRC Press

32. Zanoni S, Zavanella L (2012) Chilled or frozen? Decision strategies for sustainable food supply chains. Int J Prod Econ 140(2):731–736

33. Nyamah EY, Jiang Y, Feng Y, Enchill E (2017) Agri-food supply chain performance: an empirical impact of risk. Manage Decis

34. Singh RK, Gunasekaran A, Kumar P (2018) Third party logistics (3PL) selection for cold chain management: a fuzzy AHP and fuzzy TOPSIS approach. Ann Oper Res 267(1):531–553

35. Mor R, Singh S, Bhardwaj A, Singh L (2015) Technological implications of supply chain practices in agri-food sector: a review. Int J Supply Oper Manage 2(2):720–747
36. Kamble SS, Gunasekaran A, Sharma R (2020) Modelling the blockchain enabled traceability in agriculture supply chain. Int J Inf Manage 52:101967
37. Kaipia R, Dukovska-Popovska I, Loikkanen L (2013) Creating sustainable fresh food supply chains through waste reduction. Int J Phys Distrib Logistics Manage
38. Gold S, Kunz N, Reiner G (2017) Sustainable global agrifood supply chains: exploring the barriers. J Ind Ecol 21(2):249–260
39. Ivanov D, Dolgui A (2020) Viability of intertwined supply networks: extending the supply chain resilience angles towards survivability. A position paper motivated by COVID-19 outbreak. Int J Prod Res 58(10):2904–2915
40. Singh V, Hedayetullah M, Zaman P, Meher J (2014) Postharvest technology of fruits and vegetables: an overview. J Postharvest Technol 2(2):124–135
41. Latino M, Corallo A, Menegoli M (2018) From industry 4.0 to agriculture 4.0: how manage product data in agri-food supply chain for voluntary traceability, a framework proposed. World Acad Sci Eng Technol Int J Nutr Food Eng 12:126–130
42. Lezoche M, Hernandez JE, Díaz MDMEA, Panetto H, Kacprzyk J (2020) Agri-food 4.0: a survey of the supply chains and technologies for the future agriculture. Comput Ind 117:103187
43. Madramootoo C (ed) (2015) Emerging technologies for promoting food security: overcoming the world food crisis. Woodhead Publishing, Cambridge
44. Warfield JN (1974) Toward interpretation of complex structural models. IEEE Trans Syst Man Cybern (5):405–417
45. Mangla SK, Luthra S, Rich N, Kumar D, Rana NP, Dwivedi YK (2018) Enablers to implement sustainable initiatives in agri-food supply chains. Int J Prod Econ 203:379–393
46. Prakash S, Soni G, Rathore APS, Singh S (2017) Risk analysis and mitigation for perishable food supply chain: a case of dairy industry. BIJ 24(1):2–23
47. Raut RD, Luthra S, Narkhede BE, Mangla SK, Gardas BB, Priyadarshinee P (2019) Examining the performance oriented indicators for implementing green management practices in the Indian agro sector. J Clean Prod 215:926–943

# From the Vine to the Bottle: How Circular is the Wine Sector? A Glance Over Waste

Patricia Calicchio Berardi, Luciana Stocco Betiol, and Joana Maia Dias

**Abstract** The present chapter provides an overview of the challenges associated with the food chain, particularly in the wine sector. It starts from a global context on the food chain (Sect. 1.2), food loss, and food waste (Sect. 1.3) and stresses out circular supply chain management (Sect. 1.4). Then, circular practices in the food supply chain and the wine sector are evaluated through a systematic review and a questionnaire to companies of this sector (Sect. 2). Finally, four case studies are analyzed. In total, 47 studies highlighted the use of waste as a resource and the challenges to map data in the food supply chain. The questionnaire did not allow for a robust quantitative study. The analysis of four case studies showed that, although being aligned with the topic of sustainability, companies are unable to identify regenerative and restorative operations applicable to their activities. The final part (Sect. 3) shows that the application of the principles of circularity in the studied chain has still many constraints and opportunities that require further developments: the establishment of local arrangements integrating different agents and the alternative valorization routes for by-products to work resources optimally, thus contributing to a more deeply incorporation of circular practices.

P. C. Berardi (✉) · J. M. Dias
LEPABE, Department of Metallurgical and Materials Engineering, Faculty of Engineering, University of Porto, R. Dr. Roberto Frias, s/n, 4200-465 Porto, Portugal
e-mail: pberardi@fe.up.pt

J. M. Dias
e-mail: jmdias@fe.up.pt

P. C. Berardi · L. S. Betiol
CELOG Centro de Excelência e Logística e Supply Chain, São Paulo School of Business Administration EAESP/FGV, Rua Itapeva, 474, São Paulo, SP 01332-000, Brazil
e-mail: luciana.betiol@fgv.br

L. S. Betiol
FGVethics, Centro de Estudos em Ética, Transparência, Integridade e Compliance, São Paulo School of Business Administration, EAESP/FGV, Rua Itapeva, 474, São Paulo, SP 01332-000, Brazil

R. S. Mor et al. (eds.), *Challenges and Opportunities of Circular Economy in Agri-Food Sector*, Environmental Footprints and Eco-design of Products and Processes,
https://doi.org/10.1007/978-981-16-3791-9_9

**Keywords** Food chain · Circular supply chain management · Waste management · Wine sector · Circular economy

# 1 Introduction

## 1.1 Context

The size of the challenge in reducing waste generation in the Agri-Food chain is as complex and as necessary as the size and importance that the chain itself has in the global socioeconomic scenario. This theme has attracted attention in all spheres of the society, both internationally and locally, among governments, specialists, researchers, professionals in the sector, and from other business areas. The demand for effective and efficient use of resources by reducing food waste has an economic and a moral character. There are many difficulties that permeate this phenomenon, starting with the lack of characterization of the sector. A matter of great difficulty is accessing robust and consistent databases, with reliable data, mostly because the different methodologies and definitions do not allow comparison. The inaccurate classifications, wide variety of public policies, different rules and regulations hinder the uniformity of information. Even so, efforts are being devoted to developing platforms that will start to encourage the building of statistical information (albeit in a timid and fragile way), particularly concerning waste generation along the Agri-Food chain. However, as these information sources still lack more inputs, it is not possible to properly map waste generation in the different stages of the chain worldwide.

Considering the presented challenges and identified lack of information, the present chapter depicts an exploratory study aiming to identify circular practices being carried out in the food chain, and more particularly, in the wine chain.

## 1.2 Food Chain: Global Context

Operating to optimize the use of resources and their correct allocation is imperative for the current and future times taking into account the expected population increase and the need to fulfill associated basic demands. The world population is expected to grow from 7.2 billion in 2010 to more than 9 billion in 2050, with the associated food demand being expected to increase by 60% [83]. In order to reduce the environmental impacts resulting from this increased food demand, it is necessary to avoid unnecessary waste generation and food losses that frequently exist in the various stages of the supply chain.

According to the Food and Agriculture Organization of the United Nations (FAO) [26], one-third of all global food production is lost over different stages, with greater concentration at the consumer stage. In addition to representing a high economic cost with many inefficiencies, it is a problem that brings many environmental and social impacts of serious consequences [6, 17, 37, 60, 66], constituting an immoral action [6, 15], given that 1 billion of people globally suffers malnutrition or food insecurity [26]. Working to optimize the use of natural resources in the food chain may allow for a better balance between the variables within the food production and consumption systems, which will tend to a more resilient overall system contributing toward sustainable development in the long run.

In order to tackle the problem itself, it is necessary to understand how complex it is to deal with the food chain system, which encompasses a very wide and varied set, with different dynamics, agents (direct and indirect) and many links, from the field to the final consumer.

In fact, the food supply chain consists of many interdependent steps and operations, from field cultivation, food processing, distribution, retailing and consumer handling where direct support for storage, packaging, and transportation is required [41]. In other words, evaluating food chains means "following the product" by analyzing the material flow from the field to the final product [11]. In addition, a business relationship exists between different participants such as suppliers, producers, workers, processors, brokers, wholesalers, and retailers [44]. Such relations occur among different geographical levels, from the regional to the global scale, in which most players are made up of small and medium enterprises. When focusing on agricultural production, the predominant companies are small/family farming [28]. Another point to consider is the diversity of products included in this production system.

The Agri-Food chain includes food and beverages, with a wide range of products, such as vegetables, fruits, and animal protein of the most varied types (fish, chicken, pork, beef, among many others), with different expiration dates and varied limitations in terms of safe and appropriate consumption. For each of these products, there are specific characterizations, with multiple variables that follow regional production and commercialization norms including local laws and regulations. In other words, it is clear the complexity of aligning all these elements for an effective and efficient fight regarding the optimal use of resources and the search for a real sustainability of the world food system.

Studies show that, since 1995, trade in food and agriculture production has more than doubled in real terms. This growth is largely due to emerging and developing countries, which are the main global players in these markets, accounting for 1/3 of the global trade, according to the data from The State of Agricultural Commodity Markets [28]. This advance was only possible in the face of technological improvements, regional and international global trade policies, and infrastructure systems, which transformed productive and commercial processes, making it easier for global food and agriculture chains to grow.

When speaking specifically of agricultural production, understood as one of the main occupations in the world, we are dealing with more than 600 million farms that provide income and jobs for billions of people. According to FAO data, it is estimated that about 90% of these farms are based on family work, on small properties (less than one hectare) in developed countries, accounting for about 80% of the food produced in the world [28].

The increasing participation of the food trade and agricultural production in the global chain, due to the increase in the world population and the additional demand for food, generates both positive and negative environmental impacts. One of the negative environmental impacts lies in the fact that, associated with greater production, there is also an increase in food loss and food waste generation. Recent studies show that within the food chain, the biggest loss occurs in the fruit and vegetable chain, followed by meat, fish, seafood, and milk [2]. World Bank data indicates that 44% of waste production in the world is linked to food [45], a situation that worsened even further with the COVID-19 pandemic, especially in sectors that demand more labor in the production link [57, 81].

The concern to act strongly in fighting this great problem of high waste generation potential in the Agri-Food chain is highlighted in multiple agendas at the global and local levels, the latter having its optimum level depending on each regional context, and effective compliance to these agendas. This is what can be noted in a study conducted by the European Commission, where only 20% of the 50 largest companies in the food sector in the world had implemented, until 2017, a program aimed at food loss and food waste prevention [64].

Several global agendas seek to encourage and guide public and business policies to reduce food loss and food waste. We can temporarily identify its beginning in 2008, according to the timeline in Fig. 1.

All these movements are in line with the important UN 2030 Agenda on the Sustainable Development Goals, being essential for its progress. In parallel, it is essential to carry out an adequate diagnosis of the most diverse sectors of the economy, in which the so-called revolution/Industry 5.0 could provide new tools to work data and man and machine relationships on the various objectives that we want to achieve with the SDGs.

However, this revolution can only prosper if data on the most diverse objectives of this agenda for sustainable development is available, current, based on the same

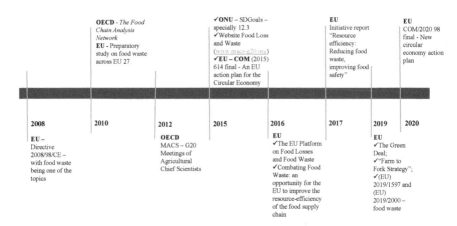

**Fig. 1** Timeline of relevant public and business policies to reduce food loss and food waste. *Source* Authors

concepts, and that can be monitored, which requires a great coordination between governments and the private sector [69], as it will be highlighted in the next sections.

## 1.3 Waste in the Food Chain

The huge amount of food wastage throughout the production and consumption processes represents a major problem in face of the inefficiency of the system. It is also necessary to consider, in addition, that a vast number of other resources are necessary for the production and transport of these foods, including land, water and energy, with many inefficiencies and improper management practices associated.

Data from the World Resources Institute (WRI) and FAO points out that food loss and food waste costs approximately $940 billion a year for the global economy [25, 49]. In other words, it is necessary to take into account not only the direct losses of resources and costs, but also all the indirect losses and costs associated with the production.

Agriculture is responsible for many environmental impacts, including those associated with land occupation, by using half of all the planet's habitable land (excluding ice and desert), namely, being also responsible for the consumption of 70% of the planet's freshwater that is destined for agricultural irrigation [39] and for 78% of global pollution of watercourses [67].

Another fact to note is that the generation of greenhouse gas emissions from the food supply chain corresponds to around 13.7 billion metric tons of carbon dioxide equivalents ($CO_2$eq), 26% of global emissions [62]. There are many studies that explore the different negative impacts derived from food losses and food waste throughout the Agri-Food supply chain [6, 13, 66, 72]. Optimizing and regenerating food production resources can thus represent a direct improvement in terms of climate change, reduction of water stress and pollution, and the regeneration of land, which can promote an increase in biodiversity, in addition, substantially contributing to other SDGs, as SDG 2, 12, 15, and 16.

In order to be able to proceed with studies on this topic, it is necessary to characterize the phenomenon. Through the literature survey, it was possible to clearly identify a barrier related to a wide range of definitions. There is no alignment among the main exponents of the theme and they are, to a large extent, concept providers, as evidenced in a study carried out by the Organization for Economic Cooperation and Development (OECD) in 2014 [6]. Teigiserova et al. [80] highlight that without clarifying the terms and their scope, it is not possible to carry out good waste management.

FAO brings two divisions to the problem: **Food Loss**—the decrease in the quantity or quality of food resulting from decisions and actions by food suppliers in the chain, excluding retailers, food service providers and consumers; and, **Food Waste**—the decrease in the quantity or quality of food resulting from decisions and actions by retailers, food service providers and consumers.[1] In another definition, the United Nations Environment Program (UNEP) in collaboration with WRI presented food loss as "food that spills, spoils, incurs an abnormal reduction in quality such as bruising or wilting, or otherwise gets lost before it reaches the consumer" while food waste refers to "food that is of good quality and fit for human consumption but that does not get consumed because it is discarded—either before or after it spoils" [33].

For the European Food Use for Social Innovation by Optimizing Waste Prevention Strategies (FUSIONS), food waste is any food, and inedible parts of food, removed from the food supply chain to be recovered or disposed of, including the following destinations: composting, crops plowed in/not harvested, anaerobic digestion, bioenergy production, cogeneration, incineration, disposal to sewer, landfill, or discarded to sea but not including food or inedible parts of food removed from the food supply chain to be sent to animal feed or bio-based material/chemistry processing [6, 24].

The same phenomenon occurs in the scientific literature where it is possible to note the lack of consensus, aggravated by the variety of frameworks. The theme of food waste is addressed in sustainability, supply chain, network research, and waste management, focusing on both food loss and food waste [15, 18], where it is highlighted that this theme must be observed in a multidisciplinary perspective [2].

Added to this is the fact that it is difficult to classify the moment when the generating event occurs and who is responsible for it, taking into account that food loss and food waste generation occur throughout all stages of the chain, all over the world, from primary production to final consumption [85]. The study by Corrado and Sala [18] highlighted this issue and showed a survey carried out between the different definitions for both reports and scientific studies.

The content under analysis is another obstacle to conceptual alignment. This is because part of the studies works with the edible portion of food, while others focus on the non-edible portion—waste.

Another factor to be considered is the metric to measure the quantities under analysis, for example, quantification, calorific value, or economic value. Most studies have worked with mass, but there are studies with dry mass and others with wet mass. A caveat made in the FAO report [27] attests that most studies are performed by physical measurement with units of analysis in tons. However, for the impact estimates (particularly the environmental ones), the measurements move to economic values, without considering the differentiation of commodities and their respective real impacts, remaining only tied to the weight. This can give bias in the analysis. It should also be noted that there are different methodologies [27], in addition to having a specific objective for each type of study. This means that the different drivers of the generation of food waste throughout the chain produce different perspectives concerning waste prevention or waste management or recovery (energy or materials).

Another key issue is the data deficiency and inconsistency. There is a great effort in trying to aggregate data on a global scale, which is a challenge, due to the great variability of observations. Among the forms of report, the data can be in single values, values in a range, mean values or mean values with a variation (dispersion).

---

[1] http://www.fao.org/platform-food-loss-waste/en/.

These values could be either in absolute terms or as percentages. Another consideration is that much of the existing literature on food loss and food waste is based on secondary data [85], and the majority produced by developed countries, which limits the sample [40].

The database on Food Losses and Food Waste (FAO STAT DATABASE),[2] created by FAO as an objective opportunity to advance in characterizing and addressing the problem of food loss and food waste in the Agri-Food chain, is a clear example of all these complexities. As much as the methodology created with the defined formulas seeks to use data from different bases, for all types of food products and from all countries, it is possible to see a certain weakness, with several gaps. There is no consistency or uniformity in the collection of these data since they originate in different media, such as reports, theses, and works published in congresses, therefore, a great variety of sources. On the other hand, of the 185 countries, only 23 reported losses in 2016 to one or more commodities. That is, the base is not sufficiently robust and still does not provide security, which interferes in the design of public and business policies for the solution, according to geographic and temporal peculiarities [2].

Among the problems pointed out for not addressing the issue of losses and waste generation and management is the non-perception of value by the holder of such resource, who often prefers to discard it for disposal and, also, because potential user does not recognize added value in the materials and thus they do not promote increased interest in the transaction [15, 49].

In order to be able to think about evaluating concrete issues in a supply chain, it is first necessary to understand the different ways of managing a supply chain and the elements and steps to take into account. This will be the focus of the next session which addresses the topic of circular supply chain management.

## 1.4   Circular Supply Chain Management

In the literature dedicated to the area of operations, the subjects related to supply chain management (SCM) have evolved to meet the demands of the global agendas, considering socio-environmental variables over time. At first, the approach was made with issues related to the environment. In the next stage, the alignment was expanded by incorporating economic, environmental, and also social factors, as a way of working on sustainability at SCM. More recently, with the pressure of legislation in some regions, and with the extension of the circular economy debate, both scientific and technical literature started to operate by introducing the circularity frameworks in SCM.

---

[2] http://www.fao.org/platform-food-loss-waste/flw-data/en/.

In a brief historical context, it is possible to identify that at the end of the 1980s Operations Management studies were strongly dedicated to operational and production factors—total quality management—with a later emphasis on just in time operations. Subsequently, the studies focused on processes of reengineering, outsourcing, in which the center was the operational and strategic harmonization of the value chain. Up to that point, there was no other role rather than the economic and operational performance of the supply chain. It was from the lean supply chain that elements such as the attempt to eliminate waste generation and the search for efficiency started to be considered as strategic decisions at SCM [20].

Since then, issues related to the environment have been incorporated into production processes, especially factors aimed at reducing negative impacts such as pollution, including the reduction of toxic elements in products and eradication of certain substances, for instance, chlorofluorocarbons (CFCs) considering their effects on ozone depletion.

The Green Economy movement has driven many legislative and organizational decisions, contributing to direct influences on SCM. Pollution prevention and cleaner production are direct reflections of what happened with the so-called Green Supply Chain Management (GSCM), aiming at an environmental approach to reduce pollution and to promote efficiency in the use of resources, without losing management perspectives of supply chain operations [78]. The companies' interest in adopting GSCM was to reduce operational costs, to integrate suppliers in the decision-making process, to adopt differentiated purchasing strategies, leading to waste reduction, substitution of materials and raw materials, reduction of overall emissions, the better use of natural resources, a more efficient development of new products, innovation creation, among others [65].

With more consolidated discussions on sustainable development and the expansion of organizational decisions through greater collaboration with sustainable production and consumption, the studies started to be dedicated to the identification of a more comprehensive concept and approach, in addition to evaluating economic results. They sought to present socio-environmental benefits both for the business, for the chain, for the society and for the environment [48]. Pollution control, recycling, reverse logistics, reprocessing, among many other issues went from a purely operational perspective to strategic assessments in which sustainable supply chain management would bring performance differentials with a greater competitive advantage [65, 75].

Furthermore, sustainable supply chain management (SSCM) started to adopt a wide review of processes and behaviors not only among the direct links of partnership, but also with all stakeholders, directly and indirectly, reaching government, financers, insurance companies, among other organizations [4]. In addition, a wide range of actions and practices has been adopted aimed at sustainability in the supply chain, such as rethinking, repairing/recovering, reconditioning, remanufacturing, repositioning with a view to extract the greatest value from resources or even enhance products/materials [36], responding to global demands for minimizing negative impacts supported by regulatory, institutional, or voluntary means [12].

As a direct reflection of the global context for the search of the conceptual evaluation of circular economy in supply chain studies, the so-called circular supply chain [7, 20] raised as an emerging area in academic literature [50].

The intention of this approach is not only to expand borders beyond the GSCM and SSCM by reducing the use of virgin resources or by operating in closed loops, but to increase the life-cycle extension of products and to stress the regenerative dimension through organizational collaboration inside and outside the productive sector to recover value and reestablish the maintenance of natural stocks [29, 30, 35].

Circular supply chain management (CSCM) demands a complete review and a new redesign of the entire value chain through an holistic and multidisciplinary perspective [10], requiring the collaboration of different supply chain actors to operate with a wide range of waste management technologies and resource recovery procedures [30] to promote the use of waste as a resource. Many elements are needed to enable the adoption and application of the circular economy in SCM: design for circularity, circular processes approach, innovation and sustainable technologies, reverse flows, industrial symbiosis, among others oriented to restore and regenerate natural resources [47].

Benefits have been pointed out regarding CSCM such as optimization of resource use, economic gains, improvements in environmental management, more efficiency in processes along the entire value chain, improving of strategies toward end-of-life extension, or even competitive advantage are some of the aspects highlighted in the developed studies [47]. However, the studies still seem to indicate an early stage in the application of the CSCM and highlight numerous barriers and difficulties in moving beyond punctual circular practices, or due to the modeling characteristics associated with the linear system, far from the principles of regeneration [30, 36, 46, 50, 68].

After presenting the general context of the food chain, waste in the food chain and circular supply chain management, the methodological approach used to analyze how the scientific community addressed the topic, as well as the results found, are presented.

## 2 Methods and Results

In order to be able to investigate more concretely the behavior of the food chain in relation to the principles of circularity, the first step of this study was to map scientific publications in the main bases that have a large number of productions, namely Web of Science, SCOPUS, Science Direct and Springer Link, for a multidisciplinary view in several scientific areas with the objective of capturing the largest number of studies and avoiding any bias.

A systematic literature review was carried out, which is a resource that greatly helps the exploration and dimensioning phase of the knowledge already produced. This is made through a structured, systematized method, which provides scientific quality, reliability and allows replication, being used to research and analyze evidence on a given topic in existing publications [23, 71, 73, 82].

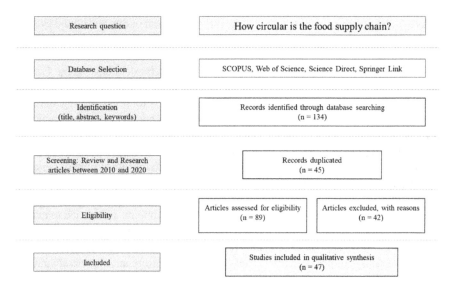

**Fig. 2** Structure of the systematic literature review process and respective results. *Source* Authors

The first step of the chosen research method is to define the research question: How circular is the food supply chain? From this question came the set of keywords used which were used as follows: **Circular Economy** and **Supply Chain** and **Food** and **Waste**. Based on a waste management perspective, the objective was to identify if the waste is being treated as a resource. The search filters for all the databases were restricted to the main topics (title, abstract, and keywords). The search was carried out in October 2020 covering the period between 2010 and 2020 and the results are summarized in Fig. 2.

The search resulted in 134 studies that appeared in scientific publications, excluding conferences proceeding and working papers. After eliminating duplicates (45), all the abstracts were examined. After careful evaluation, it was verified that 42 of the papers did not fit the scope of our research. Several were not dedicated to the food chain (e.g., Focusing on, the need for legislation for public purchases to help to reduce waste in school catering), or the object of study was not focused on waste management, but rather on other interests such as the impact of natural disasters on the food chain. These studies were disregarded given that the focus of this research is on waste management in the food chain with an emphasis on the circular economy. The articles were read and analyzed independently by the researchers, who compared the results and sought convergence based on the research question.

In total, 47 articles that met the selection criteria were considered as the final sample. It was possible to verify that the main objective of 21 of the studies was addressing waste as a material or organic resource, while 13 studied circular business models, 11 were dedicated to energy recovery from waste, and 9 declared interest in reducing waste in the food chain, as presented in Fig. 3. The vast majority did not

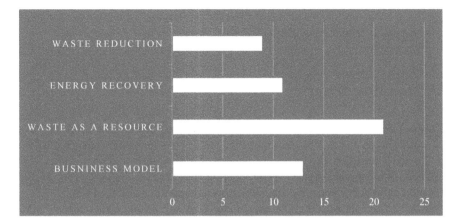

**Fig. 3** Weight of the different areas addressed in the 47 selected studies from the systematic literature review, to respond to the question how circular is the food supply chain, with focus on food, waste and circular economy. *Source* Authors

declare the subject studied in the abstract (25 papers), 13 were aimed at environmental impact studies, 7 were dedicated to economic aspects, and 5 had a social character.

Among the most used methods, of what was declared in the abstract, there is a dispersion between literature review and life-cycle assessment with 6 studies in each, others 5 carried out case studies, 4 used a survey, and 3 conducted interviews. Another 6 works used varied methods between cluster, factor analysis, matrix, and others. 17 publications did not declare the study method. There was a concentration of publications of studies that worked on circular economy in SC and food waste between the years 2019 (16) and 2020 (18). The journals that published most of the works in these subjects were: Sustainability (5), Journal of Cleaner Production (4), and Resources Conservation and Recycling (4).

As this is a study dedicated to the food supply chain, in the analysis carried out, it was possible to identify a wide range of different study objects (fresh food, milk, meat, fish, vegetables, cereals, pasta, bread), with a multiplicity of methodologies and cutouts, as pointed out in previous studies [15, 18], demonstrating, once again, how challenging it is to map and align data in the food chain sector. Concluded the analysis of this first stage, the difficulty in working holistically was evident. To move forward and better understand the behavior regarding the adoption and application of the principles of circular economy in the supply chain, we have decided to investigate a single production chain, thus focusing on one activity sector.

The chosen chain was that of wine, since wine production stands out from other Agri-Food sectors due to its long-standing historical value [1]. Wine is a product that is closely related to the local cultural history, from the cultivation of grapes that has a strong influence on regional heritage, through production itself, which can vary from more artisanal processes to fully mechanized industries with modern technologies, to the final product, which started to have not only business components varying from local consumption, but also high diversity worldwide, in addition to expanding to the aspect of promoting wine tourism [8].

## 2.1 The Wine Sector

Grapes are produced in vines (*Vitis vinifera*) being one of the largest fruit crops in the world spread in different areas of the planet [77] in which more than a half of the production goes to the production of wines [42, 52]. Considered one of the oldest productions in human history, wine production has enormous economic, social, and environmental importance in different regions around the world [1, 14, 63].

In 2018, 7.4 Mha was the total area under vines cultivation, generating 77.8 Mt of grapes where 57% was wine grape production, 36% was table grape production, and 7% was dried grape production. Worldwide, the total wine production was 292 MhL and the consumption accounted for 246 MhL, according to OIV [58]. Such production generates a significant amount of waste every year [52], where estimates point to up to 30% w/w of the material used [19]. All this amount of waste generated causes enormous impacts, not only economically but also environmentally. However, compared to other chains, the wine industry is seen as a chain with low environmental impact [1, 54].

It is an overly complex sector and with many particularities, depending on each grape variety and each manufacturer that defines the characteristics that must be highlighted in the final product. It is considered a highly regulated sector because it must remain faithful to the characteristics of each production region in order to preserve the recognition of the identity of each production area. Still, it brings a wide range of challenges considering the best practices regarding the optimal use of resources and correct waste management among the food chain processes [3, 5].

Among the main stages of production, from the grape to the final product of bottled wine, numerous singularities exist. The winemaking processes itself, by transforming the grapes into a drink that passes through different stages, until it is bottled and ready to go to the final consumer. In addition to the direct processes relevant to the production, there is a substantial impact on both transport and packaging, especially in the final phase to take the finished product to the consumer, with high waste generation and carbon footprint associated [3, 31, 61]. However, it was not the object of this study to work on the distribution and consumption stages, thus constituting a cradle to gate analysis [56].

The screening of factors that need to be analyzed in terms of optimizing the use of resources and improving waste management in the wine chain, considered thus the three main stages of the activity: viticulture, vinification, and bottling.

In the agriculture phase, it is necessary to take into account a series of elements, with all their local specificities that constitute the so-called terroir, such as soil/region, temperature, altitude, slope; the vine and its type of planting; the grape among the different qualities and varieties (white or red); whether or not there is irrigation; the types of cultivation (conventional, organic, or biodynamic), and whether manual or mechanical treatment occurs; how to treat the soil and plants with phytosanitary products; the harvest (manual or mechanical extraction). All of this represents a set of inputs (water, electricity, fertilizers/chemicals, equipment, packaging, organic materials, etc.), and in turn their respective by-products and wastes [1, 3, 5]. Different cultivation systems can also result in different amounts of waste generated for the same area and their use results in by-products and wastes [77].

Winemaking brings a combination of stages defined by the winemaker and which, according to each type of grape and the characteristics that are used in each product, vary between the different wineries. However, pressing, fermentation, stabilization, filtration, refining, and storage manufacturing processes are common.

During the production phase, many by-products and wastes are generated, mostly organic [56], including: organic solid wastes (e.g., leaves, branches, stems) from unfermented grapes, grape skins and seeds; fermented grape pomace; wine lees and filtration residues (e.g., filter plates, diatomaceous earth filter, cartridge) as a direct result of the transformation of grapes into wine [19, 42]. In addition, packaging waste, obsolete equipment, and wood (mostly related to barrels) are generated. Finally, liquid effluents result from cleaning and washing equipment [3].

In the final stage, when it is understood that the product is ready to be bottled, there is the third processing stage, highly relevant in terms of cleaning and hygiene to ensure product quality assurance [22]. In this stage, the generation of waste is strongly associated with the used materials/products such as glass, cork, plastics and cardboard, paper labels, metals, wood from boxes and pallets, and effluents derived from cleaning of the bottles [3, 43].

## 2.2 Circular Economy in the Wine Sector

Circular economy business models, with the premises of systems regeneration and restoration, go beyond the sustainability efficiency focused practices. Studies have been dedicated to identifying how much of the wine sector operates according to the principles of circularity [8, 16, 32, 51, 74]. However, what has been verified is that the search for businesses inserted in the logic of sustainable development is focused only on environmental performance and on minimizing impacts [21, 38, 54, 55, 79, 86], or in alternative means to enhance and reuse materials and nutrients to sustain other processes [9, 53, 76, 84].

Nevertheless, in addition to the scientific community, producers know that resources are increasingly scarce and, in order to reduce costs and maintain sustainable levels of operation, it is necessary to search for alternative models for the long term. The wastes generated must not only be removed from the landfill, but must be worked in optimal use, being considered valuable ingredients and important sources of nutrients for other processes adding value to the overall system [34].

This is because the by-products and wastes of this sector are mostly organic substrates that can be integrated into other chains, allowing an integral use or valorization, examples include: incorporation of skins/pomace residues in distilleries, coloring, animal feed; from the seeds it is possible to remove antioxidant and oils for food/feed applications or cosmetic industry; it is possible to reuse lees in the wine production itself or in still distilleries, and to obtain pigments and products used as nutritional supplements for food application. Many of the materials can be used as sources of bioproducts and renewable energy namely for the production of compost and biogas [19, 42, 52, 70, 77].

As previously highlighted, there is a need to establish an alignment between the different legal frameworks in order to allow for an optimal use of these resources. For example, according to Spigno et al. [77], these by-products are classified as food waste concerning FUSIONS's definition, whereas for the European Waste Framework Directive (2008/98/CE) they are considered by-products which causes direct practical effects on the destination of materials.

In order to demonstrate these challenges, we sought to build a broader picture of the adoption of circular economy practices by the wine production chain at a global level. This study initially worked on two fronts. Systematic review of the literature and application of a questionnaire with 44 questions, having as main objectives the identification of waste management practices and possible arrangements between partners which can be configured in circular flows in the supply chain. After evaluating the questionnaire, four case studies were more deeply analyzed.

Starting from the questionnaire, this collection instrument was divided into four parts, the first being the general characterization of the respondent's profile and the activities that the company develops, and the following three sections were divided by the respective viticulture, vinification, and bottling productions. In all three sections, it was asked whether there was separation of waste, quantities produced, and destinies for each type of waste. In addition, open questions were conducted to evaluate whether there was any type of sharing activities and whether the company had circular economy practices, with examples.

First, the instrument of collection was validated by a pilot group of specialists.[3] For the selection of the sample, members of The Porto Protocol[4] were identified as a potential interest group. The form was sent to the entire database by e-mail and was also available on social networks. With the title "From the vine to the bottle: how circular is the wine sector?" the questionnaire attracted a good number of stakeholders (55 respondents). However, it was not possible to perform statistical data treatment because some responses were disregarded for not having completed the form. Thus, the number of valid responses was less than necessary to carry out a robust quantitative study.

Although the attempt to work in a quantitative way was frustrated, it was possible to notice that the theme attracts interest. However, professionals seem to have

---

[3] Professors and alumni from a post-graduate enology program in northern Portugal, mostly professionals dedicated to wine production in that country.

[4] Non-profit institution that works as an open and shared platform to join efforts to fight climate changes and promote sustainability, originated in the wine sector, nowadays congregating in any other sector that is interested in the cause. It has 200 members distributed 50% in Europe and the remaining 50% around the rest of the world with more than 850 followers on social networks.

some difficulties in aligning management practices within the concepts of circular economy, being one of the challenges of the lack of segregated quantitative and qualitative information regarding all the existing by-products and wastes, independently of the company's dimension (taking into account the sample analyzed).

As an alternative, we have decided to report case studies with some of the respondent companies that presented different examples in the practices of waste management and optimization of the use of resources, or even sharing models.

Qualitative methodologies, more specifically the case study method, have been a widely used format both in studies that address the theme of circular economy in food chain [7, 59] as well as in wine sector [74]. Next, four case studies will be presented, two from the USA, and two from Portugal.

### 2.2.1 Case Studies of Circular Economy in the Wine Sector

The case studies carried out were constructed based on primary data, provided directly by the companies in relation to the production of 2019/2020, following the same line of the questionnaire. The four companies analyzed have the three activities: vineyard cultivation, winemaking, and bottling.

*Company A*

Located on the Atlantic coast of North America, this microcompany[5] has been operating since 2005 under the actual management. 3 ha of white grapes were grown, which produced 30 tons; and 3 ha of red grapes, which generated 18 tons. Of this amount, 70% under manual treatment and 30% under mechanical treatment, without irrigation, with harvesting done entirely by hand.

Despite being in a model of conventional agriculture, it is a production that seeks to look at sustainability criteria as it is inserted in a context of ecological diversity with a large area of forest, pasture, and other crops besides grapes, a system that lasts for three generations, where other crops were also explored. Of the 70 kg of organic waste generated from pruning, all of it is transformed into compost locally at a cost of 500.00 €. All waste generated in viticulture is segregated, with plastics, paper, and cardboard being reused and waste from equipment and mineral oils and lubricants being sent for recycling, with a cost of 250.00 €. As by-products, the company uses 100 tons of wood chips integrated in winemaking. In the operation, 20 tons of white grapes and 20 tons of reds were processed, each generating 100 hL of wine. Of the total waste generated, 12 tons of grape skins and seeds, 2 tons of wine lees, the entire organic load being composted locally; 4 tons of wood (barrels) were reused, and the remaining waste was segregated and sent for recycling. The cost of treating paper waste was 1200.00 € and wood treated for reuse cost 3000.00 €. Of the 200 hL of effluents generated, 100% was treated and disposed of in sequence. No data on

---

[5] According to OECD definition: Microenterprise (from 1 to 9 employees); small enterprise (from 10 to 49 employees); medium enterprise (from 50 to 249 employees); large enterprise (250 or more employees).

bottling residues were presented. Although it is not the focus of our study, it is worth mentioning that in the pandemic period, the company started offering services to share logistics with other small producers in the region and donated some products to help institutions at the community.

*Company B*

A project starting in 2018 as a microcompany located in the Dão region, in the highlands of Portugal, with only two complete harvests, has been strongly dedicated to developing a modern business, with deep roots in sustainability and circular economy, with limited preventive organic interventions. In the last harvest, 6 ha of white grapes and 16 ha of red were grown, with respective production of 25 and 75 tons, respectively. Vines are 100% organic, with 70% made by mechanical treatment. Harvesting is entirely manual. From the cultivation process, 40 tons of waste were generated, most of which (99%) were organic waste from pruning, and all were composted locally together with a small portion (4 kg) of paper and cardboard. The rest of the segregated waste is sent for recycling. The company did not generate mooring waste. This local treatment cost 1800.00 € and the company received 200.00 € for the sale of waste from equipment maintenance (transport vehicles).

In wine processing, 34 tons of white grapes were processed, 65% of which were own and 35% from third parties, plus 75 tons of red grapes, of which 79% came from own production. 180 hL of white wine and 420 hL of red wine were produced. In the transformation process, 12 tons of organic solid wastes were generated, 4 tons of grape skins and seeds, all of which 16 tons of organic solid waste composted locally. More than 14 tons of fermented grape pomace were sold as by-products, as well as the 4 tons of wine lees; 1 ton of wood was reused; of the 1 ton of paper, everything was sent for recycling, and of the 2 tons for plastic, half was for recycling and half was reused. With the sale of by-products, the company obtained revenues of 100.00 € from grape pomace and 400.00 € from lees. With regard to effluents, 900 hL were generated, a large part of which was treated in the own wastewater plant installed in the middle of the year. Thus, 20% were discarded directly before the start of wastewater plant operation, 40% were treated before discarded, and 40% were treated and reused.

In the bottling activity (40,000 bottles were processed), 550 kg of waste were generated, being 120 kg of plastic, 200 kg of paper, 200 kg of wood, 20 kg of glass, and 10 kg of metal. Cork waste was practically none. Except for wood that is reused (pallets for transportation), everything else is segregated and sent for recycling.

As circular practices, this company shared effort for composting wastes from pruning, wine production, green cleaning, and organic material with a local viticulturist. Equipment usage, logistics, and labor are shared as the company sourced a donation of 2–3 ton of chicken manure daily. Additionally, this organization shares equipment and logistics with two other wine producers that produce wine inside this winery for their brands.

*Company C*

This small business, located in Napa Valley, California in the USA, founded in the early 1970s, was one of the pioneers in organic farming, having started in 1985. It currently has half of its cultivation in organic culture and half in biodynamics, both certified. With a philosophy of natural and integrated management, it has a strong concern for the entire ecosystem, from the recovery and strengthening of the soil, diversity of plants, insects, and animals, to the incorporation of the lunar calendar. It has half a hectare of white grapes cultivation and 15 ha of red grapes with 100% mechanical treatment, with irrigation and 100% manual harvesting. The production obtained was 5 tons of white grapes and 120 tons of reds. There is separation of waste in which plastics, paper, and waste from equipment maintenance are sent for recycling. Of the organic waste from pruning, 80% was reused, 10% was composted locally, and the rest was sent to compost off-site. 2000.00 € were spent on waste management, half of which for pruning wastes.

The winemaking process resulted in 498 hL of white wine from 80 tons, of which 95% came from third parties, and 815 hL of red wine originated from 160 tons of red grapes with 80% of own grapes. 165 tons of waste were generated: 95 tons of organic solid waste, 30 tons of grape pomace, 5 tons of less, 5 tons of paper, and 2 tons of plastic. Only filtration residues and oil were not screened. Of the sorted organic waste, all are sent to off-site composting, all the wood is reused, and the other waste is recycled. With a cost of 6000.00 €, half was destined to treat organic waste; 2000.00 € were used to pay for the treatment of grape pomace, and the rest for plastic and paper. Of the 11,355 hL of effluents generated, 100% were sent for off-site treatment with costs of 20,000 €.

In the bottling process, the company generated a total of 4 tons of waste, 1 ton of plastic, 2 tons of paper and cardboard and 1 ton of glass that are destined for recycling. Wood and cork waste are internally reused, and here the company does not segregate costs because it incorporates them in the winery production side.

A point highlighted by the company was that the large generation of waste from the three activities has a higher concentration in winemaking and since it is certified both in organic production and in biodynamic production, according to the rules of these certifications, it is not possible to allocate these residues in full for both certifications. It is a company that has a holistic and integrative vision, not only of its territory, but of the entire community, for the preservation of all biodiversity, having helped to regenerate a riparian corridor. Even so, the company did not highlight circularity actions and reported not using by-products from other activities.

*Company D*

Having started its production in the early 1980s in Alentejo, Portugal, this small-sized company houses a property of multiple cultures with diversity and richness of native fauna and flora that in addition to vineyards has cork oaks, sheep grazing, and cultivation of nuts. Its principles are to preserve traditional cultures of the region and the balance of the ecosystem. Regarding the production of vines, 9 ha of white grapes and 41 ha of red grapes were cultivated by irrigation, 90% of which was organic and

10% biodynamic. The treatment of the vines was completely mechanical and the harvest was carried out 80% in manual harvesting. In total, this property harvested 47 tons of white grapes and 117 tons of red grapes. All waste of organic origin was composted locally. Plastic, paper and cardboard waste, waste from equipment maintenance and mineral oils and lubricants were sorted and all sent for recycling.

The wine production only used its own grapes, having generated a volume of 282 hL of white wine and 819 hL of red wine. The company did not report the total volume of organic waste (leaves, branches, stems), although it has indicated that all waste generated in those categories is directed for local composting. The amount of other residues was informed, with 16 tons of grape pomace and 10 tons of wine lees, both destined for energy recovery (off-site) in total. All wood waste is reused, plastic, paper and cardboard waste, and waste from equipment maintenance were sent for recycling, and filtration waste was sent for landfill disposal. The totality of 7700 hL of effluents generated was treated and discarded. Information concerning waste generation in bottling activities has not been provided. No practices for sharing or using by-products were reported.

Table 1 summarizes the main elements of each of the four cases analyzed.

There was no interest in making straightforward comparisons, given the many particularities of each production, and each region associated to those companies. However, some relevant issues can be extracted from this analysis:

- The four selected cases already represent a high relation to sustainability practices but even in companies with clear guidelines for reducing environmental impacts, waste management does not seem to be an operation fully inserted in the context of management routines, because the data was not always available and recycling appears as the prevalent end of life solution, which although relevant, represents generally a loss of value compared to other circular economy practices.
- Even when there are concerns and even active practices related to those concerns over sustainability, companies are unable to identify them as regenerative and restorative operations of the system.
- Only one of the cases had labor sharing operations and two cases highlighted the use of by-products from other activities.
- Although not being the most suitable form of recovery by the principles of circularity, only one company sent waste as raw material for energy production.
- Some by-products are produced in a relevant amount, grape pomace (ranging from 0.10 to 0.13 $t/t_{grapes\ processed}$—relatively consistent values from 3 companies) and lees (0.02–0.06 $t/t_{grapes\ processed}$) which need to be sent for treatment, with costs associated in some cases. The integrated management of such organic rich materials might thus be seen as an opportunity for future developments on circular economy practices.
- A relevant and variable amount of wastewater is produced by the companies (from 5 to 47 $hL/t_{grapes\ processed}$). This variation stands between companies without irrigation and those with irrigated cultivation. Possibly, the reported values from the companies with irrigation combine viticulture and winemaking. Either way,

**Table 1** Resume of the most relevant information gathered from the four analyzed case studies

| Description | A | B | C | D |
|---|---|---|---|---|
| Location | USA | Portugal | USA | Portugal |
| Company dimension | Microcompany | Microcompany | Small company | Small company |
| Type of viticulture | Conventional | Organic | 50% organic 50% biodynamic | 50% organic 50% biodynamic |
| Irrigation | No | No | Yes | Yes |
| *Cultivated area* (ha) | | | | |
| • White grapes | 3 | 6 | 0.5 | 9 |
| • Red grapes | 3 | 16 | 15 | 41 |
| **Total (ha)** | **6** | **22** | **15.5** | **50** |
| *Harvest production* (t) | | | | |
| • White grapes | 30 | 25 | 5 | 47 |
| • Red grapes | 18 | 75 | 120 | 117 |
| **Total (t)** | **48** | **100** | **125** | **164** |
| $t_{white\ grapes}$/ha | **10** | **4.2** | **10** | **5.2** |
| $t_{red\ grapes}$/ha | **6** | **4.7** | **8.0** | **2.8** |
| Waste from pruning | 70 kg | 40 t | Not reported | Not reported |
| *Grapes processed* (t) | | | | |
| • White grapes | 20 | 34 | 80 | 47 |
| • Red grapes | 20 | 75 | 160 | 117 |
| **Total (t)** | **40** | **109** | **240** | **164** |
| *Volume of wine* (hL) | | | | |
| • White grapes | 100 | 180 | 498 | 282 |
| • Red grapes | 100 | 420 | 815 | 819 |
| **Total (hL)** | **200** | **600** | **1313** | **1101** |
| $hL/t_{white\ grapes\ processed}$ | **5** | **5.3** | **6.2** | **6.0** |
| $hL/t_{red\ grapes\ processed}$ | **5.0** | **5.6** | **5.1** | **7.0** |
| *Waste from production* (t) | | | | |
| • Organic solid wastes (leaves, branches, stems) | Not informed | 12 | 95 | Not informed |
| • Grape skins and seeds | 12 | 4 | Not informed | Not informed |
| • Fermented grape pomace | Not informed | 14 | 30 | 16 |
| $t_{grape\ pomace/grapes\ processed\ (total)}$ | – | **0.13** | **0.13** | **0.10** |
| • Lees | 2 | 4 | 5 | 10 |
| $t_{lees/grapes\ processed\ (total)}$ | **0.05** | **0.04** | **0.02** | **0.06** |
| Wastewater (hL) | 200 | 900 | 11,355 | 7700 |
| $hL/t_{grapes\ processed\ (total)}$ | **5.0** | **8.3** | **47.1** | **47.0** |

(continued)

**Table 1** (continued)

| Waste destiny | Composting Recycling | Composting Recycling Selling | Composting Recycling | Composting Recycling Energy Landfilling |
|---|---|---|---|---|
| *Circular economy practices* | | | | |
| • Reuse | Yes | Yes | Yes | Yes |
| • Recycling | Yes | Yes | Yes | Yes |
| • Other forms of recovery | | | | Yes |
| • By-products used during operation | Yes | Yes | | |
| • Shared equipment | | Yes | | |
| • Shared logistic | Yes | Yes | | |
| • Shared workforce | | Yes | | |

*Source* Authors

results emphasize the importance to preserve resources and the relevance of developing actions to promote more circular practices in this field, focusing on the reduction of the amount of wastewater generated.

The mentioned aspects call attention to a gap that seems to exist in the wine production sector in which a huge potential for optimizing the use of resources both internally, and in systems of exchanges, shares, donations, or sales of by-products could substantially increase activities focused on redesigning the material flows. In order to allow for improvements, it is necessary not only the collaboration between multiple agents, but also to identify the different chains in the local and regional territorial scales with the potential for valorizing these resources.

# 3  Conclusions

The present chapter aimed to provide an overview of the challenges associated with the food chain, and more particularly, the wine sector, focusing on the practices concerning by-products and waste management. For that, it started by presenting a global context on the food chain, food loss and food waste and after depicts relevant issues concerning circular practices and more specifically circular supply chain management.

A more detailed study was also conducted regarding circular practices in the food supply chain, with focus on waste and more specific activities at the wine sector (including vineyard cultivation, vinification and bottling), where a systematic review was performed together with a set of 44 questions distributed through relevant platforms that gather companies of this sector from which four case studies were more deeply analyzed.

The systematic review showed that most of the 47 selected studies are directed to the use of waste as resource, through the valorization of those wastes (application of proteins in other chains, application of fibers in construction materials, bioproducts, transformation into a biofertilizer, conversion into energy, among others); however, for a food supply chain analysis those were considered inappropriate due to the different study objects and multiplicity of methodologies and cutouts, validating previously identified challenges to map and align data in the food chain sector.

The questionnaire had 55 respondents; however, statistical data treatment was not possible since the number of valid complete responses was less than necessary to carry out a robust quantitative study. The evaluation of four case studies was further performed, where quantitative and qualitative information concerning grape and wine production as well as waste and wastewater management was presented. It was verified that, although the selected companies are already aligned with the topic of sustainability, they are unable to identify regenerative and restorative operations applicable to their activities. The composting practices are transversal to the companies and some by-products are produced in relevant amounts, such as grape pomace (up to 0.13 $t/t_{grapes\ processed}$) and lees (up to 0.06 $t/t_{grapes\ processed}$) with costs for external management, highlighting an opportunity to establish integrated and more circular solutions for such resources. Wastewater is also generated in relevant amounts (from 5 to 47 $hL/t_{grapes\ processed}$) which points out the relevance of exploring additional reduction practices in this field.

Many possibilities exist to apply the principles of circularity in the studied chain, from the amount of resources not absorbed and worked in an optimal way, to the enormous potential to establish local arrangements that integrate different agents, with different valorization routes for by-products between chains; consequently allowing operation for sustainable development.

**Acknowledgements** We kindly appreciate The Porto Protocol for spreading the study to its members.

This work was financially supported by: Base Funding—UIDB/00511/2020 of the Laboratory for Process Engineering, Environment, Biotechnology and Energy—LEPABE—funded by national funds through the FCT/MCTES (PIDDAC).

# References

1. Abecassis J, Cuq B, Escudier JL, Garric G, Kondjoyan A, Planchot V, Salmon JM, de Vries H (2018) Food chains; the cradle for scientific ideas and the target for technological innovations. Innov Food Sci Emerg Technol 46(August 2017):7–17
2. Del Carmen Alamar C, Falagán N, Aktas E, Terry LA (2018) Minimising food waste: a call for multidisciplinary research. J Sci Food Agric 98(1)
3. Amienyo D, Camilleri C, Azapagic A (2014) Environmental impacts of consumption of Australian red wine in the UK. J Cleaner Prod 72:110–119
4. Andersen M, Skjoett-Larsen T (2009) Corporate social responsibility in global supply chains. Supply Chain Manage Int J

5. Arzoumanidis I, Salomone R, Petti L, Mondello G, Raggi A (2017) Is there a simplified LCA tool suitable for the agri-food industry? An assessment of selected tools. J Cleaner Prod 149:406–425
6. Bagherzadeh M, Inamura M, Jeong H (2014) Food waste along the food chain. In: OECD Food, agriculture and fisheries papers, issue 71
7. Batista L, Bourlakis M, Liu Y, Smart P, Sohal A (2018) Supply chain operations for a circular economy. Prod Plann Control 29(6):419–424
8. Berardi PC, Dias JM (2019) How has the wine sector incorporated the premises of circular economy? J Environ Sci Eng B 8(3):108–117
9. Bertran E, Sort X, Soliva M, Trillas I (2004) Composting winery waste: sludges and grape stalks. Bioresour Technol 95(2):203–208
10. Bressanelli G, Perona M, Saccani N (2019) Challenges in supply chain redesign for the circular economy: a literature review and a multiple case study. Int J Prod Res 57(23):7395–7422
11. Brunori G, Galli F, Barjolle D, Broekhuizen R, Colombo L, Giampietro M, Kirwan J, Lang T, Mathijs E, Maye D, Roest K, Rougoor C, Schwarz J, Schimitt E, Smith J, Stojanovic Z, Tisenkopfs T, Touzard J (2016) Are local food chains more sustainable than global food chains? Considerations for assessment. Sustainability 8(449)
12. Buren N, Demmers M, Heijden R, Witlox F (2016) Towards a circular economy: the role of Dutch logistics industries and governments. Sustainability 8(17)
13. Campoy-Muñoz P, Cardenete MA, Delgado MC (2017) Economic impact assessment of food waste reduction on European countries through social accounting matrices. Resour Conserv Recycl 122:202–209
14. Christ KL, Burritt RL (2013) Critical environmental concerns in wine production: an integrative review. J Cleaner Prod 53:232–242
15. Ciulli F, Kolk A, Boe-Lillegraven S (2020) Circularity brokers: digital platform organizations and waste recovery in food supply chains. J Bus Ethics 167(2)
16. Coelho MC, Pereira RN, Rodrigues AS, Teixeira JA, Pintado ME (2020) The use of emergent technologies to extract added value compounds from grape by-products. Trends Food Sci Technol 106(May):182–197
17. Commission for Environmental Cooperation (2019) Quantifying food loss and waste and its impacts
18. Corrado S, Sala S (2018) Food waste accounting along global and European food supply chains: State of the art and outlook. Waste Manage 79:120–131
19. Dávila I, Robles E, Egüés I, Labidi J, Gullón P (2017) The biorefinery concept for the industrial valorization of grape processing by-products. In: Galanakis CM (ed) Handbook of grape processing by-products. Academic Press, Cambridge
20. De Angelis R, Howard M, Miemczyk J (2018) Supply chain management and the circular economy: towards the circular supply chain. Prod Plann Control 29(6):425–437
21. Delmas MA, Grant LE (2014) Eco-labeling strategies and price-premium: the wine industry puzzle. Bus Soc 53(1)
22. Englezos V, Rantsiou K, Cravero F, Torchio F, Giacosa S, Río Segade S, Gai G, Dogliani E, Gerbi V, Cocolin L, Rolle L (2019) Minimizing the environmental impact of cleaning in winemaking industry by using ozone for cleaning-in-place (CIP) of wine bottling machine. J Cleaner Prod 233:582–589
23. Esposito B, Sessa MR, Sica D, Malandrino O (2020) Towards circular economy in the agri-food sector. A systematic literature review. Sustainability 12(18)
24. European Commission (2017) Food use for social innovation by optimizing food waste recovery strategies. In: Innovation strategies in the food industry: tools for implementation
25. FAO (2018) El estado de la seguridad alimentaria y la nutrición en el mundo 2017. Fomentando la resiliencia en aras de la paz y la seguridad alimentaria
26. FAO (2019) Food loss and food waste. In: Food loss and food waste. http://www.fao.org/food-loss-and-food-waste/flw-data
27. FAO (2019) The state of food and agriculture 2019. Moving forward on food loss and waste reduction. Food and Agriculture Organization of the United Nations

28. FAO (2020) The state of agricultural commodity markets 2020. In: The state of agricultural commodity markets 2020. Food and Agriculture Organization of the United Nations
29. Farooque M, Zhang A, Liu Y (2019) Barriers to circular food supply chains in China. Supply Chain Manage 24(5):677–696
30. Farooque M, Zhang A, Thürer M, Qu T, Huisingh D (2019) Circular supply chain management: a definition and structured literature review. J Cleaner Prod 228:882–900
31. Ferrara C, De Feo G (2018) Life cycle assessment application to the wine sector: a critical review. Sustainability (Switzerland) 10(2)
32. Fiore E, Stabellini B, Tamborrini P (2020) A systemic design approach applied to rice and wine value chains. The case of the InnovaEcoFood project in piedmont (Italy). Sustainability 12(21):1–28
33. Flanagan K, Robertson K, Hanson C (2019) Reducing food loss. World Resources Institute, p 130
34. Galanakis CM (2017) In: Galanakis CM (ed) Handbook of grape processing by-products. Sustainable solutions. Academic Press, Cambridge
35. Geissdoerfer M, Vladimirova D, Evans S (2018) Sustainable business model innovation: a review. J Cleaner Prod 198:401–416
36. Ghisellini P, Ulgiati S (2019) Circular economy transition in Italy. Achievements, perspectives and constraints. J Cleaner Prod 243(September):118360
37. Goodwin L (2019) Time's up for food waste! European Commission. http://library1.nida.ac.th/termpaper6/sd/2554/19755.pdf
38. Gouveia C, Liberato MLR, DaCamara CC, Trigo RM, Ramos AM (2011) Modelling past and future wine production in the Portuguese Douro valley. Clim Res 48(2–3):349–362
39. Gruère G, Shigemitsu M, Crawford S (2020) Agriculture and water policy changes: stocktaking and alignment with OECD and G20 recommendations. OECD Food, Agriculture and Fisheries Papers
40. Guo X, Broeze J, Groot JJ, Axmann H, Vollebregt M (2020) A worldwide hotspot analysis on food loss and waste. Associated greenhouse gas emissions, and protein losses. Sustainability 12(18)
41. Hao L, Li Y, Gong P, Xiong W (2019) Copyright. In: Godoi FC, Bhandari BR, Prakash S, Zhang M (eds) Fundamentals of 3D food printing and applications. Elsevier, Amsterdam, p iv
42. Hogervorst JC, Miljić U, Puškaš V (2017) Extraction of bioactive compounds from grape processing by-products. In: Galanakis CM (ed) Handbook of grape processing by-products. Academic Press, Cambridge
43. Iannone R, Miranda S, Riemma S, De Marco I (2016) Improving environmental performances in wine production by a life cycle assessment analysis. J Cleaner Prod 111:172–180
44. Jarosz L (2000) Understanding agri-food networks as social relations. Agric Hum Values 17:279–283
45. Kaza S, Yao LC, Bhada-Tata P, Van Woerden F (2018) What a waste 2.0: a global snapshot of solid waste management to 2050. https://openknowledge.worldbank.org/handle/10986/30317
46. Kiefer CP, González PDR, Carrillo-Hermosilla J (2018) Drivers and barriers of eco-innovation types for sustainable transitions: a quantitative perspective. Bus Strategy Environ 28(1):155–172
47. Lahane S, Kant R, Shankar R (2020) Circular supply chain management: a state-of-art review and future opportunities. J Cleaner Prod 258:120859
48. Linton JD, Klassen R, Jayaraman V (2007) Sustainable supply chains: an introduction. J Oper Manage 25(6):1075–1082
49. Lipinski B, Hanson C, Lomax J, Waite R, Searchinger T (2013) Reducing food loss and waste (Installment 2 of creating a sustainable food future)
50. Mangla SK, Luthra S, Mishra N, Singh A, Rana NP, Dora M, Dwivedi Y (2018) Barriers to effective circular supply chain management in a developing country context. Prod Plann Control 29(6):551–569
51. Manniello C, Statuto D, Di Pasquale A, Picuno P (2020) Planning the flows of residual biomass produced by wineries for their valorization in the framework of a circular bioeconomy. In: Lecture notes in civil engineering, vol 67, pp 295–303

52. Maroun RG, Rajha HN, Vorobiev E, Louka N (2017) Emerging technologies for the recovery of valuable compounds from grape processing by-products. In: Galanakis CM (ed) Handbook of grape processing by-products. Academic Press, Cambridge, pp 155–181
53. Martínez Salgado MM, Ortega Blu R, Janssens M, Fincheira P (2019) Grape pomace compost as a source of organic matter: evolution of quality parameters to evaluate maturity and stability. J Cleaner Prod 216:56–63
54. Merli R, Preziosi M, Acampora A (2018) Sustainability experiences in the wine sector: toward the development of an international indicators system. J Cleaner Prod 172:3791–3805
55. Navarro A, Puig R, Fullana-i-Palmer P (2017) Product vs corporate carbon footprint: some methodological issues. A case study and review on the wine sector. Sci Total Environ 581–582:722–733
56. Neto B, Dias AC, Machado M (2013) Life cycle assessment of the supply chain of a Portuguese wine: from viticulture to distribution. Int J Life Cycle Assess 18(3):590–602
57. OECD (2020) Food supply chains and COVID-19: impacts and policy lessons. https://read.oecd-ilibrary.org/view/?ref=134_134305-ybqvdf0kg9&title=Food-Supply-Chains-and-COVID-19-Impacts-and-policy-lessons
58. OIV, International Organisation of Vine and Wine (2020) 2019 Statistical report on world vitiviniculture
59. Paes LAB, Bezerra BS, Deus RM, Jugend D, Battistelle RAG (2019) Organic solid waste management in a circular economy perspective—a systematic review and SWOT analysis. J Cleaner Prod 239:118086
60. Papargyropoulou E, Lozano R, Steinberger JK, Wright N, Ujang ZB (2014) The food waste hierarchy as a framework for the management of food surplus and food waste. J Cleaner Prod 76:106–115
61. Ponstein HJ, Ghinoi S, Steiner B (2019) How to increase sustainability in the finish wine supply chain? Insights from a country of origin based greenhouse gas emissions analysis. J Cleaner Prod 226:768–780
62. Poore J, Nemecek T (2018) Reducing food's environmental impacts through producers and consumers. Science 360(6392):987–992
63. Pretorius IS (2000) Tailoring wine yeast for the new millennium: novel approaches to the ancient art of winemaking. Yeast 16(8):675–729
64. Principato L, Ruini L, Guidi M, Secondi L (2019) Adopting the circular economy approach on food loss and waste: the case of Italian pasta production. Resour Conserv Recycl 144(December 2018):82–89
65. Rao P, Holt D (2005) Do green supply chains lead to competitiveness and economic performance? Int J Oper Prod Manage 25(9):898–916
66. Ribeiro I, Sobral P, Peças P, Henriques E (2018) A sustainable business model to fight food waste. J Cleaner Prod 177:262–275
67. Ritchie H (2020) Environmental impacts of food production. Our World in Data. https://ourworldindata.org/environmental-impacts-of-food
68. Ritzén S, Sandström GÖ (2017) Barriers to the circular economy—integration of perspectives and domains. Procedia CIRP 64:7–12
69. Sachs J, Schmidt-Traub G, Kroll C, Lafortune G, Fuller G, Woelm F (2020) The sustainable development goals and COVID-19
70. Sánchez-Gómez R, Alonso GL, Salinas MR, Zalacain A (2017) Reuse of vine-shoots wastes for agricultural purposes. In: Galanakis CM (ed) Handbook of grape processing by-products. Academic Press, Cambridge, pp 79–104
71. Sassanelli C, Rosa P, Rocca R, Terzi S (2019) Circular economy performance assessment methods: a systematic literature review. J Cleaner Prod 229:440–453
72. Scherhaufer S, Moates G, Hartikainen H, Waldron K, Obersteiner G (2018) Environmental impacts of food waste in Europe. Waste Manage 77:98–113
73. Schmeisser B (2013) A systematic review of literature on offshoring of value chain activities. J Int Manage 19(4):390–406

74. Sehnem S, Ndubisi NO, Preschlak D, Bernardy RJ, Santos Junior S (2020) Circular economy in the wine chain production: maturity, challenges, and lessons from an emerging economy perspective. Prod Plann Control 31(11–12):1014–1034

75. Seuring S, Muller M (2008) From a literature review to a conceptual framework for sustainable supply chain management. J Cleaner Prod 16:1699–1710

76. Silva V, Igrejas G, Falco V, Santos TP, Torres C, Oliveira AMP, Pereira JE, Amaral JS, Poeta P (2018) Chemical composition, antioxidant and antimicrobial activity of phenolic compounds extracted from wine industry by-products. Food Control 92:516–522

77. Spigno G, Marinoni L, Garrido G (2017) State of the art in grape processing by-products. In: Galanakis CM (ed) Handbook of grape processing by-products: sustainable solutions. Academic Press, Cambridge

78. Srivastava SK (2007) Green supply-chain management: a state-of-the-art literature review. Int J Manage Rev 9(1):53–80

79. Szolnoki G (2013) A cross-national comparison of sustainability in the wine industry. J Cleaner Prod 53:243–251

80. Teigiserova DA, Hamelin L, Thomsen M (2020) Towards transparent valorization of food surplus, waste and loss: clarifying definitions, food waste hierarchy, and role in the circular economy. Sci Total Environ 706:136033

81. The World Bank (2020) Atlas of sustainable development goals 2020: from world development indicators. https://datatopics.worldbank.org/sdgatlas/goal-12-responsible-consumption-and-production/#footnote2

82. Tranfield D, Denyer D, Smart P (2003) Towards a methodology for developing evidence-informed management knowledge by means of systematic review. Br J Manage 14(3):207–222

83. United Nations Environment Programme (2017) Food and food waste. Sustainable Food Systems. https://www.unenvironment.org/explore-topics/resource-efficiency/what-we-do/sustainable-lifestyles/food-and-food-waste

84. Xu R, Ferrante L, Briens C, Berruti F (2009) Flash pyrolysis of grape residues into biofuel in a bubbling fluid bed. J Anal Appl Pyrol 86(1):58–65

85. Xue L, Liu G, Parfitt J, Liu X, Van Herpen E, Stenmarck Å, O'Connor C, Östergren K, Cheng S (2017) Missing food, missing data? A critical review of global food losses and food waste data. Environ Sci Technol 51(12):6618–6633

86. Zambon I, Colantoni A, Cecchini M, Mosconi EM (2018) Rethinking sustainability within the viticulture realities integrating economy, landscape and energy. Sustainability 10(2)

# Carbon Footprint of Raw Milk and Other Dairy Products

**Rishabh Sahu and Tripti Agarwal**

**Abstract** Dairy and livestock sector is a significant contributor of anthropogenic greenhouse gas emissions. Carbon footprint (CF) is commonly used to indicate the greenhouse gas (GHG) emissions ($CO_2$ equivalent) at various life cycle stages of a product. Studies undertaken globally on CF of dairy products were reviewed, reported CF values for various products are summarized and important contributing factors are discussed. In various studies undertaken globally, CF of dairy products has been calculated by using different international standards and methodologies like ISO 14040, 14044 and 14067, publicly available specification (PAS 2050). Most of the studies have used functional units such as kilogram greenhouse gas emissions per kilogram of fat-and-protein-corrected-milk (FPCM) and energy correlated milk (ECM). Direct emissions of carbon dioxide ($CO_2$), methane ($CH_4$) and nitrous oxide ($N_2O$) from on-farm livestock production and indirect $CO_2$ and $N_2O$ emissions related to inputs on the farm have generally been considered in various studies. Enteric methane ($CH_4$) has been reported as the major source of dairy farm emissions, followed by manure management, fertilizer production and its application. Processed milk products were found to have higher CF value as compared to the unprocessed milk. Various mitigation strategies have been suggested for emission reduction from dairy farms for example balanced feed rations and concentrates according to animal body requirements during lactation period, reducing use of nitrogen based fertilizers and increasing efficiency in application during crop production, use of biogas in place of cow dung, anaerobic digestion (AD) and efficient manure management.

**Keywords** Carbon footprint · Life cycle assessment (LCA) · Milk · Dairy products · Mitigation

R. Sahu · T. Agarwal (✉)
Department of Agriculture and Environmental Sciences, National Institute of Food Technology
Entrepreneurship and Management (NIFTEM), Kundli, Haryana, India

© The Author(s), under exclusive license to Springer Nature Singapore Pte Ltd. 2021    177
R. S. Mor et al. (eds.), *Challenges and Opportunities of Circular Economy in Agri-Food Sector*, Environmental Footprints and Eco-design of Products and Processes,
https://doi.org/10.1007/978-981-16-3791-9_10

# 1  Introduction

Warming of the Earth is unambiguous. Anthropogenic greenhouse gas (GHG) emissions, due to industrialization, have resulted in a significant increase in methane ($CH_4$), carbon dioxide ($CO_2$), nitrous oxide ($N_2O$) and other greenhouse gases in the atmosphere. Earth's temperature has been recorded to rise by 0.85° since the 1880s. It has an extensive impact on natural systems; influencing water resources, melting of ice reserves, shifting of geographical range of marine, freshwater and many terrestrial species, their migration pattern and seasonal activity, increase and decrease in temperature in extreme levels. Many unprecedented changes have been recorded over each of the last three decades. The temperature of the Earth's surface is predicted to rise beyond the temperatures during the twenty-first century under all estimated emission frameworks and result in long-lasting and irreversible changes to climate components. Limiting the rise in global average temperature under 2 °C above pre-industrial levels and consistent effort to check the temperature rise to 1.5 °C would reduce the GHG emission to greater extent. Accordingly, atmospheric GHG concentrations must be limited to 550 ppm Carbon dioxide equivalents [21]. Mitigation and adaptation with effective decision making and analysis tools are complementary approaches to manage and reduce climate change [15].

## 1.1  Impact of Dairy on Climate

Livestock sector contribute significantly, i.e., ~ 14.5% to the total anthropogenic greenhouse gas emissions (7.1 Gt $CO_2$eq per year of 49 Gt $CO_2$eq per year) globally. The dairy sector accounts for ~ 3% of the global GHG emissions [18]. As per the national inventories, in all developing countries, GHG emissions from livestock sector account for ~ 9% of their gross GHG emissions but surpasses 20% of the gross GHG emission in more than one-third of developing countries. Methane gas, which has a higher warming potential than Carbon dioxide, constitutes approximately 44% of the emissions from livestock sector [10].

Contribution of dairy cattle and beef to the global livestock sector emission was found to be the highest (74% of livestock emission globally) [1]. Currently, the contribution of enteric fermentation is ~ 32–40% to GHG emission from agriculture globally and ~ 90% of emission from enteric fermentation is from cattle and buffaloes. This implies that one ton of methane and nitrous oxide heat the earth 28 and 265 times more than the Carbon dioxide, respectively, over a 100 year time horizon. GHG emissions from the dairy sector were found to increase by 18% between 2005 and 2015 due to substantial growth in milk production and increase in consumer demand. However, dairy farming emission per unit of product were found to have a reducing trend at the rate of 1% per year since 2005 [5]. The GHG emissions during the life cycle of a product/process/facility is expressed in $CO_2$ equivalent considering the IPCC-100 year global warming potential (GWP) for nitrous oxide, methane and carbon dioxide which are 265, 28 and 1, respectively [17].

## 1.2 Demand Pattern for Dairy Products

About 6 billion people (> 80% of the world population) consume milk and dairy products daily. According to the IFPRI report, there is a growth in milk consumption by 0.2 and 3.3% in developed and developing countries, respectively, from 1993 to 2020. Growth of milk production is 0.4% in developed countries and 3.2% in developing countries [4]. According to the United Nations, the world population is predicted to rise from the current 7.6–8.6 billion by 2030, 9.8 billion by 2050 and exceed 11.2 billion by 2100. This is expected that the demand of calories from animal origin will double by 2050 [11]. High GHG emissions are anticipated from livestock sector due to forthcoming demand for livestock products [32]. This provides opportunity and challenges to the dairy sector to fulfill global demand with nutritious and sustainable food [5].

## 2 Methodology for Carbon Footprint Calculation

Existing methodologies for carbon footprint calculation are as follows [13]:

- ISO 14044 (2006):
  It defines requirements and gives guidelines for life cycle analysis of product including goal and scope of life cycle assessment (LCA), product life cycle inventory analysis, product life cycle impact assessment, life cycle interpretation, reporting and critical review and limitations.
- ISO 14067: Carbon Footprint of Product:
  This standard is used to compute the GHG emissions from organizations and their activities and specifies requirements and guidance for studies to calculate the product's actual carbon footprint and partial carbon footprint based on LCA defined in ISO 14040 and 14044.
- PAS 2050 (Publicly Available Standard): 2011:
  PAS 2050 defines requirements for the evaluation of the greenhouse gas emissions from the life cycle of products according to the LCA principles and techniques (i.e., ISO 14040/44). Requirements are defined for recognizing the system boundary and data requirements for analysis and the result calculation. It also includes the six GHGs covered under the Kyoto protocol.
- ISO 14064:
  ISO 14064-1:2006 defines requirements and principles for reporting and quantification of GHG emission at organization level, ISO 14064-2:2006 at project level, ISO 14064-3:2006 can be applied for both organizational and project level

quantification whereas ISO 14069 GHG is guidance for the application of ISO 14064.

Intergovernmental panel on climate change (IPCC) formed by the world meteorological organization and united nation environment program, is the leading body for climate change assessment. IPCC [14] approach is used for GHG evaluation of the dairy sector. IPCC guidelines is based on following three approach

Tier 1

This is a simplified approach in which default emission factors either from detailed tier 2 methodology or by literature reviews are used. Tier 1 method is used where detailed data is not available or where the enteric fermentation is not a major category.

Tier 2

It involves more complex methodology, requires detailed country specific data for specific livestock categories, tier 2 approach is used where a huge portion of emission results from livestock of countries total emission.

This approach requires to define animals, management circumstances, diet quality and animal productivity to evaluate more accurately diet intake to evaluate methane emission from enteric fermentation as well as determination of nitrogen and manure excretion rate to facilitate the consistency and accuracy of $N_2O$ and $CH_4$ emission from manure management.

Tier 3

Some countries, where livestock emission is a major concern, may use the tier 3 approach. This approach requires developing sophisticated models for diet composition, concentration, feed availability, seasonal variation and mitigation strategy.

Functional unit (FU) and allocation method are two of the main criteria which decide the final value of the CF. GHG emissions (kg $CO_2$eq) per kilogram of fat and protein corrected milk (FPCM) is the most commonly used FU for milk. However, emissions per kilogram of energy corrected milk (ECM) have also been reported in few studies. Economic allocation is used in a number of studies reporting CF of dairy products. However, as per ISO guidelines, system expansion is recommended than the allocation wherever possible.

# 3    Carbon Footprint of Dairy Products

Carbon footprint of dairy products has been reported from different parts of the world. Most of the studies have reported CF of raw milk using LCA methodology/PAS 2050 with a cradle to farm-gate system boundary. GHG emissions are most commonly reported in kilogram of $CO_2$eq/kg of fat and protein corrected milk (FPCM). Details of CF footprint studies on milk and other dairy products are given in Table 1.

**Table 1** Carbon footprint of milk and other dairy products reported from different countries

| S. No. | Place | Product | CF | Scope | Method | References |
|---|---|---|---|---|---|---|
| 1. | Australia | Milk | 0.39–1.35 kg of $CO_2$eq/kg of ECM | Cradle to farm gate | Integrated farm system model (IFSM) | Sejian et al. [26] |
| 2. | Canada western province | Milk | 0.93 kg of $CO_2$eq/L of milk | Cradle to factory gate | ISO standards 14040 series:2006 | Vergé et al. [28] |
| | Canada eastern province | Milk | 1.12 kg of $CO_2$eq/L of milk | | | |
| | Canada | Cheese | 5.3 kg of $CO_2$eq/L of milk | | | |
| | | Butter | 7.3 kg of $CO_2$eq/L of milk | | | |
| | | Milk powder | 10.1 kg of $CO_2$eq/L of milk | | | |
| 3. | China | Skim milk powder (SMP) | 4.72 kg of $CO_2$eq per FU | Cradle to grave | PAS 2050:2008 | Xu et al. [33] |
| | | Fluid milk | 0.46 kg of $CO_2$eq per FU | | | |
| | | Yogurt | 0.31 kg of $CO_2$eq per FU | | | |
| 4. | Denmark | Fresh dairy products | 1.1 kg $CO_2$eq/kg | Farm to customer | (ISO 2006), IDF [12] | Flysjö et al. [6] |
| | | Butter and butter blends | 8.1 kg $CO_2$eq/kg | | | |
| | | Cheese | 6.5 kg $CO_2$eq/kg | | | |
| | | Milk powder and whey based products | 7.4 kg $CO_2$eq/kg | | | |
| | | Other, very broad category | 1.2 kg $CO_2$eq/kg | | | |
| 5. | India | Milk | 1.7 kg $CO_2$-eq/kg FPCM | A cradle to farm gate | ISO 14044 2006, IPCC [14] | Garg et al. [7] |

(continued)

**Table 1** (continued)

| S. No. | Place | Product | CF | Scope | Method | References |
|--------|-------|---------|-----|-------|--------|-----------|
| 6. | India | Milk | 5.9 kg $CO_2$eq kg$^{-1}$ dry wt. | Production, processing, transportation and preparation | IPCC [16] | Pathak et al. [22] |
| | | Curd | 7.4 kg $CO_2$eq kg$^{-1}$ dry wt. | | | |
| | | Lassi | 6.1 kg $CO_2$eq kg$^{-1}$ dry wt. | | | |
| | | Butter | 1.2 kg $CO_2$eq kg$^{-1}$ dry wt. | | | |
| 7. | Iran, Tehran | Packaged fluid milk | 1.57 kg $CO_2$eq/kg FPCM milk at farm gate and 1.73 kg $CO_2$eq/L of pasteurized milk packaged in a plastic pouch | Cradle to milk processing gate | Attributional LCA: ISO 14040:2006 | Daneshi et al. [3] |
| 8. | Ireland | Milk | 0.87–1.72 kg of $CO_2$eq/kg of FPCM Mean 1.11 kg of $CO_2$eq/kg of FPCM | Cradle to farm gate | PAS 2050 | O'brien et al. [20] |
| 9. | Ireland | Milk | 1.20 kg of $CO_2$eq/kg of FPCM | Cradle to farm gate | PAS 2050, BIS, 2011, IPCC [14] | O'Brien et al. [19] |
| 10. | Ireland | Milk | 1.09–1.93 kg $CO_2$eq/kg | Cradle to farm gate | LCA, ISO 2006, IDF [12] | Rice et al. [24] |
| | | | 1.09–1.82 kg $CO_2$eq/ECM | | | |
| 11. | Ireland | Fat filled powder | 1.65 kg $CO_2$eq/kg solids | Farm gate to processor gate | – | Yan and Holden [34] |
| | | Skimmed milk powder | 1.40–1.70 kg $CO_2$eq/kg solids | | | |
| | | Butter | 0.41–0.62 kg $CO_2$eq/kg solids | | | |
| 12. | Ireland | Milk | 1.3–1.5 kg $CO_2$eq/kg of ECM per year | Cradle to farm gate | IPCC guidelines | Casey and Holden [2] |

(continued)

**Table 1**  (continued)

| S. No. | Place | Product | CF | Scope | Method | References |
|---|---|---|---|---|---|---|
| 13. | Italy | Milk | 1.11 kg $CO_2$eq/kg FPCM | Cradle to farm gate | ISO/TS 14067 (2014), tier 1 and 2 approach given by IPCC [14] | Vida et al. [29] |
| 14. | Kenya, central (382 farms) | Milk | 2.19 and 3.13 kg of $CO_2$eq/kg of FPCM | Cradle to farm gate | Attributional LCA method | Wilkes et al. [31] |
| 15. | Portugal | UHT milk | 1.74 kg of $CO_2$eq/kg of packaged energy-corrected milk (ECM)-at dairy factory gate | Cradle to factory gate | Life cycle assessment (LCA) | González-García et al. [8] |
|  |  | Raw milk at farm gate | 0.74 kg of $CO_2$eq/kg of FPCM at dairy farm gate |  |  |  |
| 16. | Spain, Catalonia | Yogurt production | 1.94 kg $CO_2$eq/kg of yogurt | Cradle to factory gate | ISO/TS 14067 (2013) | Vasilaki et al. [27] |
| 17. | Sub-tropical regions | Milk | 1.06 kg $CO_2$eq/kg of ECM | Cradle to farm gate | IPCC guidelines | Ribeiro-Filho Id et al. [23] |
| 18. | Developed dairy regions | Milk | 1.3–1.4 kg $CO_2$eq/kg FPCM | Cradle to farm gate | IPCC, GLEAM model (FAO) | FAO and GDP [5] |
|  | Developing dairy regions | Milk | 4.1–6.7 kg $CO_2$eq/kg FPCM |  |  |  |
| 19. | New Zealand | Goat milk | 0.90 kg of $CO_2$eq/kg of FPCM | Cradle to farm gate | ISO, 2006a (14040) and ISO, 2006b (14044) | Robertson et al. [25] |

*Source* Author

## 3.1  Milk

A cradle to farm-gate study was conducted in Ireland to calculate Carbon Footprint of milk from grass-based dairy farms ($n = 124$) using attributional LCA method. The CF of milk ranged from 0.87 to 1.72 kg of $CO_2$eq/kg of FPCM and a mean of $1.11 \pm 0.13$ kg of $CO_2$eq/kg of FPCM. On-farm GHG emissions contributed 80% to the total CF. A significant variation was found between carbon footprint of milk from different farms which was particularly attributed to farm N efficiency which indicate that improving farm efficiency and performance by adopting management practices can mitigate GHG emission of milk [20]. The average carbon footprint of milk in smallholder dairy farms situated in central Kenya varied between 2.19 and 3.13 kg $CO_2$eq/kg FPCM. The highest contribution to the total GHG emissions was found to be from enteric fermentation ($CH_4 - 55.5\%$). Feed production and transport, and manure management contributed 31.6% and 12.6%, respectively to the total CF. Significant variations in the GHG emissions were found among different farms with respect to the feeding systems (i.e., grazing, zero-grazing and mixed feeding systems). CF of farms with grazing only feeding systems was found to be significantly higher than the farms with zero-grazing and mixed systems. However, the difference was not observed when the uncertainty in input parameters was considered [31]. Casey and Holden [2] conducted a study in Ireland to determine greenhouse gas emission of milk production. The average GHG emission at the farm-gate was calculated as 1.50 kg of $CO_2$eq/kg of ECM per year and 1.3 kg of $CO_2$eq/kg per year with economic allocation between meat and milk. Out of total emissions, enteric fermentation accounted for 49%, dung management 11%, concentrate feed 13%, electricity and diesel consumption 5% and fertilizer 21% to the total GHG emissions. They also found that a reduction of 14–18% in carbon footprint can be achieved if the cows are reared in an extensive farm management, 14–26% reduction by elimination of non-milking cows and 28–33% reduction by combining both the options [2]. Another study was conducted on nationally representative 221 dairy farms (grass-based) in Ireland to examine the relationship between CF of milk and financial performance of the farms. CF of milk ranged from 0.60 to 2.13 kg of $CO_2$eq with an average of 2.13 kg of $CO_2$eq per kg of FPCM. Regression analysis revealed that CF is reduced and farm profit is increased when the milk production per hectare per cow and the grazing season is increased. However, CF and farm profit were adversely affected when increased milk production was achieved by an increase in concentrate feeding [20]. Life cycle assessment as per ISO 14040 and 14044 was done in a traditional dairy farming system (1368 animals) having the facility of anaerobic digestion (AD) and photovoltaic electricity generation systems. The Carbon footprint of farm was found to be 1.11 kg $CO_2$eq/kg FPCM. Inclusion of AD was found to reduce the CF by 0.26 kg $CO_2$eq/kg FPCM. However, contribution of photovoltaic (PV) system was recorded as negligible because of small dimensions of the technology [30]. In a study of dairy farms in Anand, Gujarat, India, the average Carbon footprint of buffalo milk was calculated as 2.7, 2.5 and 3.0 kg $CO_2$eq/kg FPCM on digestibility, economic and mass basis, respectively, whereas for cow milk CF was 2.0, 1.9 and 2.3 kg $CO_2$eq/kg FPCM, respectively [7]. CF of dairy production was assessed in sub-tropical climate zones in which cows may graze on tropical and temperate pasture. Three different feed scenarios viz. 100% total mixed ration ad libitum, 75 and 50% total mixed ration and access to the tropical or temperate pastures during lactation. The CF of three scenarios was found to be similar with an average of 1.06 kg $CO_2$eq/kg ECM. Decreasing the TMR intake and including the grazing on the pastures was found to potentially reduce the GHG emissions to a minor extent [23]. A cradle to farm-gate study was conducted to calculate CF of indoor ($n = 3$) and outdoor goat farms ($n = 2$) in New Zealand. The average CF for indoor farms per ha was higher i.e. 11.05 t of $CO_2$eq than per ha CF for the outdoor farms (5.38 t of $CO_2$eq). However, CF per kg of FPCM was not significantly different for indoor (0.81 kg) and outdoor farms (1.03 kg). Methane from enteric fermentation contributed the highest to the CF. Use of supplementary feed material was found to be an important contributor to the GHG emission. The study concluded that manure management system and supplementary feed can substantially affect the carbon footprint [25]. Life cycle assessment methodology was used to identify major environmental hotspots for packaged ultra-high temperature (UHT) milk (whole milk, partially and fully skimmed milk) and cocoa milk in Portugal by applying a cradle to factory gate approach. It was found that the production at the farm level is the main hotspot among several categories, mainly due to the emissions from manure management, fertilizer production and enteric fermentation. Whereas significant on-site emissions from the dairy factory were from energy requirements and packaging material production for the product [8]. Carbon footprint (CF) of packaged fluid milk in Tehran was estimated to be 1.57 kg $CO_2$eq/kg FPCM upto farm-gate level and 1.73 kg $CO_2$eq/L of pasteurized milk packaged in a plastic pouch. Enteric fermentation contributed the highest (30%), followed by electricity (14%), diesel (8.9%), manure emissions (8.8%) and transportation (8.6%) to the CF of milk product [3].

A study was conducted in Ireland on 54 farms in which a new functional unit, base price-adjusted milk (BPAM), was proposed for raw milk at the farm gate to better capture the economic function of milk. It was found that kilogram BPAM was significantly correlated with kg ECM for the sampled farms. The average GHG emission of milk ranged from 1.09 to 1.93 kg $CO_2$eq/kg with no quality correction and the average GHG of milk ranged from 1.09 to 1.82 kg $CO_2$eq/kg ECM. It was found that excessive biological contamination in the raw milk could increase GHG emission by > 200%. The authors concluded that the studies focused on farmers and processors should consider using BPAM as a functional unit to capture more properties of the milk than just milk solids [24].

## 3.2   Other Dairy Products

Carbon footprints (CFs) of various types of yogurts were estimated using LCA methodology in a Spanish dairy plant of La Fageda. The total carbon footprint was found to be 1.94 kg $CO_2$eq for 1 kg of yogurt production. Raw milk and milk-based ingredients contributed 80–96% to the $CO_2$eq in addition to major contribution to all the impact categories. Packaging, pasteurization and refrigerators accounted for 70% of the energy consumption in the plant. However it is to be noted that the GHG emission varies depending on the region and the production process [27]. Life cycle assessment was done to assess cumulative energy demand and carbon footprint of three types of products (fat filled powder (FFP), skimmed milk powder (SMP) and butter) from four companies. The carbon footprint for butter ranged from 0.41 to 0.62 kg $CO_2$eq/kg solids, for SMP from 1.40 to 1.70 kg $CO_2$eq/kg solids and for fat filled powder (FFP) CF was 1.65 $CO_2$eq/kg solids. The study recommended using the site-specific data for each industry [34]. A study was conducted in a dairy company, Arla Foods, Denmark to develop a method for assessing the carbon footprint of five major categories of dairy products, from cradle to wholesale or retail, using LCA tool SimaPro7. Fresh dairy products (FDP) (crème fraiche, cream, consumer milk and yogurt) were found to emit 1.1 kg $CO_2$eq/kg, butter and butter blends (BSM) (butter and blend products that are a mixture of vegetable oil and butter) 8.1 kg $CO_2$eq/kg, cheese (all cream cheese, mold cheese, white cheese and yellow cheese) emitted 6.5 kg $CO_2$eq/kg, whey based products and milk powder (whey protein concentrates, skimmed milk powder, whole milk powder and other whey based products) emitted 7.4 kg $CO_2$eq/kg and other product groups (everything not classified in any of the other categories for example cooking products custard and various desserts) accounted for 1.2 kg $CO_2$eq/kg. Raw milk and purchased dairy products accounted for the largest share of GHG emission. Butter and butter blends had slightly higher share than average share of non-dairy raw materials and ingredients. Whey based products and milk powder were found to have substantial energy use per kg of product, while fresh dairy product and 'other' had a comparatively larger impact from packaging. Transportation contributed comparatively small to the emission from all final products [6]. In a life cycle assessment study, CF of cheddar cheese was found to be 14 kg $CO_2$eq/kg of cheddar. Anaerobic digestion of cheese whey was done for energy recovery which led to 2% reduction in CF of cheese [9].

A study was conducted to calculate the CF of main Canadian dairy products using cradle to factory gate system boundary. On-farm activities contributed 90% to the total GHG emissions for the studied dairy products. Percentage contribution of different production and processing steps was calculated for fluid milk and yogurt. Farm production was found to contribute 86.9 and 72.2% to the fluid milk and yogurt, respectively. On the other hand, processing contributed 6.5 and 16.8% to the GHG emissions from fluid milk and yogurt, respectively. Two different functional units (FU) viz. $CO_2$eq/kg of product and per kilogram of protein content were used to present the CF. CF of yogurt was found to be 55% higher than the fluid milk when per kilogram of product FU was used, but the difference was only 23% when per kilogram of protein content was used as a FU. For cheese and cream, CF was 5 and 2 kg of $CO_2$eq/kg of product, respectively. However, CF per kg of protein was 22 and 83 kg of $CO_2$eq which is several-fold higher than the CF presented in per kg of product. A critical difference was observed in the CF for butter with the two FUs due to very low protein per unit mass of butter. CF per kg of product was 7 kg of $CO_2$eq, while per kg of protein was 729 kg of $CO_2$eq [28]. In a study on Indian food items, CF was found to be 5.9 kg $CO_2$eq $kg^{-1}$ dry wt. for milk, 7.4 kg $CO_2$eq $kg^{-1}$ dry wt. for curd, 6.1 kg $CO_2$eq $kg^{-1}$ dry wt. for lassi and 1.2 kg $CO_2$eq $kg^{-1}$ dry wt. for butter. However, CF based on the fresh wt. was 0.76 kg $CO_2$eq $kg^{-1}$ dry wt. for milk, 0.74 kg $CO_2$eq $kg^{-1}$ dry wt. for curd, 0.34 kg $CO_2$eq $kg^{-1}$ dry wt. for lassi and 0.10 kg $CO_2$eq $kg^{-1}$ dry wt. for butter [22].

# 4 Mitigation Measures

As per the Global Dairy Platform and FAO [5], taking measures in the following main areas can help the dairy sector in reducing the GHG emissions: (1) carbon capturing and sequestration, (2) efficiency improvement; and (3) linking the dairy production to the circular bio-economy. Strategies such as nutrient recovery through anaerobic digestion, transforming manure into nutrient-rich products could be an alternative of synthetic fertilizers [5]. Research, regulations, policies and incentives could be other ways to support low-carbon choices. The control of foot and mouth disease has been reported to result in reduction in emission intensity by a significant amount [35].

# References

1. Caro D, Davis SJ, Bastianoni S, Caldeira K (2014) Global and regional trends in greenhouse gas emissions from livestock. Clim Change 126(1–2):203–216. https://doi.org/10.1007/s10584-014-1197-x
2. Casey JW, Holden NM (2005) Analysis of greenhouse gas emissions from the average Irish milk production system. Agric Syst 86(1):97–114. https://doi.org/10.1016/j.agsy.2004.09.006
3. Daneshi A, Esmaili-Sari A, Daneshi M, Baumann H (2014) Greenhouse gas emissions of packaged fluid milk production in Tehran. https://doi.org/10.1016/j.jclepro.2014.05.057
4. Delgado C, Rosegrant M, Steinfeld H, Ehui S, Courbois C (2001) Livestock to 2020: the next food revolution. Outlook Agric 30(1):27–29. https://doi.org/10.5367/000000001101293427
5. FAO and GDP (2018) Climate change and the global dairy cattle sector—the role of the dairy sector in a low-carbon future, Rome, 36 pp. License: CC BY-NC-SA-3.0 IGO
6. Flysjö A, Thrane M, Hermansen JE (2014) Method to assess the carbon footprint at product level in the dairy industry. Int Dairy J 34(1):86–92. https://doi.org/10.1016/j.idairyj.2013.07.016
7. Garg MR, Phondba BT, Sherasia PL, Makkar HPS (2016) Carbon footprint of milk production under smallholder dairying in Anand district of Western India: a cradle-to-farm gate life cycle assessment. Anim Prod Sci 56(3):423. https://doi.org/10.1071/AN15464
8. González-García S, Castanheira ÉG, Dias AC, Arroja L (2013) Using life cycle assessment methodology to assess UHT milk production in Portugal. Sci Total Environ 442:225–234. https://doi.org/10.1016/j.scitotenv.2012.10.035
9. Gosalvitr P, Cuellar-Franca R, Smith R, Azapagic A (2019) Energy demand and carbon footprint of cheddar cheese with energy recovery from cheese whey. Energy Procedia 161. https://doi.org/10.1016/j.egypro.2019.02.052
10. Gupta R, Dasgupta A (2020) Milk will drive methane emissions in India. Clim Change 161(4):653–664. https://doi.org/10.1007/s10584-020-02715-4
11. Havlík P, Valin H, Herrero M, Obersteiner M, Schmid E, Rufino MC, Mosnier A, Thornton PK, Böttcher H, Conant RT, Frank S, Fritz S, Fuss S, Kraxner F, Notenbaert A (2014) Mercator research institute on global commons and climate change. Int Cent Trop Agric 10829. https://doi.org/10.1073/pnas.1308044111
12. IDF (2010) A common carbon footprint approach for dairy: the IDF guide to standard lifecycle assessment methodology for the dairy sector. Bull Int Dairy Fed 445:1–40
13. IDF (2015) A common carbon footprint approach for dairy: the IDF guide to standard lifecycle assessment methodology for the dairy sector. Bull Int Dairy Fed 479/2015
14. IPCC (2006) Emissions from livestock and manure management, vol 4, Chap 10. Contributing authors: Hatfield JL, Johnson DE, Bartram D, Gibb D, Martin JH
15. IPCC (2014) Chapter climate change 2014 synthesis report summary for policymakers summary for policymakers

16. IPCC (2007) Climate change 2007: the physical science basis. Contribution of Working Group I to the fourth assessment report of the Intergovernmental Panel on Climate Change (Eds S Solomon, D Qin, M Manning, Z Chen, M Marquis, KB Averyt, M Tignor, HL Miller), pp 129–234. Cambridge University Press, Cambridge, UK

17. IPCC (2013) Climate change 2013: the physical science basis. Contribution of Working Group I to the fifth assessment report of the Intergovernmental Panel on Climate Change (EdsTF Stocker, D Qin, GK Plattner, M Tignor, SK Allen, J Boschung, A Nauels, Y Xia, V Bex, PM Midgley), pp 649–740. Cambridge University Press, Cambridge, UK

18. Opio C, Gerber P, Mottet A, Falcucci A, Tempio G, Macleod M, Vellinga T, Henderson B, Steinfeld H (2013) Greenhouse gas emissions from ruminant supply chains - a global life cycle assessment. FAO, Rome, Italy

19. O'Brien D, Hennessy T, Moran B, Shalloo L (2015) Relating the carbon footprint of milk from Irish dairy farms to economic performance. https://doi.org/10.3168/jds.2014-9222

20. O'brien D, Brennan P, Humphreys J, Ruane E, Shalloo L (2014) An appraisal of carbon footprint of milk from commercial grass-based dairy farms in Ireland according to a certified life cycle assessment methodology. https://doi.org/10.1007/s11367-014-0755-9

21. Pandey D, Agrawal M (2014) Carbon footprint estimation in the agriculture sector. Springer, Singapore, pp 25–47. https://doi.org/10.1007/978-981-4560-41-2_2

22. Pathak H, Jain N, Bhatia A, Patel J, Aggarwal PK (2010) Carbon footprints of Indian food items. Agr Ecosyst Environ 139(1–2):66–73. https://doi.org/10.1016/j.agee.2010.07.002

23. Ribeiro-Filho Id HMN, Civiero M, Kebreab E (2020) Potential to reduce greenhouse gas emissions through different dairy cattle systems in subtropical regions. Plos One. https://doi.org/10.1371/journal.pone.0234687

24. Rice P, O'Brien D, Shalloo L, Holden NM (2019) Defining a functional unit for dairy production LCA that reflects the transaction between the farmer and the dairy processor. Int J Life Cycle Assess 24(4):642–653. https://doi.org/10.1007/s11367-018-1486-0

25. Robertson K, Symes W, Garnham M (2015) Carbon footprint of dairy goat milk production in New Zealand. J Dairy Sci 98:4279–4293. https://doi.org/10.3168/jds.2014-9104

26. Sejian V, Prasadh RS, Lees AM, Lees JC, Al-Hosni YAS, Sullivan ML, Gaughan JB (2018) Assessment of the carbon footprint of four commercial dairy production systems in Australia using an integrated farm system model. Carbon Manage 9(1):57–70. https://doi.org/10.1080/17583004.2017.1418595

27. Vasilaki V, Katsou E, Pons S, Col On J (2016) Water and carbon footprint of selected dairy products: a case study in Catalonia. https://doi.org/10.1016/j.jclepro.2016.08.032

28. Vergé XPC, Maxime D, Dyer JA, Desjardins RL, Arcand Y, Vanderzaag A (2013) Carbon footprint of Canadian dairy products: calculations and issues. J Dairy Sci 96(9):6091–6104. https://doi.org/10.3168/jds.2013-6563

29. Vida E, Eurosia D, Tedesco A, Tedesco DEA (2017) The carbon footprint of integrated milk production and renewable energy systems—a case study. Sci Total Environ 609:1286–1294. https://doi.org/10.1016/j.scitotenv.2017.07.271

30. Vida E, Tedesco DEA (2017) The carbon footprint of integrated milk production and renewable energy systems—a case study. Sci Total Environ 609:1286–1294. https://doi.org/10.1016/j.scitotenv.2017.07.271

31. Wilkes A, Wassie S, Odhong' C, Fraval S, Van Dijk S (2020) Variation in the carbon footprint of milk production on smallholder dairy farms in central Kenya. https://doi.org/10.1016/j.jclepro.2020.121780

32. Wilkes A, Reisinger A, Wollenberg E, van Dijk S (2017) Measurement, reporting and verification of livestock GHG emissions by developing countries in UNFCCC: current practices and opportunities for improvement. In: CCAFSReport No. 17. CCAFS, Wageningen, The Netherlands

33. Xu CC, Huang J, Chen F (2013) The application of carbon footprint in agri-food supply chain management: case study on milk products. Adv Mater Res 807–809:1988–1991. https://doi.org/10.4028/www.scientific.net/AMR.807-809.1988

34. Yan M, Holden NM (2018) Life cycle assessment of multi-product dairy processing using Irish butter and milk powders as an example. J Clean Prod 198:215–230. https://doi.org/10.1016/j.jclepro.2018.07.006
35. York L, Heffernan C, Rymer C (2017) A comparison of policies to reduce the methane emission intensity of smallholder dairy production in India. Agr Ecosyst Environ 246:78–85. https://doi.org/10.1016/j.agee.2017.05.032

# Valorization of By-Products from Food Processing Through Sustainable Green Approaches

Deepak Kumar, Md. Shamim, Santosh K. Arya,
Mohammad Wasim Siddiqui, Deepti Srivastava, and Shilpa Sindhu

**Abstract** The food processing sector is a very prominent section of the international market and increased up to approximately $4.5 trillion by 2024 (Businesswire in Global food processing market report, 2019: trends, forecasts, and competitive analysis (2013–2024), 2019 [1]). The increasing demand for healthy fruits and vegetables, organic processed foods, nutraceuticals and functional foods, seafood with packaged food products like ready-to-eat, and frozen processed foods is expected to drive the market growth after pandemic COVID-19 incidence. During processing and handling of horticultural crops, generate large volumes of wastes annually and become a major concern to the whole world. Fruits and vegetables have many bioactive compounds (polyphenols, flavonoids, organic acids, aroma, and flavoring agents, etc.) which have a positive impact on human health because of anti-inflammation, anti-allergic, anti-cancer, anti-atherogenic, and antioxidants properties, depending upon extraction their methods from wastes, efficacy, and bioavailability to the body. Applications of conventional (soxhlet, maceration, percolation, hydro-distillation) and green techniques (solid-phase, supercritical, accelerated solvent, microwave, and ultrasound extractions) for the valorization of horticultural wastes' conversion in many bioactive components are followed by the potentiality, scalability, and sustainability of the

D. Kumar (✉) · S. K. Arya
R&D Division, Nextnode Bioscience Pvt. Ltd., Opposite GEB Office, Kadi Kalol Road, Kadi, Gujarat 384440, India

Md. Shamim
Department of Molecular Biology and Genetic Engineering (PBG), Dr. Kalam Agricultural College, Kishanganj, Bihar Agricultural University, Sabour, Bhagalpur, Bihar 813210, India

M. W. Siddiqui
Department of Food Science and Post-Harvest Technology, Bihar Agricultural University, Sabour, Bhagalpur, Bihar 813210, India

D. Srivastava
Integral Institute of Agricultural Science and Technology, Integral University, Dasauli, Lucknow, Uttar Pradesh 226021, India

S. Sindhu
Schools of Management, The Northcap University, Gurgaon, Haryana 122017, India

extraction process and highlight the concept of the circular economy as "Waste to Wealth."

**Keywords** Horticultural wastes · Bioactive compounds · Green technologies · Valorization · Sustainability

# 1  Introduction

The demand for fruits and vegetables has expanded prominently due to changing dietary habits and the growing world population. Moreover, horticultural commodities take a crucial bit of human life and are used as the most important food commodity [2]. The fruit and vegetable waste is generated during the processing and improper handling and remains underutilized for industries due to lack of appropriate processing technologies essential for their efficient valorization [3]. The wastes are acquired from the horticultural crop produce processing sectors comprise core, rind, peels, pods, shell, pomace, rag, stones, and seeds, etc. [4, 5]. The prior studies revealed that around ~ 60 metric ton waste is generated during all processing steps [5, 6]. Asian countries (mainly China, India, and the Philippines) are the largest shareholders in generating horticultural wastes [7].

The waste and losses of fruit and vegetables are a serious issue and observe during all phases of harvesting, processing, grading, storage, handling at home before or after preparation, as well as at transport to warehouses, and supply chain of marketing, respectively [2, 8]. These generated horticultural processed wastes have many phenolics, flavonoids, terpenes, essential oils, aromatic, and flavoring agents. Hussain et al. [2] and Sagar et al. [6] critically explained in their studies showed that valorization of underutilized horticultural wastes generates profit, reduces liability, and supports the economy of the concerned industry as well as the country.

Food processing waste can be exploited to isolate potential phenolics, functional molecules which might be explored in the pharmaceutical, textile, food, and beverage industries. Moreover, proper use of horticultural waste may provide a benchmark for solving the environmental complications by adopting a green approach as well as improving human health through functional foods and nutraceuticals substances (carotenoids, phenols, vitamins, dietary fibers, and other pigments, among several others) as such. Hence, there are requirements of strategies for their processing and valorization of wastes to enable sustainable utilization and reduce the environmental burden, respectively. Moreover, horticultural wastes valorization would be helpful in saving natural resources and limiting the judicious and environmental issues [9]. This chapter explores the benefits of fruit and vegetable wastes as a starting point for valorization and their uses in the circular economy.

## 2 Global Production, Losses and Decay of Horticultural Produce

The global fruit and vegetable industrial market is growing steadily by nearly 49% over the last ten years. As per the data of the Food and Agricultural Organization (FAO), 872 million metric tons (MMT) of fruits and 588 MMT of vegetable production were observed globally in 2018–2019 [10, 11]. China was the largest fruit and vegetable producing country in 2018–2019, and followed by India, Brazil, and the United State of America (USA) in the world [11, 12].

As per the FAO report, it was indicated that post-harvest losses reached majorly up to 40% for fruit and vegetables, 30% for dairy and fishes, and 20% for cereals [13, 14]. Moreover, annually it is estimated that nearly one-third of the total food produced is waste or decay in the world which had to cost around $685 billion US Dollars [13, 14]. The losses and wastes in horticultural produce generated because of pre-and-post operational activities and consumer preferences for quality. A huge amount of horticultural wastes are produced during processing, including roots, skin, leaves, pulp, pomace, seeds, and others [15]. Besides that, it is estimated that one-fourth portion of the horticultural commodity is a waste part for primary consumers, and the rest other generated during conversion in value-added food products [4, 16, 17]. It is also seen that many horticultural commodities cannot be utilized as fresh and require primary processing to get concerning products such as tea, turmeric, etc. [6, 8, 18]. The highest loss and decay in horticultural crops were observed for citrus fruits 48–55% (54.28 MMT), for apple fruits 25–30% (21.81 MMT) and followed by grapes for 15–25% (11.57 MMT) globally in 2018–2019 [19, 20].

In concern of wastage, papaya fruits' processing generates 32% unusable pulp, 6.5% of seeds, and about 53% of the final product. Similarly, the processing of mangoes generates about 18% of inoperable pulp, 11% of peels, 13.5% of seeds, and 58% of the concerned products. Sixteen percent of waste is generated during the peeling of mandarins. Furthermore, pineapple processing produces about 15% of the top, 14% of peels, as waste [6]. The other examples for the nature of waste generated from fruit and vegetable are illustrated in Table 1.

## 3 Application of Horticultural Wastes in Circular Economy

At the industrial level, there is a requirement of an alternate approach to managing the horticultural wastes to minimize the high volume wastage as a garbage lot. Therefore, there are requirements for green, recyclable, and sustainable approaches to the effective usage of this garbage in converting to value-added products that support the economies of industries as well as countries. Horticultural wastes have potential bioactive molecules such as phenolics, organic acids, enzymes, dietary fibers, flavor, and aroma chemicals, which fulfill the concept of the circular economy. In the current scenario, European countries alone generate around 88 MT of organic

**Table 1** Global production of horticultural commodity (fruits and vegetables), waste, and typical decay during processing (2018–2019)

| Horticultural commodity | Global production (MMT) | Nature of waste | Global waste and decay (MMT) | Decay (%) | Literature cited |
|---|---|---|---|---|---|
| Apple fruit | 87.24 | Pomade, peel, seeds | 21.81 | 25–30 | Wang et al. [20], Bhushan et al. [21], https://www.statista.com/ |
| Banana fruit | 116.41 | Peel, fruit stalk | 20.95 | 18–35 | Chakraborty et al. [22], https://www.statista.com/ |
| Citrus fruits (oranges, lime/lemons, etc.) | 113.10 | Peel, seeds, rag | 54.28 | 48–55 | Bisht et al. [19], Joshi and Sharma [23], https://www.statista.com/ |
| Dragon fruit | 11.98 | Rind, skin, seeds | 2.30 | 25–48 | Jalgaonkar et al. [24], https://www.globenewswire.com/ |
| Grapes | 77.14 | Seeds, stem, skin, pomace | 11.57 | 15–25 | González-Centeno et al. [25], Llobera and Canellas [26], https://www.statista.com/ |
| Guava fruit | 6.75 | Core, peel, seeds | 0.81 | 12–25 | Kanwal et al. [27], Gupta and Joshi [28], Food Outlook [29] |
| Jackfruit | 3.70 | Rind, seeds | 1.67 | 45–80 | Akter and Haque [30], Saxena et al. [31] |
| Mango | 55.85 | Peel, stone | 22.34 | 40–45 | Mitra et al. [32], https://www.statista.com/ |
| Papaya | 13.74 | Rind, seeds | 1.37 | 10–30 | Han et al. [33], Parni and Verma [34], https://www.statista.com/ |
| Passion fruit | 1.46 | Skin, seeds | 0.58 | 40–50 | Sagar et al. [6], Almeida et al. [35], Food Outlook [29] |
| Pineapple | 28.18 | Core, skin | 8.45 | 30–55 | Choonut et al. [36], Nunes et al. [37], https://www.statista.com/ |

(continued)

**Table 1** (continued)

| Horticultural commodity | Global production (MMT) | Nature of waste | Global waste and decay (MMT) | Decay (%) | Literature cited |
|---|---|---|---|---|---|
| Onion | 99.97 | Outer skin, leaves | 24.99 | 25–30 | Sharma et al. [38], https://www.statista.com/ |
| Peas | 22.77 | Peel, shells | 9.10 | 40–45 | Sagar et al. [6], Verma et al. [39], https://www.statista.com/ |
| Potato | 370.43 | Peel | 55.56 | 15–40 | Sepelev and Galoburda [40], https://www.statista.com/ |
| Tomato | 180.77 | Core, skin, seeds | 18.07 | 10–20 | Løvdal et al. [41], https://www.statista.com/ |
| Rambutan | 1.38 | Skin, seeds | 0.69 | 50–65 | Food Outlook [29], Issara et al. [42], Sagar et al. [6] |
| Cauliflower | 26.91 | Leaves, stems | 12.10 | 45–50 | Pankar and Bornare [43], https://www.statista.com/ |
| Beetroot | 588.27 | Peels, parings, stalks | 58.82 | 10 | Seremet et al. [44], Costa et al. [45], https://www.tridge.com/ |

*MMT* Million metric tons

horticultural wastes that are mostly dumped as garbage lots without any utilization [46]. Additionally, the development of new processes and approaches is very crucial to improve the upstream and downstream recovery of value-added bioactive compounds from wastes, which are all based on the concept of circular economy [47]. Sustainable green processing and biotechnological methods might be exploited for efficient isolation of bioactive metabolites from these horticultural wastes and find futuristic industrial applications as such.

Horticultural wastes' valorization plays an important role into the supply chain through reusing the material and returning back to processing and allowing circular economic growth, along with contributing minimal negative environmental effects [48]. Moreover, Agri-Food industries-based wastes and by-products can be used in producing low cost (fuel, organic manure) and costly value-added products (flavoring agents, organic acids, dietary fibers, enzymes, flavonoids, vitamins, and a source of prebiotics, etc.), respectively [5, 49]. Additionally, horticultural wastes utilization in the conversion of valuable products is a zero waste concept [50, 51]. Moreover, this concept shows another notion of the circular economy "Taste the waste," respectively [2].

# 4 Valorization of Horticultural Wastes

The valorization approach is based on circular economy and translation of fruit and vegetable waste which remains underutilized to value-added products without disturbing the environment and ecosystem [2]. As per the available literature, functional bioactive compounds can be extracted from Agri-Food wastes [52]. Mostly, bioactive compounds possess antioxidant, cardio-protective, anti-tumor, anti-obesity, antiviral and antibacterial activities and proved health beneficial properties, etc. [53, 54]. The green approaches are required to obtain bioactive such as polyphenolic compounds, pigments, dietary fibers, fatty acids, and organic acids, etc., from food processing industrial wastes [50]. The powder extracted from carrot, banana, mango kernel, orange peels, and pomace of pineapple is used in making medicine, health supplement, cosmetic products, flavoring agents, and perfumeries products, etc. [55]. In many available reports, dietary fibers extracted from wastes of apple skin, orange pomace, mango peels, and carrot pomace have been used as a functional food ingredients in many bakery products like cakes, biscuits, wheat rolls, and buns, etc. [56].

# 5 Extraction Methods of Bioactive Compounds

Extraction plays a basic step to isolate the targeted bioactive molecules from the horticultural wastes. In the current scenario, many approaches (conventional or novel) for extractions like steam extractions, hydro-distillation, solid-phase extraction, pressurized liquid method, electro-permeabilization treatment, accelerated solvent extraction, microwave irradiation method, ultrasound extraction, high-voltage electric discharge (HVED), and some modified solvent systems are applied for the isolation of bioactive compounds [57]. The make-up and quality of obtained bioactive compounds from agro wastes were based on extraction techniques (viz. solvent extraction, drying method, and intensity of treatment) as well as processing conditions [58]. Various treatment and extraction methods also affect the structure and concentration of bioactive metabolites. Solvent extraction mediated processes can damage the thermolabile-based bio-molecules, whereas accelerated solvent extraction (ACE) can lead to low yield of bioactive molecules and high cost [57, 59]. The modified wet-milling technique was used for high purity fiber extraction from agro-waste [57]. Further, novel processing techniques like solvent extraction, electro-permeabilization treatment, accelerated solvent extraction (ACE), microwave irradiation extractions, and high-voltage electric discharge (HVED) have many benefits and certain drawbacks. But, the application of these sustainable, innovative, and green extraction approaches supports quality, less environmental pollution, and appropriate handling aspects to workers during processing [6].

## 5.1 Conventional Extraction Methods

The conventional techniques are considered classical methods because they have been part of the extraction of bioactive compounds for a long time. These techniques are mainly based upon applied heat/or power and solvent extraction or their combination. There are mainly three conventional techniques, namely soxhlet extraction [60], maceration [61], percolation [62], decoction [63], and hydro-distillation [64].

Out of these, the soxhlet technique has been much acclaimed and popular and mostly used for achieving the oils from plant-based materials [65, 66]. N-hexane-based soxhlet extractions of agarwood leaves gave the maximum oil yield containing phytol, hexadecanoic acid compared to other polar and nonpolar solvents [67]. Guthrie et al. [68] have investigated the yield of polyphenolic compounds from the peel of kiwifruit through water extraction. The total phenolics and flavonoids were found superior to the ethanol extraction method.

The maceration technique is very popular for a long time and used in the preparation of herbs-based tonics in herbal industries. It is a low-cost technique and became popular to extract essential oils and bioactive compounds but not suitable for high yield extractions [61]. Safdar et al. [69, 70] have extracted polyphenols by maceration and ultrasound-assisted extraction (UAE) method and found the highest polyphenols (67.58 GAE/g of extract) in UAE method with ethanol (80%) and least (18.66 GAE/g of extract) polyphenolics with ethyl acetate (100%) maceration technique. Similarly in the case of kinnow peel, maximum polyphenols (32.48 mg GAE/g extract) have isolated with 80% ethanol through the UAE method and low phenolic acid (8.64 mg GAE/g extract) with 80% ethyl acetate by maceration method [69, 70].

Percolation is an extraction method and comparatively superior to maceration because of using constant feeding of fresh solvent to replace saturated solvent [71]. Fu et al. [62] used 55% alcohol to soak medicinal raw material and then percolated for extraction of the active component which was reliable and satisfactory in quality. Moreover, Zhang et al. [72] extracted the fucoxanthin with percolation and refluxing methods from *Undaria pinnatifida* and observed purity which was higher with the percolation method than refluxing technique.

The decoction is an old method to extract tannin, flavonoids, and alkaloids, but it has a substantial amount of hydrophilic impurities. Moreover, this method is not being recommended for the isolation of heat-sensitive or volatile compounds. Arina and Harisun [73] have extracted tannin from *Quercus infectoria* (Manjakani) by decoction method. The extracts had contained the highest tannin content (2233 mg/g) and antioxidant activity (93.42%).

The reflux extraction method is comparatively more methodical than maceration and decoction, which require less solvent and time for extraction. However, it contains some limitations to extract heat-labile components from natural products.

Kongkiatpaiboon and Gritsanapan [74] used *Stemona collinsiae* root for extraction of bio-insecticide (didehydrostemofoline) with 70% ethanol refluxing which showed the highest yield compared to other solvents. Zhang [75] has extracted maximum yields of puerarin and Baicalin by reflux method using 60% ethanol as a solvent than the decoction technique.

Hydro-distillation or/ steam distillation as the name suggests is a water-based classical technique in which plant source samples are hydrated to isolate aromatic oils and various functional molecules. This technique is not suitable for heat-labile bioactive molecules because most are degraded at high temperatures [6]. Shakir & Salih [76] had extracted essential oil from different citrus fruits (orange, lemon, and mandarin) peels by steam distillation and microwave-assisted steam distillation. The results showed that microwave-assisted steam distillation gave a higher yield, shorter extraction time, and purity than steam distillation. Similarly, Fakayode and Abobi [77] optimized the extraction process for oil and pectin from orange (*Citrus sinensis*) peels using steam distillation with response surface approach. The extracted yield of essential oil and pectin was ranged from 0.57–3.24% and 12.93–29.05%, respectively. The applications and limitations of conventional methods are described in Table 2.

### 5.1.1 Novel Extraction Methods

The extractions of bioactive products through conventional methods have characterized difficulty in low extraction recovery, obtaining high purity, time-consuming, and huge losses of costly solvent in process. Therefore, novel techniques have emerged to tackle the limitations of conventional methods. Microwave irradiation extraction (MIE) technique is based upon its heating principles utilizing microwaves (electric field and magnetic field) [82]. Microwave irradiation extraction showed many benefits over other techniques including high extract amount, decreased equipment size, and temperature gradient to extract polyphenolics [6, 83]. The microwave irradiation method has the efficacy to isolate the bioactive compounds with reduced consumption of organic solvent [84]. Dorta et al. [85] extracted and found the maximum antioxidants from mango peel compared to traditional solvent extraction (TE).

Enzyme-assisted extraction is a handy and novel approach to isolate bound polyphenolics, flavonoids, terpenes, and oils with prominent yield from food-agro wastes [86, 87]. Li et al. [88] have extracted polyphenolics from the peels of five citrus family members (mandarin, orange, grapefruit, and two lemons) with the highest yield through celluzyme MX among different enzyme treatments. Similarly, phenolic compounds from grape waste were reported higher and best with novoferm enzyme treatment among the other enzymes like pectinex, and celluclast, respectively [6].

The solvent of liquid–liquid extraction approach is a prominent method to separate two completely mixed liquids from each other through mass transfer. The concerned solvent contains significant affinity and selectivity for more than one completely mixed feed solution. In this technique, one extract contains the desired product, and the second feed solution having less solute known as the raffinate [89]. This technique

**Table 2** Conventional extraction methods, their applications, limitations for targeting bioactive compounds

| Targeted bioactive compounds | Technique | Applications | Limitations | Literature cited |
|---|---|---|---|---|
| Lipid/fat extraction | Soxhlet extraction | – Classical technique <br> – Suitable technique for non-volatile oils | – Not ecofriendly <br> – Time-consuming | Azmir et al. [65], Gracis-Sales et al. [66], Soxhlet [60] |
| Polyphenols, anthocyanins | Maceration | – Extraction of thermolabile components | – Long extraction time <br> – Low extraction efficiency | Albuquerque et al. [78], Jovanović et al. [79] |
| Alkaloids | Percolation | – More prominent technique than maceration <br> – Hydrophilic and hydrophobic solvents | – Requirement of large volume of solvents <br> – Time-consuming technique | Fu et al. [62], Zhang et al. [72] |
| Flavonoid glycosides, tannins, alkaloids | Decoction | – Water is used as solvent <br> – Moderate time is required than percolation | – Not suitable for thermolabile compounds | Arina and Harisun [73], Barba et al. [63], Zhang and Su [80] |
| Oil and bioactive compounds | Steam distillation | – Simplest method to extract volatile oils <br> – Small-scale structure is required for functioning the industry | – Sensitive to heat-labile compounds <br> – Slow and time-consuming process | Sagar et al. [6], Shakir and Salih [76], Azmir et al. [65], Vankar [64] |
| Natural products from herbs | Reflux extraction | – Potent extraction method than percolation/or maceration | – Not used for thermolabile compounds <br> – Time saving than other conventional methods | Kongkiatpaiboon and Gritsanapan [81], Zhang [75] |

is very useful for the beverage industry to extract polyphenolics, limonoids, and flavonoids from selected horticultural wastes. The best results can be availed through a precise selection of the suitable solvent for the extraction process because solvent's selectivity has an impact on the quality and yield of phenolic compounds [6, 8].

Likewise enzyme-assisted extraction, solid-phase extraction is also based on leaching concepts. In this method, solid material repeatedly passes through a solvent and increases or changes the diffusion coefficients (solid material imbibe) and concentration gradient by mass transfer operation [90]. Solid-phase extraction is suitable for isolating important phytochemicals, polyphenols from fruits and vegetables, sucrose from beet, protein fractions from oil seed, hydrocolloids from algae, and limonoids from neem kernels, respectively [2].

Supercritical fluid extraction method is a novel extraction method which separates and extracts bioactive components through supercritical fluid/or solvent. The critical point is lower than the temperature and pressure of the supercritical fluid. Moreover, many bioactive components have an adequate solubility in supercritical fluid because of having similar density to an ordinary solvent. Hernandez et al. [91] used the supercritical fluid extraction method to extract bioactive compounds from the cocoa husk and reported high phenolics (35.11 EAG mg/g extract), flavan-3-ols (12.89 EEP mg/g extract), and carotenoid content (64.35 mg/g extract). Castro-Vergas et al. [92] have used different solvents with supercritical fluid extraction methods to extract phenolic fractions from guava seeds.

The ultrasound-assisted extraction technique is a rapid and efficient green approach to extract bioactive components from different substances. This technique follows the principle of ultrasonic cavitation and is mostly used to isolate the botanical extracts. Maran et al. [93] extracted bioactive compounds from *Nephelium lappaceum* L. fruit peel by ultrasound-assisted extraction technique under optimum conditions. The highest content of total anthocyanin (10.26 mg/100 g), phenolics (552.64 mg GAE/100 g), and flavonoid (104 mg RE/100 g) was found at 20 W (ultrasound power), 20 min (extraction time), and 1:18.6 g/ml (solid–liquid ratio). Additionally, Almusallam et al. [94] have used ultrasound-assisted extraction technology with response surface methodology to maximize the yield of total phenolic compounds from date palm spikelets.

Accelerated solvent extraction is a novel green method which is broadly used in isolation of different bioactive molecules from natural substances. The principle of this technique is based upon applying pressure and heat to solvent and samples for achieving speed and efficacy. Paes et al. [95] have used accelerated solvent and supercritical extraction methods to extract bioactive compounds from blueberry residues. Total polyphenol content and anthocyanin were reported higher in both accelerated solvent and supercritical extraction methods compared to the solvent extraction method. The other extraction techniques are briefly depicted with advantages and limitations in Table 3.

## 5.2  *Extraction of Dietary Fiber from Horticultural Wastes*

Dietary fiber is the non-digestible compound that was first defined by Hipsley [109]. The dietary fibers constitute the inner linings of plant cells and a very important part of the human diet. The definition and concept of dietary fiber included hemicellulose, lignin, and cellulose. Moreover, dietary fiber was further redefined by Trowell [110] as a roughage bioactive compound mainly based on digestion resistance. Dietary fiber is very essential for lively diets that are directly or indirectly, associated with cardiovascular diseases, diabetes, hypertension, obesity, and stroke disorders [2, 8]. Dietary

**Table 3** Novel extraction methods, their applications, and limitations for targeting bioactive compounds

| Targeted bioactive compounds | Technique | Applications | Limitations | Literature cited |
|---|---|---|---|---|
| Polyphenolic compounds | Solvent extraction | – Simple technique for liquid polyphenoilc molecules estimation<br>– Protecting the polyphenolic compounds from thermal degradation | – Cost intensive<br>– Degradation rate of phenolics is high due to external and internal factors<br>– Uses of expensive and hazardous solvents in process | Garcia-Salas et al. [66], Espinosa-Alonso et al. [96] |
| Photochemicals of medicinal plants | Solid-phase extraction | – Handling is easy<br>– Separation and potentiality in extraction are higher than solvent extraction | – Cost intensive<br>– Suitable and for hydrophilic active molecules<br>– Not recommended highly evaporable compounds | Abd-Talib et al. [90], Vuckovic [97] |
| Volatile and triterpenic compounds | Supercritical fluid extraction (SFE) | – Gives better mass transfer<br>– Time saving and ecofriendly<br>– Suitable for volatile compounds<br>– Recycling and reusing can be possible in SFE, in result low wastage | – Cost intensive and thermodynamic complicated system<br>– Polar system cannot be dissolved<br>– Not suitable for pharmaceuticals samples | Geeta et al. [98], Abbas et al. [99] |
| Phytosterols | Electro-permeabilization treatment extraction | – Continuous mode and processing capacity of raw material are higher to other methods<br>– Improved yield within short processing time<br>– Ecofriendly and less energy labile | – Highly sensitive method and require proper optimization of all operationally parameters for extraction | Patras et al. [100], Puertolas et al. [101] |

(continued)

202                                          D. Kumar et al.

**Table 3** (continued)

| Targeted bioactive compounds | Technique | Applications | Limitations | Literature cited |
|---|---|---|---|---|
| Phytochemical extraction from agro-industrial waste | Accelerated solvent extraction | – Recommended for bioactive compounds extraction from solid-phase<br>– Suitable to hydrophilic active molecules<br>– Less solvent and time required for extraction | – Not suitable for high recovery of bio-molecule | Dobias et al. [102], Klejdus et al. [103] |
| Extractions of bounded phytochemicals and oils | Enzyme-assisted extraction (EAE) | – Ecofriendly because water is used in process<br>– Favorable technique for extraction of cell wall entrapped molecules<br>– Titre amount or recovery rate is maximum to other methods | – Not suitable to oil rich bounded phytochemicals | Niranjan and Hanmoungjai [104], Puri et al. [105] |
| Rapid extractions of bioactive compounds (polyphenols) | Microwave irradiation extraction | – Quality labile process<br>– High recovery of targeted compound in less time | – Expensive because of high cost of apparatus and equipments<br>– Complicated in operation, required highly skilled person<br>– Unfit for heat-labile compounds | Phaiphan et al. [83], Sagar et al. [6], Sticher [106] |
| Carotenoids, lipids, chlorophyll, phenolic compounds | Ultrasound-assisted extraction (UAE) | – High yield<br>– Minimum time<br>– Energy and power saving | – Process parameters should be in optimum condition for extraction | Yahya et al. [107], Barba et al. [108], Azmir et al. (2015) |

fibers (soluble and/or insoluble) with regards to extraction methods are manufactured through chemical methods, enzymatic, and microbial fermentations, respectively. To date, many methods are proposed for the extraction of dietary fibers and bioactive metabolites from plant-based resources [2].

Insoluble (IDF) and soluble dietary fibers (SDF) are the main fractions of total dietary fiber on the basis of water solubility. The insoluble dietary fibers consist prominently hemicellulose and cellulose, respectively in its composition, while in the case of soluble fibers include gums (monomers of sugars from beans and legumes), pectin (sugars from legumes, whole grains, and pulpy fruits like papaya, guava, etc.), and mucilage (cactus, okra, aloe vera, succulent and aquatic plants) [2, 6].

In the juice industry, apple pomace is the main waste material that contains prominent amounts of edible fiber. The dietary fiber was higher in apple peel (0.91–1.79% FW) compared to pomace [6, 111]. Grapes pomace was also high in both soluble and insoluble fibers, but prominently consist pectins (small portions), hemicellulose, and cellulose [112]. Mango wastes/or by-products have also possessed high amounts of dietary fibers (40.6–72.5%) among other fruit wastes [113]. Moreover, Ashoush and Gadallah [114] observed the crude fiber in mango peel (9.33 g/100 g dried mango peel) and kernels (0.2 g/100 dried mango kernels) which were used as dietary fibers (10% organoleptic acceptance) indifferent bakery products like cookies, muffins, rolls, and cakes [2]. The dietary fibers amount in lemon peel and pulp was significantly observed 1.5 times higher in pulp compared to peel wastes and optimal for human consumption [115].

The dietary fiber content was found higher in inner layers of the onion (68.3%@ dry weight basis) and the lowest in the outer skin (66.3% dry weight basis), respectively [2, 6]. Potato peel and potato solid wastes were observed as good sources of crude fibers (6.1–12.5% in the peel, 2.7–3.5% in potato solid waste) [249], respectively. The total dietary fibers in different horticultural wastes are elaborated in Table 4 [2].

## 5.3 Extraction of Important Bioactive Compounds from Horticultural Wastes

Phenolics and bioactive molecules are the largest class of isoprenoids family which attributes essence (aroma), the nutritional quality of horticultural crops, and biological functions [142, 143]. Additionally, polyphenolic compounds are divided into various classes such as tannins, lignans, and the prominent major class flavonoids (subclass: flavanonols, flavonols, flavones, anthocyanidins, and flavonoids) among other classes [6, 8, 144].

Lee and Wrolstad [145] investigated the distribution of anthocyanin and polyphenols in blueberries wastes (skins, flesh, and seeds) and found that skins were rich in antioxidant activity, while other wastes (flesh, seeds) have varying amounts of

**Table 4** Dietary (insoluble and soluble) fiber in different horticultural waste

| Type of horticultural waste | TDF (%) | IDF (%) | SDF (%) | Literature cited |
|---|---|---|---|---|
| Apple peels | 0.91–1.79 | 0.46–1.39 | 0.40–0.43 | Patocka et al. [111], Sagar et al. [6], Gorinstein et al. [116] |
| Apple pomace | 26.5–42.5 | 21.8–33.5 | 5.9–7.5 | Wang et al. [20], Ktenioudaki et al. [117] |
| Apricot seeds | 27–35 | 11.5–14.8 | – | Fatima et al. [118], Seker et al. [119] |
| Banana peel | 42.8–50.25 | 29.9 | 12.9 | Budhalakoti [120], Wachirasiri et al. [121] |
| Carrot pomace | 64–70 | 45–50 | 13–24 | Hussain et al. [2] |
| Cauliflower waste | 3.11–65 | – | – | Stojceska et al. [122], Femenia et al. [123] |
| Dates seeds | 57.87–92.4 | – | – | Sagar et al. [6], Elleuch et al. [124] |
| Garlic husk | 62.23–85.6 | 58.07–76.48 | 4.16–9.12 | Liurong et al. [125], Kallel et al. [126] |
| Grapes pomace | 77.2–77.9 | 68.4–73.5 | 3.7–9.5 | Llobera and Canellas [26], Valiente et al. [127] |
| Grape seeds | 35–40 | – | – | Ma and Zhang [128], Bagchi et al. [129] |
| Green chilli peel and seeds | 80.41 | – | – | Sagar et al. [6], Mckee and Latner [130] |
| Lemon peel | 14.0 | 9.0–9.04 | 4.0–5.0 | Rafiq et al. [131], Gorinstein et al. [116] |
| Mango peel | 40.6–72.5 | 27.8–49.5 | 12.8–23.0 | Ajila and Rao [113] |
| Kiwifruit pomace | 25.8–28.0 | 18.7–19.0 | 7.1–10.0 | Soquetta et al. [132] |
| Orange pulp and peel | 35.0–57.0 | 47.6–54.0 | 9.41–22.0 | Chau and Huang [133], Grigelmo-Miguel and Martin-Belloso [134] |
| Pea hulls | 91.5 | 87.4 | 4.1 | Hussain et al. [2], Ralet et al. [135] |
| Peach pomace | 35.8–54.5 | 26.1–35.5 | 9.7–19.1 | Gracia-Amezquita et al. [136], Pagan and Ibarz [137] |
| Potato peel | 5.0–26.6 | 2.4–24.0 | 0.48–2.6 | Camire et al. [138], Liu et al. [139] |
| Pear pomace | 36.1–43.9 | 22.0–36.3 | 14.1–19.1 | Gracia-Amezquita et al. [136], Pagan and Ibarz [137] |
| Tomato pomace and peel | 50.0–86.2 | 25.0–71.8 | 14.3–25.0 | Gracia-Amezquita et al. [136], Navarro-González et al. [140] |

(continued)

**Table 4** (continued)

| Type of horticultural waste | TDF (%) | IDF (%) | SDF (%) | Literature cited |
|---|---|---|---|---|
| Raspberry pomace | 53.0–77.5 | 50.0–75 | 1.7–2.5 | Hussain et al. [2], Gorecka et al. [141] |

– Data not available, *TDF* total dietary fiber, *IDF* insoluble dietary fiber, *SDF* soluble dietary fiber

cinnamic acid, flavonol glycosides along with anthocyanins, respectively. Sivagurunathan et al. [146] have reported polyphenolic content in the range of 0.14–0.24 g/100 ml in samples of cashew apple wastes through GC-MS. In another study, Andrade et al. [147] also extracted polyphenolic compounds (1975.64 mg/100 g) from cashew apple waste. Process yield and quality of total phenolics were observed higher under supercritical fluid extraction (1.738%) and lower in an ethyl alcohol solvent (0.380%)-based extraction from guava seeds [148]. Lezoul et al. [149] have used different solvents and extraction methods to extract polyphenolics, and flavonoids from gooseberry wastes and observed the highest content of polyphenolics in roots (1576.95 mg GA/100 g dry mass), followed by leaves (1340 mg GA/100 g dry mass and flowers (940 mg GA/100 g dry mass), respectively. Anthocyanins namely cyaniding 3-O-arabinoside, peonidin-3-O-galactoside, peonidin-3-O-glucoside, and Peonidin 3-O-arabinose were extracted from cranberry press wastes [150]. Similarly, Jurikova et al. [151] studied the European cranberry, which was very rich in polyphenols (12.4–207.3 mg/100 g fresh weight), quercetin (0.52–15.4 mg/100 g fresh leaves), and proanthocyanidins (1.5–5.3 mg/100 g fresh leaves) compounds, respectively.

Martinez-Fernandez et al. [152] have recovered polyphenolics (22.48–32.87 mg/g dry peel), alkaloids (450–610 mg/kg dry peel), and polysaccharides (35.7% wt) from potato peel samples. Five colorants anthocyanins molecules were extracted from black carrot pomace through thermosensation in which cyanidin-3 xyloside-galactoside-glucoside-ferulic acid was maximum (60.85–74.22 mg/L) and followed by cyaniding-3-xyloside-galactoside (49.56–70.12 mg/L), respectively [153]. Moreover, Seremet et al. [44] observed the highest betacyanin content (9.80 mg/g dry material basis) from red beetroot under infusion method. There are very diverse variations among different horticultural wastes which are illustrated in Table 5.

## 5.4 Bioconversion of Horticultural Wastes in Enzymes, Flavor, and Organic Acids

### 5.4.1 Industrial Enzymes

Amylases are very important among all enzyme groups which are further divided into $\alpha$-amylase, $\beta$-amylase, and glucoamylase, respectively. Amylase can be produced

**Table 5** Polyphenolics extracted from horticultural wastes

| Horticultural commodity and waste part | Polyphenolics/antioxidants molecules | Literature cited |
|---|---|---|
| Apple pomace, peels, and leaves | Flavonoids, proanthocyanides, quercetin, catechins, phloretin glycosides, chrologenic acid, cryptochlorogenic acid, 3-ydroxyphlorizin, triterpenes | Patocka et al. [111], Sudha et al. [154], Schieber et al. [155] |
| Acerola peels | Vitamin C, phenolics, and anthocyanins | Wadhwa et al. [8], Sancho et al. [156] |
| Banana peels, bract, and leaves | Carotenoids (xanthophylls, laurate, palmitate or caprate), anthocyanidins (pelargonidin, petunidin, malvidin, peonidin, delphinidin), cyaniding, polyphenols (flavonoids and terpenoids) | Pathak et al. [157], Marie-Magdeleine et al. [158], Subagio et al. [159], Pazmino-Duran et al. [160] |
| Beetroot pomace, leaves | Betalains, anthocyanins, polyphenolics, betacyanin | Seremet et al. [44], Domínguez et al. [161] |
| Blueberry skins and leaves | Cinnamic acid, caffeic acid, polyphenolics, anthocyanins | Teleszko and Wojdyło [162], Lee and Wrolstad [145] |
| Broccoli and cabbage by-products | Phenolic acids, flavonoids, vitamin C, glucosinolates | Wadhwa et al. [8], Papaioannou and Liakopoulou-Kyriakides [163], Dominguez-Perles et al. [164] |
| Carrot pomace | Carotene ($\alpha$ and $\beta$), hydroxycinnamic derivatives like dicaff eoylquinic and chlorogenic acid | Agcam et al. [153], Zhang and Hamauzu [165], Schieber et al. [155] |
| Citrus juice, peels, pulp, and seeds | Limonoids-highly oxygenated triterpenoid, eriocitrin, hesperidin, and hesperidin | Anticona et al. [166], Gomez-mejia et al. [167] |
| Cashew apple waste | Organic acid, vitamin c, polyphenols, unsaturated fatty acids, antioxidants, minerals | Andrade et al. [147], Sivagurunathan et al. [146] |
| Cranberry press residues, fruits, and levaces | Cyanidin-3-O-arabinoside, Peonidin-3-O-glucoside, Peonidin 3-O-arbinoside, myricetin-3-xylopiranoside, dimethoxymyricetin-hexoside, methoxyquercetin-pentoside, Quercetin | Jurikova et al. [151], Sagar et al. [6], Klavis et al. [150], Teleszko and Wojdyło [162] |
| Date palm spikelets, pits, seeds | Antioxidants, rutin, catechin, anthocyanins, polyphenolics | Almusallam et al. [94], Djaoudene et al. [168] |

(continued)

**Table 5** (continued)

| Horticultural commodity and waste part | Polyphenolics/antioxidants molecules | Literature cited |
|---|---|---|
| Grapes seeds, pomace, and skin | Procyanidins, anthocyanins, catechin, stilbenes, flavonol glycosides, epicatechin, epigallocatechin, picatechin gallate, dietary fiber | Ferri et al. [169], Gulcu et al. [170], Babbar et al. [171] |
| Gooseberry peels, roots, leaves, stem, flowers | Polyphenolic, flavonoids compounds | Lezoul et al. [149], Poltanov et al. [172] |
| Guava pulp, seeds, leaves, peels | Melanin, polyphenolic compounds, dietary fiber | Permadi and Nugroho [173], Castro-Vergas et al. [148], Jimenez-Escrig et al. [174] |
| Kiwifruit peel | Polyphenols, flavonoids, caffeic, and protocatechuic acid | Guthrine et al. [175], Sagar et al. [6], Wijngaard et al. [176], Mattila et al. [177] |
| Litchi pericarp, seed, and bark | Flavonoids, anthocyanin (cyaniding-3-rutinoside), phenolics (procyanidins, epicatechin, steroids, vitamin C, sesquiterpenes) | Rosales et al. [178], Wadhwa et al. [8] |
| Longan seeds | Antioxidants, polyphenolic compounds | Rakariyatham et al. [179], Panyathep et al. [180] |
| Mango peel, seed kernels, pulp, and ripe peel | Gallates, ellagic acid, flavonol glycosides, gallotannins, syringic acid, quercitin, mangiferin pentoside, carotenoids ($\beta$-carotene), vitamins, gallic acid, enzymes, and dietary fibers | Varakumar et al. [181], Ajila et al. [4] |
| Purple star apple peel | Ellagic acid, myricetin, sinapic caffeic, ferulic acid, gallic acid, gallotannins, gallates | Moo-Huchin et al. [182] |
| Onion skin and waste | Quercetin 3, quercetin 40-O-monoglucoside, 40-O-diglucoside | Wadhwa et al. [8], Kim and Kim [183] |
| Potato peel | Polyphenols, glycoalkaloids, Polysaccharides | Martinez-Fernandez et al. [152], Sagar et al. [6], Choi et al. [184] |
| Persimmon peel | Gallic acid, p-coumaric acid, protocatechuic, catechin, proanthocyanidins, ascorbic acid, epigallocatechin, catechin gallate | Wadhwa et al. [8], Bubba et al. [185] |
| Red beet peel | Cyclodopa glucoside derivatives, tryptophane | Seremet et al. [44], Sagar et al. [6], Kujala et al. [186] |
| Rambutan peels | Corilagin, ellagic acid, geraniin | Boyano-Orozco et al. [187], Monrroy et al. [188] |

(continued)

**Table 5** (continued)

| Horticultural commodity and waste part | Polyphenolics/antioxidants molecules | Literature cited |
|---|---|---|
| Tomato pomace, skin | Lycopene, sterols, carotenes, tocophereols, flavonoids, terpenes, and ascorbic acid | Szabo et al. [189, 190], Wadhwa et al. [8] |

through solid-state fermentation through microorganisms which is an environmentally safe process. Sharanappa et al. [191] have utilized fruit waste for α-amylase production through *Aspergillus niger* under solid-state fermentation and reported the highest amylase activity after 48 h (PH 6.0 and temp, 25 °C) of incubation. Similarly, Rizk et al. [192] reported maximum activity of α-amylase (168,215 U/g) from fruit wastes using *A. niger*, which was isolated from mango peels. Moreover, Msarah et al. [193] used two *Bacillus* sp. Namely *Bacillus licheniformis* (HULUB1) and *Bacillus subtilis* (SUNGB2) to produce α-amylase from food wastes and reported that SUNGB2 had higher enzyme activity (22.140 U/mL) than *B. licheniformis* (18.15 U/mL), respectively.

Cellulases are mainly applied in the food processing sector (such as making sensory products and extracting phenolic compounds) and classified in β-D-glucosidase, exo-1,4-β-glucanase, and endo-1,4-β-D-glucanase. Food wastes of horticultural produce (such as grape pomace, and banana peel) are a good source of cellulose and are used for cellulase enzyme production by microbial fermentation at the commercial level [194]. Mrudula and Murugammal [195] recorded CMCase and FPase activities 8.90 U/g dry mycelial bran (DMB) and 3.57 U/g DMB by using *A. niger* under submerged and solid-state fermentation from coir waste. The extracellular cellulase was 14.7 fold more than submerged fermentation. Gordillo-Fuenzalida et al. [196] used manufacturing food wastes for cellulase production by *Trichoderma* sp. and reported the maximum activity in culture supernatant to rest other cellulase enzymes. Furthermore, Sinjaroonsak et al. [197] have used *Streptomyces thermocoprophilus* (TC13W) for the production of cellulase and xylanase and reported maximum activities of 925 Unit/g APEFB (pretreated empty fruit bunch) and 1796 U/g APEFB under optimum conditions (PH 6.5, Tem 40 °C, and 120 h incubation time), respectively. Additionally, Mostafa et al. [198] the reported maximum yield of cellulase (8.11 U/ml), β-glucosidase (2.75 U/ml), and FPase (4.55 U/ml) through *Paenibacillus alvei* under submerged fermentation. The complete description is laid down in Table 6.

Invertase is a very prominent enzyme in the sugar industry that is used for invert sugar production. Moreover, invertase enzymes make the product fresh and soft during storage [240]. Fruit peel waste was considered a good source for the production of higher levels of invertase under solid-state fermentation conditions. Horticultural waste, sucrose, glucose, and fructose were mixed with *A. niger* to produce invertase [6]. Among all carbon sources, fructose was found superior as a carbon source in extracellular invertase production to rest others. Uma et al. [203] have optimized the culture conditions for achieving the high levels of invertase through *A. flavus*

**Table 6** Applications of horticultural wastes for production of enzymes, flavor, and organic acids using microorganisms

| Horticultural wastes | Product | Applied microorganism | Literature cited |
|---|---|---|---|
| *Enzymes* | | | |
| Banana waste, potato peels, cabbage waste, coconut waste, date waste, loquat kernels, mango kernel, cassava waste, orange waste | Amylases | *Bacillus* spp., *Bacillus subtilis*, *Aspergillus niger, Bacillus licheniformis, Pseudomonas* sp., *Aspergillus oryzae, Penicillium expansum, Fusarium solani, Streptomyces* sp. | Msarah et al. [193], Rizk et al. [192], Wadhwa et al. [8], Said et al. [199], Dabhi et al. [200] |
| Date palm waste, banana waste, cabbage waste, kinnow waste, mango peel, palm kernel cake, pea peel waste, and vegetable waste | Cellulases | *Paenibacillus alvei, Streptomyces thermocoprophilus, Trichoderma* sp., *Aspergillus niger* | Sinjaroonsak et al. [197], Mostafa et al. [198], Gordillo-Fuenzalida et al. [196], Mrudula and Murugammal [195] |
| Peels of pineapple, pomegranate, orange peel waste, Banana peel, sapota | Invertases | *Penicillium* sp., *Aspergillus niger, Aspergillus flavus* | Nehed and Atalla [201], Mehta and Duhan [202], Uma et al. [203] |
| Orange peel waste, banana waste, pineapple waste, strawberry waste, lemon peel, cashew apple, grape pomace | Pectinases | *Fusarium* spp., *Aspergillus* spp., *Penicillium chrysogenum, Bacillus* Spp., *Aspergillus niger, Aspergillus foetidus, Aspergillus awamori, Penicillium citrinum* | Sudeep et al. [204], El Enshasy et al. [205], Sagar et al. [6] |
| Faba bean waste, pineapple peel, tomato waste, melon peels, orange peel, watermelon rind, banana peel | Xylanases | *Aspergillus oryzae* MN894021, *Trichoderma koeningi, Trichoderma* spp. *Aspergillus fumigatus, Chaetomium globosum* | Atalla and El Gamal [206], Atalla et al. [207], Zehra et al. [208], Bandikari et al. [209] |

(continued)

**Table 6** (continued)

| Horticultural wastes | Product | Applied microorganism | Literature cited |
|---|---|---|---|
| Coconut cake, mahua cake, lemon peel, banana skin | Lipases | *Aspergillus niger, Lasiodiplodia theobromae, Colletotrichum gloeosporioides, Chaloropsis thielarioides* | Selvakumar and Sivashanmugam [210], Venkatesagowda et al. [211], Kumar and Kanwar [212], Parihar [213] |
| Bark of cocoa beans, grape seeds, kiwifruit waste, orange peel, potato peels, apple pomace | Laccases | *Marasmius* sp., *Trametes hirsute, Pleurotus* sp. | Ramadiyanti et al. [214], Botella et al. [215], Couto et al. [216], Krishna and Chandrasekaran [217] |
| Corn stalk, organic solid waste | Glucanases | *Fusarium oxysporum*, Anaerobic digestion with anaerobic microorganism | Tian et al. [218], Panagiotou et al. [219] |
| *Flavor* | | | |
| Sugar beet pulp waste, pineapple cannery waste | Vanillin and vanillic acid | *Pycnoporus cinnabarinus, Aspergillus niger* | Laufenberg et al. [220], Lesage-Meessen et al. [221] |
| Carrot pomace | Isoamyl acetate (banana) | *Bacillus licheniformis, Ceratocystis fimbriata* | Rao and Sobha [222], Torres et al. [223], Fischbach et al. [224] |
| Carrot pomace | Ethyl butyrate (pineapple) | *Burkholderia multivorans, Ceratocystis fimbriata* | Rao and Sobha [222], Dandavate et al. [225], Laufenberg et al. [220] |
| Olive press cake | δ-Decalactone, γ-decalactone | *Ceratocystis moniliformis, Pityrosporum ovale* | Sagar et al. [6], Laufenberg et al. [226] |
| *Organic acids* | | | |
| Pineapple peel, papaya peel | Acetic acid | *Saccharomyces cerevisiae* + *Acetobacter aceti* | Malenica and Bhat [227], Sagar et al. [6], Vikas and Mridul [228] |

(continued)

and obtained 5.8 fold more yield recovery. Likewise, Nahed and Atalla [201] used fungal strain *Penicillium* sp. and *Trichoderma virdie* for invertase production from agricultural wastes and reported maximum invertase enzyme activity (1.98 U/ml) in 5.0% w/v orange peel waste enriched medium. The huge demand used to be required

**Table 6** (continued)

| Horticultural wastes | Product | Applied microorganism | Literature cited |
|---|---|---|---|
| Banana peel, pineapple peel, apple pomace, date wastes | Citric acid | *Aspergillus niger, Yarrowia lipolytica* | Priscilla and Gnaneel [229], Wadhwa et al. [8], Prabha and Rangaiah [230], Acourene and Ammouche [231], Dhillon et al. [232] |
| Pineapple peel, orange and pomegranate waste, radish, and carrot waste | Ferulic acid | *Aspergillus niger* | Ray [233], Wadhwa et al. [8], Tilay et al. [234] |
| Sweet corn waste, sapota waste, banana peel, papaya peel, potato peel, orange peel | Lactic acid | *Lactobacillus casei, Rhizopus oryzae* MTCC 8784, *Lactobacillus plantarum, Lactobacillus delbrucekii* | Mora-Villalobos et al. [235], Panda and Ray [49], Krishnakumar [236], Kumar and Shivakumar [237], Jawad et al. [238], Afifi [239] |

in food and allied industries to make candies, confectionery, jam, and pharmaceuticals products, respectively.

El Enshasy et al. [205] optimized pectinase production under submerged fermentation by *A. niger* and reported 10% higher yield with pectinase activity (109.63 U/ml) at PH 5.5 than uncontrolled condition. The results of pectinases activity were reported superior at 5.8 PH (72.3 U/ml), 30 °C temperature (75.4 U/ml), and 0.5% substrate concentration (112.0 U/ml) using *A. niger* under submerged fermentation. The main applications of pectinase were to be in the beverages industry (wines, and juices) for purification, concentration, and extraction [6].

Xylanase enzyme is generally used to extract plant oils and starch from plant-based materials. Krishna [241] applied xylanase to create different textures and thicknesses in baked products. Zehra et al. [208] studied the production of xylanase and pectinase using *Aspergillus fumigatus* under submerged and solid-state fermentation and observed that addition of xylan (0.25%) or pectin (1.0%) in growth media induced the optimum production of both enzymes. Furthermore, Atalla and El Gamel [206] used pomegranate waste as a substrate for producing the xylanase by *Chaetomium globosum* and reported maximum xylanase activity (1398 U/ml) after seven days of incubation.

Selvakumar and Sivashanmugam [210] used organic solid waste for optimization of lipase production and found the maximum activity of lipase (57.45 U/ml) in organic waste which was digested by the anaerobic microorganism. Moreover, this partially purified lipase gave an 88.63% yield of biodiesel from palm oil. Laccase enzyme has a potential role in the detoxification of wastes of papers, pesticides, herbicides, and effluents. Ramadiyanti et al. [214] have produced laccase enzymes from the bark of cocoa beans using *Marasmius* sp. and reported the highest laccase activity (2.89 U/L) after 5 days of incubation. In glucanase production, Tian et al. [218] used food waste under solid-state fermentation using *A. niger* and achieved the highest extracellular endoglucanase activity (17.38 U/g dry substrate) under solid-state fermentation. Sharma et al. [242] have tea waste for the production of tannase enzyme by *A. niger* under SSF and reported the highest tannase (1.87 U/g dry substrates) at optimum condition (temp 30 °C, incubation time 96 h, PH 5.0). Further, tannase was used for detannification of guava juice (59.25% after 60 min). Regards to protease production, Prakasham et al. [243] produced alkaline protease using chickpea along with different food-agro wastes as a carbon source by *Bacillus* spp. under submerged fermentation. Elumalai et al. [244] optimized protease enzyme production from groundnut cake under submerged fermentation along with light-emitting diodes by using *B. subtilis* B22. Under these conditions, the highest protease activity (335 U/mL) was observed at optimum conditions. Similarly, Thakrar et al. [245] used different agro-industrial wastes for alkaline protease production by *Nocardiopsis alba* OM-4 under SSF and reported the highest protease activities in protein rich wastes like pulse flour, vegetables pills, green gram, etc., compared to other wastes.

### 5.4.2 Aromas and Flavoring Agents

Aromas and flavoring agents are very important ingredients of food processing and cosmetic industries which can be extracted from various horticultural wastes which are the good origin of bioactive molecules related to flavor and aromas. The demand and market of fragrances, aromas, and flavors have increased day by day because of the requirement of natural, safe, and familiar sources after pandemic COVID-19 incidence [246].

Vanillic acid is the parent precursor of Vanillin (4-hydroxy-3-methoxybenzaldehyde) and main flavoring agent, pharmaceutical, cosmetic, and food industries [246]. Fermented pods of vanilla orchids (*Vanilla planifolia*) are generally used for extraction of natural vanillin flavoring compound [15]. In the current scenario, different alternative methods including modern biology have also been used for the production of vanillin using microorganisms. In this line, pineapple peel waste was investigated for ferulic acid, which is a precursor of vanillic acid [6, 8]. For biosynthesis of vanillin, two ways process was adopted in which *A. niger* converted the horticultural waste into ferulic acid and further in vanillin with the help of Basidiomycetes (*Pycnoporus cinnabarinus*, or *Phanerochaete chrysosporium*), respectively [247].

Another pineapple flavoring compound "ethyl butyrate" was produced from apple pomace using the microorganism, *Ceratocystis fimbriata* under solid-state fermentation. Similarly, olive press cake was used as a substrate to produce coconut flavor component "δ-decalactone" through a solid-state bioconversion method using the help of *Ceratocystis moniliformis* [248, 249].

### 5.4.3 Organic Acids

Fruits and vegetable pomace, sugarcane bagasse, cassava bagasse, and molasses as a carbon source under solid-state fermentation by *A. niger*, has been used for manufacturing citric acid industrially [250]. Many scientists used sugary substrates and *Aspergillus* strains for citric acid production such as date extract/molasses, respectively [231, 251, 252]. Moreover, Iralapati and Kummari [253] used different fruit peels (mango, banana, orange) for the production of citric acid through *A. niger* under submerged fermentation. The highest concentration of citric acid was found in orange (21.76 g/kg), followed by banana (14.08 g/kg) and mango (3.2 g/kg), respectively.

Lactic acid is a prominent usable organic acid for both food and non-food industries [8]. Similarly, Mudaliyar et al. [254] observed lactic acid concentration (63.33 g/L of fermentation media) with mango peel submerged fermentation by *Lactobacillus casei*. Moreover, Vuanyuan et al. [255] used fruit and vegetable waste for high amounts of lactic acid (10–20 g/L) by *Bifidobacterium* under homo and heterofermentative condition.

Ferulic acid is a plant-based antioxidant that is primarily used in the cosmetic and food industries. It is most abundantly found as hydroxycinnamic acid in the inner linings of plant tissues. Ray et al. [233] and Tilay et al. [234] have extracted the ferulic acid from peels of pineapple, orange, and pomegranate, respectively. Li et al. [256] reported ferulic acid as the main polyphenolics in pineapple peels. Ferulic acid has antioxidant properties and is beneficial for sperm viability and motility in individuals and protects the sperm membranes by reduction of peroxidative damage [8].

Acetic acid is organic acid systematically called as ethanoic acid and produced through bioconversion of sugary wastes such as molasses, beetroot, carrot, and white radish leafage waste. Jin et al. [257] used carrot leafage as the substrate to produce acetic acid in high yield using hydrothermal two-stage process. This approach for organic acid production through vegetable waste serves two aims/or purposes (i) decreasing the pollution problem, (ii) reduction the price of raw substrate, and valorization of horticultural wastes through the recycling process [8].

## 6   Conclusion

The fruits and vegetable wastes are generated due to a lack of proper processing operations and irrelevant handling. The prior literature and studies show that horticultural wastes are abundant in bioactive molecules including edible fibers, organic acids, flavor, and enzymes that can be used in the cosmetic, agriculture, and pharmaceutical sectors. The findings have also shown that escalated usage of these bioactive compounds (fiber, lycopene, sterols, carotenes, flavonoids, and ascorbic acid) as food supplements for curing the disease. In the current scenario, there is a wide range of dietary bioactive molecules designed according to the global market demands. Except that, there exists a big void to use these bioactive molecules' extractions from horticultural food wastes, valorization, and application in different industries. These existing gaps might be fulfilled through adopting green technologies (such as supercritical fluid extraction, solid-phase extraction, etc.) for extracting bioactive molecules from horticultural wastes without degradation the quality and optimum recovery as such.

The scenario of global demand requires new technologies, operational strategies, and business models to minimize the losses of horticultural crop produces. There is urgent requirement of integration for all drivers of production, post-harvest operations, supply chains, better markets, and governmental policies to achieve systematic and sustainable growth in this sector. Moreover, public and private sectors have to join hands and should start a global initiative for valourization of horticultural wastes through the "Save Food" campaign worldwide especially in Asian countries.

## References

1. Businesswire (2019) Global food processing market report, 2019: trends, forecasts, and competitive analysis (2013–2024). ResearchAndMarkets.com. https://www.businesswire.com/news/home/20190904005488/en/Global-Food-Processing-Market-Report-2019-Trends-Forecasts-and-Competitive-Analysis-2013-2024---ResearchAndMarkets.com
2. Hussain S, Jõudu I, Bhat R (2020) Dietary fiber from underutilized plant resources—a positive approach for valorization of fruit and vegetable wastes. Sustainability 12:5401
3. Elik A, Yanik DK, Istanbullu Y, Guzelsoy NA, Yavuz A, Gogus F (2019) Strategies to reduce post-harvest losses for fruits and vegetables. Strategies 5:29–39
4. Ajila CM, Aalami M, Leelavathi K, Rao UP (2010) Mango peel powder: a potential source of antioxidant and dietary fiber in macaroni preparations. Innov Food Sci Emerg Technol 11:219–224
5. Saini A, Panesar PS, Bera MB (2019) Valorization of fruits and vegetables waste through green extraction of bioactive compounds and their nano-emulsions based delivery system. Bioresour Bioprocess 6:26
6. Sagar NA, Pareek S, Sharma S, Yahia EM, Lobo MG (2018) Fruit and vegetable waste: bioactive compounds, their extractions and possible utilization. Compr Rev Food Sci Food Saf 17:512–531. https://doi.org/10.1111/1541-4337.12330
7. FAO (2017) FAO statistics data 2014. Available from: www.fao.org/faostat/en/_data. Accessed 26 June 2017
8. Wadhwa M, Bakshi MPS, Makkar HPS (2015) Wastes to worth: value added products from fruit and vegetable wastes. CAB Rev 10(043_2015)
9. Putnik P, Kovacevic DB, Jambrak AZ, Barba FJ, Cravotto G, Binello A, Lorenzo JM, Shpigelman A (2017) Innovative "Green" and novel strategies for the extraction of bioactive added value compounds from citrus wastes—a review. Molecules 22:680. https://doi.org/10.3390/molecules22050680
10. Bizvibe (2020). https://www.bizvibe.com/blog/global-fruit-industry-factsheet
11. Statista (2020) https://www.statista.com/statistics/264662/top-producers-of-fresh-vegetables-worldwide/
12. Statista (2020) https://www.statista.com/statistics/264065/global-production-of-vegetables-by-type/
13. FAO (2019) Global initiative on food loss and waste reduction
14. FAO (2019) http://www.fao.org/partnerships/private-sector/stories/story/en/c/1239069/
15. Panouille M, Ralet MC, Bonnin E, Thibault JF (2007) Recovery and reuse of trimmings and pulps from fruit and vegetable processing. In: Waldron K (ed) Handbook of waste management

and co-product recovery in food processing. Woodhead Publishing Limited, Cambridge; CRC Press, Boca Raton, pp 417–447

16. Ajila CM, Bhat SG, Rao UP (2007) Valuable components of raw and ripe peels from two Indian mango varieties. Food Chem 102:1006–1011

17. Laufenberg G, Schulze N, Waldron K (2009) A modular strategy for processing of fruit and vegetable wastes into value-added products. In: Waldron KW (ed) Handbook of waste management and co-product recovery in food processing. Woodhead Publishing Limited, New York, pp 286–353

18. Miljkovic D, Bignami G (2002) Nutraceuticals and methods of obtaining nutraceuticals from tropical crops. U.S. Patent Application No. 10/067,569

19. Bisht TS, Sharma SK, Rawat L, Chakraborty B, Yadav V (2020) A novel approach towards the fruit specific waste minimization and utilization: a review. J Pharmacognosy Phytochem 9(1):712–722

20. Wang X, Kristo E, LaPointe G (2019) The effect of apple pomace on the texture, rheology and microstructure of set type yogurt. Food Hydrocolloids 91:83–91

21. Bhushan S, Kalia K, Sharma M, Singh B, Ahuja PS (2008) Processing of apple pomace for bioactive molecules. Crit Rev Biotechnol 28:285–296

22. Chakraborty C, Mukherjee A, Banerjee B, Mukherjee S, Bandyopadhyay K (2017) Utilization of banana peel and pulp as functional ingredient in product development: a review. Int J Eng Res Sci Technol 6(1):137–148

23. Joshi VK, Sharma SK (2011) Food processing waste management, treatment & utilization technology. New India Publication Agency, New Delhi, pp 1–30

24. Jalgaonkar K, Mahawar MK, Bibwe B, Kannaujia P (2020) Postharvest profile, processing and waste utilization of dragon fruit (*Hylocereus* spp.): a review. Food Rev Int. https://doi.org/10.1080/87559129.2020.1742152

25. González-Centeno MAR, Jourdes M, Femenia A, Simal S, Rosselló C, Teissedre P-L (2013) Characterization of polyphenols and antioxidant potential of white grape pomace byproducts (*Vitis vinifera* L.). J Agric Food Chem 61:11579–11587

26. Llobera A, Canellas J (2007) Dietary fibre content and antioxidant activity of Manto Negro red grape (*Vitis vinifera*): pomace and stem. Food Chem 101:659–666

27. Kanwal N, Randhawa MA, Iqbal Z (2016) A review of production, losses and processing technologies of guava. Asian J Agric Food Sci 4(2):96–101

28. Gupta K, Joshi VK (2000) Fermentative utilization of waste from food processing industry. In: Joshi VK (ed) Postharvest technology of fruits and vegetables: handling, processing, fermentation and waste management. Indus Pub Co., New Delhi, pp 1171–1193

29. Food Outlook, FAO (2018) Biannual report on global food markets. http://www.fao.org/3/I8080e/I8080e.pdf

30. Akter F, Haque MA (2019) Jackfruit waste: a promising source of food and feed. Ann Bangladesh Agric 23(1):91–102

31. Saxena A, Bawa AS, Raju PS (2011) Jackfruit (*Artocarpus heterophyllus* Lam.). In: Yahia EM (ed) Postharvest biology and technology of tropical and subtropical fruits. Woodhead Publishing Limited, Cambridge, pp 275–298

32. Mitra SK, Pathak PK, Devi HL, Chakraborty I (2013) Utilization of seed and peel of mango. Acta Hortic 992:593–596

33. Han Z, Park A, Su WW (2018) Valorization of papaya fruit waste through low-cost fractionation and microbial conversion of both juice and seed lipids. RSC Adv 8:27963

34. Parni B, Verma Y (2014) Biochemical properties in peel, pulp and seeds of *Carica papaya*. Plant Arch 14:565–568

35. Almeida JM, Lima VA, Giloni-Lima PC, Knob A (2015) Passion fruit peel as novel substrate for enhanced β-glucosidases production by *Penicillium verruculosum*: potential of the crude extract for biomass hydrolysis. Biomass Bioenerg 72:216–226

36. Choonut A, Saejong M, Sangkharak K (2014) The production of ethanol and hydrogen from pineapple peel by *Saccharomyces cerevisiae* and *Enterobacter aerogenes*. Energy Proc 52:242–249

37. Nunes MCN, Emond JP, Rauth M, Dea S, Chau KV (2009) Environmental conditions encountered during typical consumer retail display affect fruit and vegetable quality and waste. Postharvest Biol Technol 51:232–241
38. Sharma K, Mahato N, Nile SH, Lee ET, Lee YR (2016) Economical and environmentally-friendly approaches for usage of onion (*Allium cepa* L.) waste. Food Funct 7:3354–3369
39. Verma N, Bansal MC, Kumar V (2011) Pea peel waste: a lignocellulosic waste and its utility in cellulase production by *Trichoderma reesei* under solid state cultivation. BioResources 6(2):1505–1519
40. Sepelev I, Galoburda R (2015) Industrial potato peel waste application in food production: a review. Res Rural Dev 1:130–136
41. Løvdal T, Droogenbroeck BV, Eroglu EC, Kaniszewski S, Agati G, Verheul M, Skipnes D (2019) Valorization of tomato surplus and waste fractions: a case study using Norway, Belgium, Poland, and Turkey as examples. Foods 8:229
42. Issara U, Zzaman W, Yang TA (2014) Rambutan seed fat as a potential source of cocoa butter substitute in confectionary product. Int Food Res J 21:25–31
43. Pankar SA, Bornare DT (2018) Studies on cauliflower leaves powder and its waste utilization in traditional product. Int J Agric Eng 11(Special issue):95–98
44. Seremet D, Durgo K, Jokí S, Hudek A, Cebin AV, Mandura A, Jurasovi J, Komes D (2020) Valorization of banana and red beetroot peels: determination of basic macrocomponent composition, application of novel extraction methodology and assessment of biological activity in vitro. Sustainability 12:4539. https://doi.org/10.3390/su12114539
45. Costa APD, Hermes VS, Rios AO, Flores SH (2017) Minimally processed beetroot waste as an alternative source to obtain functional ingredients. J Food Sci Technol 54(7):2050–2058
46. Imbert E (2017) Food waste valorization options: opportunities from the bioeconomy. Open Agric 2:195–204. https://doi.org/10.1515/opag-2017-0020
47. Campos DA, Gómez-García R, Vilas-Boas AA, Madureira AR, Pintado MM (2020) Management of fruit industrial by-products—a case study on circular economy approach. Molecules 25:320
48. Ghisellini P, Cialani C, Ulgiati S (2016) A review on circular economy: the expected transition to a balanced interplay of environmental and economic systems. J Clean Prod 114:11–32
49. Panda SK, Ray RC (2015) Microbial processing for valorization of horticultural wastes. In: Shukla LB, Pradhan N, Panda S, Mishra BK (eds) Environmental microbial biotechnology. Springer International Publishing, New Delhi, pp 203–221
50. Banerjee J, Singh R, Vijayaraghavan R, MacFarlane D, Patti AF, Arora A (2017) Bioactives from fruit processing wastes: green approaches to valuable chemicals. Food Chem 225:10–22
51. Elia V, Gnoni MG, Tornese F (2017) Measuring circular economy strategies through index methods: a critical analysis. J Clean Prod 142:2741–2751
52. Ben-Othman S, Jõudu I, Bhat R (2020) Bioactives from agri-food wastes: present insights and future challenges. Molecules 25:510
53. Takshak S (2018) Bioactive compounds in medicinal plants: a condensed review. SEJ Pharm Nat Med 1:13–35
54. Yahia EM (2010) The contribution of fruit and vegetable consumption to human health. In: De La Rosa LA, Alvarez-Parrilla E, González Aguilar GA (eds) Fruit and vegetable phytochemicals—chemistry and human health, 2nd edn. Wiley-Blackwell, Hoboken, pp 3–51
55. Akozai A, Alam S (2018) Utilization of fruits and vegetable waste in cereal based food (cookies). Int J Eng Res Technol 7:383–390
56. Sahni P, Shere DM (2018) Utilization of fruit and vegetable pomace as functional ingredient in bakery products: a review. Asian J Dairy Food Res 37:202–211
57. Maphosa Y, Jideani VA (2016) Dietary fiber extraction for human nutrition—a review. Food Rev 32:98–115
58. Fuentes-Alventosa JM, Rodríguez-Gutiérrez G, Jaramillo-Carmona S, Espejo-Calvo J, Rodríguez-Arcos R, Fernández-Bolaños J, Guillén-Bejarano R, Jiménez-Araujo A (2009) Effect of extraction method on chemical composition and functional characteristics of high dietary fibre powders obtained from asparagus by-products. Food Chem 113:665–671

59. Ma MM, Mu TH (2016) Effects of extraction methods and particle size distribution on the structural, physicochemical, and functional properties of dietary fiber from deoiled cumin. Food Chem 194:237–246
60. Soxhlet F (1879) Die gewichtsanalytische Bestimmung des Milchfettes. Dingler's Polytech J 232:461–465
61. Khoddami A, Wilkes MA, Roberts TH (2013) Techniques for analysis of plant phenolic compounds. Molecules 18:2328–2375
62. Fu M, Zhang L, Han J, Li J (2008) Optimization of the technology of ethanol extraction for Goupi patch by orthogonal design test. Zhongguo Yaoshi (Wuhan, China) 11(1):75–76
63. Barba FJ, Zhu Z, Koubaa M, Sant'Ana AS, Orlien V (2016) Green alternative methods for the extraction of antioxidant bioactive compounds from winery wastes and by-products: a review. Trends Food Sci Technol 49:96–109
64. Vankar PS (2004) Essential oils and fragrances from natural sources. Resonance 9:30–41
65. Azmir J, Zaidul ISM, Rahman MM, Sharif KM, Mohamed A, Sahena F, Jahurul MHA, Ghafoor K, Norulaini NAN, Omar AKM (2013) Techniques for extraction of bioactive compounds from plant materials: a review. J Food Eng 117:426–436
66. Garcia-Salas P, Morales-Soto A, Segura-Carretero A, Fernández-Gutiérrez A (2010) Phenolic-compound-extraction systems for fruit and vegetable samples. Molecules 15:8813–8826
67. Lee NY, Yunus MAC, Idham Z, Ruslan MSH, Aziz AHA, Irwansyah N (2017) Extraction and identification of bioactive compounds from agarwood leaves. IOP Conf Ser Mater Sci Eng 162:012028. https://doi.org/10.1088/1757-899X/162/1/012028
68. Guthrie F, Wang Y, Neeve N, Quek SY, Mohammadi K, Baraoutian S (2020) Recovery of phenolic antioxidants from green kiwifruit peels using subcritical water extraction. Food Bioprod Process 122:136–144
69. Safdar MN, Kausar T, Jabbar S, Mumtaz A, Ahad K, Saddozai AA (2017) Extraction and quantification of polyphenols from kinnow (Citrus reticulate L.) peel using ultrasound amd maceration techniques. J Food Drug Anal 25(3):488–500
70. Safdar MN, Kausar T, Nadeem M (2017) Comparison of ultrasound and maceration techniques for the extraction of polyphenols from mango peel. J Food Process Preserv 41(4):e13028. https://doi.org/10.1111/jfpp.13028
71. Zhang QW, Lin LG, Ye WC (2018) Techniques for extraction and isolation of natural products: a comprehensive review. Chin Med 13:20. https://doi.org/10.1186/s13020-018-0177-x
72. Zhang H, Wang W, Fu ZM, Han CC, Song Y (2014) Study on comparison of extracting Fucoxanthin from Undaria pinnatifida with percolation extraction and refluxing methods. Zhongguo Shipin Tianjiaji 9:91–95
73. Arina MZI, Harisun Y (2019) Effect of extraction temperature on tannin content and antioxidant activity of Quercus infectoria (Manjakani). Biocatal Agric Biotechnol 19:e101101. https://doi.org/10.1016/j.bcab.2019.101104
74. Kongkiatpaiboon S, Gritsanapan W (2013) Optimized extraction for high yield of insecticidal didehydrostemofoline alkaloid in Stemona collinsiae root extracts. Ind Crops Prod 41:371–4
75. Zhang L (2013) Comparison of extraction effect of active ingredients in traditional Chinese medicine compound preparation with two different method. Heilongjiang Xumu Shouyi 9:132–133
76. Shakir IK, Salih SJ (2015) Extraction of essential oils from citrus by-products using microwave-steam distillation. Iraqi J Chem Petrol Eng 16(3):11–22
77. Fakayode OA, Abobi KE (2018) Optimization of oil and pectin extraction from orange (Citrus sinensis) peels: a response surface approach. J Anal Sci Technol 9:20. https://doi.org/10.1186/s40543-018-0151-3
78. Albuquerque BR, Prieto MA, Barreiro MF, Rodrigues A, Curran TP, Barros L, Ferreira ICFR (2017) Catechin-based extract optimization obtained from Arbutus unedo L. fruits using maceration/microwave/ultrasound extraction techniques. Ind Crops Prod 95:404–415
79. Jovanović AA, Đorđević VB, Zdunić GM, Pljevljakušić DS, Šavikin KP, Godevac DM, Bugarski BM (2017) Optimization of the extraction process of polyphenols from Thymus serpyllum L. herb using maceration, heat- and ultrasound-assisted techniques. Sep Purif Technol 179:369–380

80. Zhang Z, Su Z (2000) Recovery of taxol from the extract of *Taxus cuspidata* callus cultures with Al$_2$O$_3$ chromatography. J Liq Chromatogr Relat Technol 23(17):2683–2693

81. Kongkiatpaiboon S, Gritsanapan W (2013) Optimized extraction for high yield of insecticidal didehydrostemofoline alkaloid in *Stemona collinsiae* root extracts. Ind Crops Prod 41:371–374

82. Pare JJ, Bélanger JM, Stafford SS (1994) Microwave-assisted process (MAPTM): a new tool for the analytical laboratory. Trends Anal Chem 13:176–184

83. Phaiphan A, Churat S, Doungta T, Wichalin P, Khanchai W, Penjumras P (2020) Effects of microwave and ultrasound on the extraction of pectin and its chemical characterization of banana (*Musa sapientum* L.) peels. Food Res 4(6):2030–2036

84. Zhang HF, Yang XH, Wang Y (2011) Microwave assisted extraction of secondary metabolites from plants: current status and future directions. Trends Food Sci Technol 22:67

85. Dorta E, Lobo MG, González M (2013) Improving the efficiency of antioxidant extraction from mango peel by using microwave-assisted extraction. Plant Foods Hum Nutr 68:190–199

86. Rosenthal A, Pyle DL, Niranjan K (1996) Aqueous and enzymatic processes for edible oil extraction. Enzyme Microb Technol 19:402–420

87. Sharma A, Khare SK, Gupta MN (2002) Enzyme-assisted aqueous extraction of peanut oil. J Am Oil Chem Soc 79:215–821

88. Li BB, Smith B, Hossain MM (2006) Extraction of phenolics from citrus peels: II. Enzyme-assisted extraction method. Sep Purif Technol 48:189–196

89. Muller E, Berger R, Blass E, Sluyts D, Pfennig A (2008) Liquid–liquid extraction equipment. Ullman's encyclopedia of industrial chemistry (online version). Wiley-VCH Verlag GmbH & Co. KGaA, Weinheim

90. Abd-Talib N, Mohd-Setapar SH, Khamis AK (2014) The benefits and limitations of methods development in solid phase extraction: mini review. J Teknol 69:69–72

91. Hernandez SMP, Estevez JJ, Giraldo LJL, Mendo CJM (2019) Supercritical extraction of bioactive compounds from Cocoa husk: study of the main parameters. Rev Fac Ing Univ Antioquia 91:95–105

92. Castro-Vergas H, Baumann W, Ferreira S, Parada-Alfonso F (2019) Valorization of papaya (Carcia papaya L.) agroindustrial waste through the recovery of phenolic antioxidants by supercritical fluid extraction. J Food Sci Technol 56:3055–3066. https://doi.org/10.1007/s13197-019-03795-6

93. Maran JP, Manikandan S, Nivetha CV, Dinesh R (2017) Ultrasound assisted extraction of bioactive compounds from *Nephelium lappaceum* L. fruit peel using control composite face centered response surface design. Arab J Chem 10(1):S1145–S1157

94. Almusallam IA, Ahmed IAM, Babiker EE, Juhami FYA, Fadima GJF, Osma MA, Maiman SAA, Ghafoor K, Alqab HAS (2021) Optimization of ultrasound assisted extraction of bioactive properties from date palm (*Phoenix dactylifera* L.) spikelets using response surface methodology. LWT 140:110816. https://doi.org/10.1016/j.lwt.2020.110816

95. Paes J, Dotta R, Martinez J (2013) Extraction of phenolic compounds from blueberry (*Vaccinium myrtillus* L.) residues using supercritical CO$_2$ and pressurized water. In: III Iberoamerican conference on supercritical fluids. Cartagena de Indias, Colombia, pp 1–10

96. Espinosa-Alonso LG, Lygin A, Widholm JM, Valverde ME, Paredes-Lopez O (2006) Polyphenols in wild and weedy Mexican common beans (*Phaseolus vulgaris* L.). J Agric Food Chem 54:4436–4444

97. Vuckovic D (2013) High-throughput solid-phase microextraction in multi-well-plate format. Trends Anal Chem 45:136–153

98. Geeta HP, Srinivas G, Champawat PS (2020) Supercritical fluid extraction of bioactive compounds from bioresource: a review. Int J Curr Microbiol App Sci 9(4):559–566

99. Abbas KA, Mohamed A, Abdulamir AS, Abbas HA (2008) A review on supercritical fluid extraction as new analytical method. Am J Biochem Biotechnol 4:345–353

100. Patras A, Choudhary P, Rawson A (2017) Recovery of primary and secondary plant metabolites by pulsed electric field treatment. In: Miklavčič D (eds) Handbook of electroporation. Springer, Cham. https://doi.org/10.1007/978-3-319-32886-7_182

101. Puertolas E, Luengo E, Álvarez I, Raso J (2012) Improving mass transfer to soften tissues by pulsed electric fields. Fundamentals and applications. Annu Rev Food Sci Technol 3:263–282
102. Dobias P, Pavlíkova P, AdamM EA, Benova B, Ventura K (2010) Comparison of pressurised fluid and ultrasonic extraction methods for analysis of plant antioxidants and their antioxidant capacity. Open Chem 8:87–95
103. Klejdus B, Kopecký J, Benešová L, Vacek J (2009) Solid-phase/supercritical-fluid extraction for liquid chromatography of phenolic compounds in freshwater microalgae and selected cyanobacterial species. J Chromatogr A 1216:763–771
104. Niranjan K, Hanmoungjai P (2004) Enzyme-aided aquous extraction. In: Dunford NT, Dunford HB (eds) Nutritionally enhanced edible oil processing. AOCS Press, Champaign. ebook ISBN: 978-1-4398-2227-2
105. Puri M, Sharma D, Barrow CJ (2012) Enzyme-assisted extraction of bioactives from plants. Trends Biotechnol 30:37–44
106. Sticher O (2008) Natural product isolation. Nat Prod Rep 25:517–554
107. Yahya NA, Wahab RA, Xine TLS, Hamid MA (2019) Ultrasound-assisted extraction of polyphenols from pineapple skin. AIP Conf Proc 2155:020002-1–020002-5. https://doi.org/10.1063/1.5125506
108. Barba FJ, Parniakov O, Pereira SA et al (2015) Current applications and new opportunities for the use of pulsed electric fields in food science and industry. Food Res Int 77:773–798
109. Hipsley EH (1953) Dietary "fibre" and pregnancy toxaemia. Med J Aust 2(9):341–342
110. Trowell H (1976) Dietary fibre redefined. Lancet 307(7966):1129
111. Patocka J, Bhardwaj K, Klimova B, Nepovimora E, Wu Q, Landi M, Kuca K, Valis M, Wu W (2020) *Malus domestica*: a review on nutritional features, chemical composition, traditional and medicinal value. Plants 9:1408. https://doi.org/10.3390/plants9111408
112. Kammerer D, Claus A, Schieber A, Carle R (2005) A novel process for the recovery of polyphenols from grape (*Vitis vinifera* L.) pomace. J Food Sci 70:C157–C163
113. Ajila CM, Rao UJSP (2013) Mango peel dietary fibre: Composition and associated bound phenolics. J Funct Foods 5(1):444–450
114. Ashoush IS, Gadallah MGE (2011) Utilization of mango peels and seed kernels powders as sources of phytochemicals in biscuit. World J Dairy Food Sci 6:35–42
115. Russo M, Bonaccorsi I, Torre G, Saró M, Dugo P, Mondello L (2014) Underestimated sources of flavonoids, limonoids and dietary fibre: availability in lemon's by-products. J Funct Foods 9:18–26
116. Gorinstein S, Martín-Belloso O, Park YS, Haruenkit R, Lojek A, Cíz M, Caspi A, Libman I, Trakhtenberg S (2001) Comparison of some biochemical characteristics of different citrus fruits. Food Chem 74:309–315
117. Ktenioudaki A, O'Shea N, Gallagher E (2013) Rheological properties of wheat dough supplemented with functional by-products of food processing: brewer's spent grain and apple pomace. J Food Eng 116:362–368
118. Fatima T, Bashir O, Gani G, Tashooq BA, Jan N (2018) Nutritional and health benefits of apricots. Int J Unani Integr Med 2(2):05–09
119. Seker IT, Ozboy-Ozbas O, Gokbulut I, Ozturk S, Koksel H (2010) Utilization of apricot kernel flour as fat replacer in cookies. J Food Process Preserv 34:15–26
120. Budhalakoti N (2018) Evaluation of β-carotene content and antioxidant activity of banana peels and banana peel extracted insoluble dietary fibres. Int J Agric Environ Biotechnol 11(5):781–789
121. Wachirasiri P, Julakarangka S, Wanlapa S (2009) The effects of banana peel preparations on the properties of banana peel dietary fibre concentrate. Songklanakarin J Sci Technol 31:605–611
122. Stojceska V, Ainsworth P, Plunkett A, Ibanoglu E, Ibanoglu S (2008) Cauliflower by-products as a new source of dietary fibre, antioxidants and proteins in cereal based ready-to-eat expanded snacks. J Food Eng 87(4):554–563
123. Femenia A, Robertson JA, Waldron KW, Selvendran RR (1998) Cauliflower (*Brassica oleracea* L.), globe artichoke (*Cynara scolymus*) and chicory witloof (*Cichorium intybus*) processing by-products as sources of dietary fibre. J Sci Food Agric 77:511–518

124. Elleuch M, Besbes S, Roiseux O, Blecker C, Deroanne C, Drira NE, Attia H (2008) Date flesh: chemical composition and characteristics of the dietary fibre. Food Chem 111:676–682
125. Liurong H, Wenxue Z, Jing C, Zibo L (2019) Antioxidant and physicochemical properties of soluble dietary fiber from garlic straw as treated by energy-gathered ultrasound. Int J Food Prop 22(1):678–688
126. Kallel F, Driss D, Chaari F, Belghith L, Bouaziz F, Ghorbel R, Chaabouni SE (2014) Garlic (*Allium sativum* L.) husk waste as a potential source of phenolic compounds: influence of extracting solvents on its antimicrobial and antioxidant properties. Ind Crops Prod 62:34–41
127. Valiente C, Arrigoni E, Esteban RM, Amado R (1995) Grape pomace as a potential food fiber. J Food Sci 60:818–820
128. Ma ZF, Zhang H (2017) Phytochemical constituents, health benefits, and industrial applications of grape seeds: a mini-review. Antioxidants 6:71
129. Bagchi D, Bagchi M, Stohs SJ, Ray SD, Sen CK, Preuss HG (2002) Cellular protection with proanthocyanidins derived from grape seeds. Ann N Y Acad Sci 957:260–270
130. McKee LH, Latner TA (2000) Underutilized sources of dietary fiber: a review. Plant Foods Hum Nutr 55:285–304
131. Rafiq S, Kaul R, Sofi S, Bashir N, Nazir F, Nayik GA (2018) Citrus peel as a source of functional ingredient: a review. J Saudi Soc Agric Sci 17:351–358
132. Soquetta MB, Stefanello FS, Da Mota Huerta K, Monteiro SS, Da Rosa CS, Terra NN (2016) Characterization of physiochemical and microbiological properties, and bioactive compounds, of flour made from the skin and bagasse of kiwi fruit (*Actinidia deliciosa*). Food Chem 199:471–478
133. Chau CF, Huang YL (2003) Comparison of the chemical composition and physicochemical properties of different fibers prepared from the peel of *Citrus sinensis* L. Cv. Liucheng. J Agric Food Chem 51:2615–2618
134. Grigelmo-Miguel N, Martin-Belloso O (1997) Dietary fiber as a by-product of orange fruit extraction. In: Book of abstracts. Institute of Food Technologists annual meeting. Institute of Food Technologists, Orlando, FL, USA, 13–14 June 1997, p 39
135. Ralet MC, Della VG, Thibault JF (1993) Raw and extruded fibre from pea hulls. Part I: Composition and physico-chemical properties. Carbohydr Polym 20:17–23
136. Gracia-Amezquita LE, Tejada-Ortigoza V, Serna-Saldivar SO, Welti-Chanes J (2018) Dietary fiber concentrates from fruits and vegetables by-products: processing, modification, and application as functional ingredients. Food Bioprocess Technol 11:1439–1463
137. Pagan J, Ibarz A (1999) Extraction and rheological properties of pectin from fresh peach pomace. J Food Eng 39:193–201
138. Camire ME, Violette D, Dougherty MP, Mclaughlin MA (1997) Potato peel dietary fiber composition: effects of peeling and extrusion cooking processes. J Agric Food Chem 3:1404–1408
139. Liu Q, Tarn R, Lynch D, Skjodt NM (2007) Physicochemical properties of dry matter and starch from potatoes grown in Canada. Food Chem 105:897–907
140. Navarro-González I, García-Valverde V, García-Alonso J, Periago MJ (2011) Chemical profile, functional and antioxidant properties of tomato peel fiber. Food Res Int 44(5):1528–1535
141. Gorecka D, Pachołek B, Dziedzic K, Górecka M (2010) Raspberry pomace as a potential fiber source for cookies enrichment. Acta Sci Polonorum Technol Alimentaria 9:451–461
142. Ignat I, Volf I, Popa VI (2011) A critical review of methods for characterisation of polyphenolic compounds in fruits and vegetables. Food Chem 126:1821–1835
143. Lapornik B, Prosek M, Wondra AG (2005) Comparison of extracts prepared from plant by-products using different solvents and extraction time. J Food Eng 71:214–222
144. Robbins RJ (2003) Phenolic acids in foods: an overview of analytical methodology. J Agric Food Chem 51:2866–2887
145. Lee J, Wrolstad RE (2006) Extraction of anthocyanins and polyphenols from blueberry processing waste. J Food Sci. https://doi.org/10.1111/j.1365-2621.2004.tb13651.x

146. Sivagurunathan P, Sivasankari S, Muthukkaruppan SM (2011) Determination of phenolic compounds in red variety of cashew apple (*Anacardium occidentale* L.) by GC-MS. J Basic Appl Biol 5(1&2):227–231

147. Andrade RAM, Macial MIS, Santos AMP, Melo EA (2015) Optimization of extraction process of polyphenols from cashew apple agro-industrial residues. Food Sci Technol Campinas 35(2):354–360

148. Castro-Vergas H, Rodriguez-Varela L, Ferrreira SRR, Parada-Alfonso F (2010) Extraction of phenolic fraction from guava seeds (*Psidium guajava* L.) using supercritical carbon dioxide and co-solvent. J Supercrit Fluids 51(3):319–324

149. Lezoul NEH, Belkadi M, Habibi F, Guillen F (2020) Extraction processes with several solvents on total bioactive compounds in different organs of three medicinal plants. Molecules 25:4672. https://doi.org/10.3390/molecules25204672

150. Klavins L, Kviesis J, Klavins M (2017) Comparison of methods of extraction of phenolic compounds from American cranberry (Vaccinium macrocarpon L.) press residues. Agron Res 15(S2):1316–1329

151. Jurikova T, Skrovankova S, Mleck J, Balla S, Snopek L (2019) Bioactive compounds, antioxidant activity, and biological effects on European cranberry (*Vaccinium oxycoccos*). Molecules 24:24. https://doi.org/10.3390/molecules24010024

152. Martinez-Fernandez JS, Seker A, Davaritouchaee M, Gu X, Chen S (2021) Recovering valuable bioactive compounds from potato peels with sequential hydrothermal extraction. Waste Biomass Valorization 12:1465–1481. https://doi.org/10.1007/s12649-020-01063-9

153. Agcam E, Akyildiz A, Balasubramaniam VM (2017) Optimization of anthocyanins extractions from black carrot pomace with thermosonication. Food Chem 235:461–470

154. Sudha ML, Baskaran V, Leelavathi K (2007) Apple pomace as a source of dietary fiber and polyphenols and its effect on the rheological characteristics and cake making. Food Chem 104:686–692

155. Schieber A, Stintzing FC, Carle R (2001) By-products of plant food processing as a source of functional compounds—recent developments. Trends Food Sci Technol 12:401–413

156. Sancho SO, da Silva ARA, de Sousa Dantas AN, Magalhaes TA, Lopes GS, Rodrigues S (2015) Characterization of the industrial residues of seven fruits and prospection of their potential application as food supplements. J Chem. Available from: URL: https://doi.org/10.1155/2015/264284

157. Pathak PD, Mandavgane SA, Kulkarni BD (2017) Fruit peel waste: characterization and its potential uses. Curr Sci 113(3):444–454

158. Marie-Magdeleine C, Boval M, Philibert L, Borde A, Archiméde H (2010) Effect of banana foliage (*Musa paradisiaca*) on nutrition, parasite infection and growth of lambs. Livestock Sci 131:234–239

159. Subagio A, Morita N, Sawada S (1996) Carotenoids and their fatty-acid esters in banana peel. J Nutr Sci Vitaminol 42:553–566

160. Pazmino-Duran EA, Giusti MM, Wrolstad RE, Gloria MBA (2001) Anthocyanins from banana bracts (*Musa × paradisiaca*) as potential food colorants. Food Chem 73:327–332

161. Domínguez R, Munekata PES, Pateiro M, Maggioline A, Bohrer B, Lorenzo JM (2020) Red Beetroot, a potential source of natural additives for the meat industry. Appl Sci 10:8340. https://doi.org/10.3390/app10238340

162. Teleszko M, Wojdyło A (2015) Comparison of phenolic compounds and antioxidant potential between selected edible fruits and their leaves. J Funct Foods 14:736–746

163. Papaioannou EH, Liakopoulou-Kyriakides M (2012) Agro-food wastes utilization by *Blakeslea trispora* for carotenoids production. Acta Biochim Pol 59:151–153

164. Dominguez-Perles R, Moreno DA, Carvajal M, Garcia-Viguera C (2011) Composition and antioxidantcapacity of a novel beverage produced with green tea and minimally-processedbyproducts of broccoli. Innovative Food Sci Emerg Technol 12(3):361–368. https://doi.org/10.1016/j.ifset.2011.04.005

165. Zhang D, Hamauzu Y (2004) Phenolic compounds and their antioxidant properties in different tissues of carrots. Food Agric Environ 2:95–100

166. Anticona M, Blesa J, Frigola A, Esteve MJ (2020) High biological value compounds extraction from citrus waste with non-conventional methods. Foods 9:811. https://doi.org/10.3390/foo ds9060811
167. Gomez-Mejia E, Rosales-Conrado N, Leon-Gonzalez ME, Madrid Y (2019) Citrus peels waste as a source of value added compounds: extraction and quantification of bioactive polyphenols. Food Chem 295:285–299
168. Djaoudene O, Lopez V, Casedas G, Les F, Schisano C, Bey MB, Tenore GC (2019) *Phonex dactylifera* L., seeds: a by products as a source of bioactive compounds with antioxidant and enzyme inhibiting properties. Food Funct 19:4953–4965
169. Ferri M, Vannini M, Ehrnell M, Eliassan L, Xanthakis E, Monari S, Sisti L, Marchese P, Celli A, Tassoni A (2020) From winey waste to bioactive compounds and new polymeric biocomposites: a contribution to the circular economy concept. J Adv Res 24:1–11
170. Gulcu M, Uslu N, Ozcan MM, Gokmen F, Ozcan MM, Tijana B, Gezgin S, Dursun N, Gecgel U, Ceylan DA, Lemiasheuski V (2019) The investigation of bioactive compounds of wine, grape juice, and boiled grape juice wastes. J Food Process Preserv 43(1):e13850
171. Babbar N, Oberoi HS, Uppal DS, Patil RT (2011) Total phenolic content and antioxidant capacity of extracts obtained from six important fruit residues. Food Res Int 44:391–396
172. Poltanov EA, Shikov AN, Dorman HJ, Pozharitskaya ON, Makarov VG, Tikhonov VP et al (2009) Chemical and antioxidant evaluation of Indian gooseberry (*Emblica officinalis* Gaertn., syn. *Phyllanthus emblica* L.) supplements. Phytotherapy Res 23:1309–1315
173. Permadi EE, Nugroho LH (2019) The effect of orange, pineapple, and guava waste extract on phenolic content in green betel (*Piper betle* L.). In: 1st International conference on bioinformatics, biotechnology, and biomedical engineering (BioMIC 2018). AIP conference proceedings, pp 020016-1–020016-8. https://doi.org/10.1063/1.5098421
174. Jimenez-Escrig A, Rincon M, Pulido R, Saura-Calixto F (2001) Guava fruit (*Psidium guajava* L.) as a new source of antioxidant dietary fibre. J Agric Food Chem 49:5489–5493
175. Guthrie F, Wang Y, Neeve N, Quek SY, Mohammadi K, Baroutian S (2020) Recovery of phenolic antioxidants from green kiwifruit peel using subcritical water extraction. Food Bioprod Process 122:136–144. https://doi.org/10.1016/j.fbp.2020.05.002
176. Wijngaard HH, Roßle C, Brunton N (2009) A survey of Irish fruit and vegetable waste and by-products as a source of polyphenolic antioxidants. Food Chem 116:202–207
177. Mattila P, Hellstrom J, Torronen R (2006) Phenolic acids in berries, fruits, and beverages. J Agric Food Chem 54:7193–7199
178. Rosales MP, Jimenez RE, Ramos MOA, Hernandez BMT, Garcia OF, Salgado CMP, Lopez CMS (2019) Phenolic compounds in the pulp, pericarp and seeds of Litchi (*Litchi chinensis* Sonn.) in different stages of maturation. Nutr Food Sci Int J 8(3). NFSIJ.MS.ID.555736
179. Rakariyatham K, Liu X, Liu Z, Wu S, Shahidi F, Zhou D, Zhu B (2020) Improvement of phenolic contents and antioxidants activities of Laongan (*Dimocarpus longan*) peel extracts, by enzymatic treatment. Waste Biomass Valorization 11:3987–4002
180. Panyathepa A, Chewonarina T, Taneyhillb K, Vinitketkumnuen U (2013) Antioxidant and anti-matrix metalloproteinases activities of dried longan (*Euphoria longana*) seed extract. Sci Asia 39:12–18
181. Varakumar S, Kumar YS, Reddy OVS (2011) Carotenoid composition of mango (*Mangifera indica* L.) wine and its antioxidant activity. J Food Biochem 35:1538–1547
182. Moo-Huchin VM, Moo-Huchin MI, Estrada-León RJ, Cuevas-Glory L, Estrada-Mota IA, Ortiz-Vázquez E, Betancur-Ancona D, Sauri-Duch E (2015) Antioxidant compounds, antioxidant activity and phenolic content in peel from three tropical fruits from Yucatan, Mexico. Food Chem 166:17–22
183. Kim SJ, Kim GH (2006) Quantification of quercetin in different parts of onion and its DPPH radical scavenging and antibacterial activity. Food Sci Biotechnol 15:39–43
184. Choi SH, Kozukue N, Kim HJ, Friedman M (2016) Analysis of protein amino acids, non-protein amino acids and metabolites, dietary protein, glucose, fructose, sucrose, phenolic, and flavonoid content and antioxidative properties of potato tubers, peels, and cortexes (pulps). J Food Compos Anal 50:77–87

185. Bubba MD, Giordani E, Pippucci L, Cincinelli A, Checchini L, Galvan P (2009) Changes in tannins ascorbic acid and sugar contents in astringent persimmons during on-tree growth and ripening and in response to different postharvest treatments. J Food Compos Anal 22:668–677
186. Kuala T, Loponen J, Pihlaja K (2001) Betalains and phenolics in red beetroot (*Beta vulgaris*) peel extracts: extraction and characterisation. Z Naturforsch C 56:343–348
187. Boyano-Orozco L, Gallardo-Velazquez T, Meza-Marquez OG, Osorio-Revilla G (2020) Microencapsulation of Rambutan peels extract by spray drying. Foods 9:899. https://doi.org/10.3390/foods9070899
188. Monrroy M, Arauz O, Garcia JR (2020) Active compound identification in extracts of *N. lappaceum* peel and evaluation of antioxidant capacity. J Chem. Article ID 4301891. https://doi.org/10.1155/2020/4301891
189. Szabo K, Cătoi AF, Vodnar DC (2018) Bioactive compounds extracted from tomato processing by-products as a source of valuable nutrients. Plant Foods Hum Nutr 73:268–277. https://doi.org/10.1007/s11130-018-0691-0
190. Szabo K, Diaconesa Z, Cotai A, Vodnar DC (2019) Screening of ten tomato varieties processing waste for bioactive components and their related antioxidant and antimicrobial activities. Antioxidants 8:292. https://doi.org/10.3390/antiox8080292
191. Sharanappa A, Wani KS, Patil P (2011) Bioprocessing of food industrial waste for alpha amylase production by solid state fermentation. Int J Adv Biotechnol Res 2(4):473–480
192. Rizk MA, El-kholany EA, Abo-Mosalum EMR (2019) Production of alpha amylase by *Aspergillus niger* isolated from mango kernel. Middle East J Appl Sci 9(1):134–141
193. Msarah MJ, Ibrahim I, Hamid AA, Aqma WS (2020) Optimisation and production of alpha amylase from thermophilic *Bacillus* spp. and its application in food waste biodegradation. Heliyon 6:e04183
194. Norsalwani TT, Norulaini NN (2012) Utilization of lignocellulosic wastes as a carbon source for the production of bacterial cellulases under solid state fermentation. Int J Environ Sci Dev 3:136–140
195. Mrudula S, Murugammal R (2011) Production of cellulase by *Aspergillus niger* under submerged and solid state fermentation using low waste as a substrate. Braz J Microbiol 42:1119–1127
196. Gordillo-Fuenzalida F, Echeverria-Vega A, Cuadrus-Orellana S, Faundez C, Kahne T, Morales-Vera R (2019) Cellulases production by *Trichoderma* sp. using food manufacturing wastes. Appl Sci 9:4419. https://doi.org/10.3390/app9204419
197. Sinjaroonsak S, Chaiyaso T, H-Kittikun A (2020) Optimization of cellulase and xylanase productions by *Streptomyces thermocoprophilus* TC13W using low cost pretreated oil palm empty fruit bunch. Waste Biomass Valorization 11:3925–3936. https://doi.org/10.1007/s12 649-019-00720-y
198. Mostafa YS, Alamri SA, Hashem M, Nafady NA, AboElyousr KAM, Mohamed ZA (2020) Thermostable cellulase biosynthesis from *Paenibacillus alvei* and its utilization in lactic acid production by simultaneous saccharification and fermentation. Open Life Sci 15:185–197
199. Said A, Leila A, Kaouther D, Sadia B (2014) Date wastes as substrate for the production of α-amylase and invertase. Iran J Biotechnol 12:41–49
200. Dabhi BK, Vyas RV, Shelat HN (2014) Use of banana waste for the production of cellulolytic enzymes under solid substrate fermentation using bacterial consortium. Int J Curr Microbiol Appl Sci 3:337–346
201. Nehad EA, Atalla SMM (2020) Production and immobilization of invertase from *Penicillium* sp. using orange peel waste as substrate. Equpt Pharm J 19:103–112
202. Mehta K, Duhan JS (2014) Production of invertase from *Aspergillus niger* using fruit peel waste as a substrate. Intl J Pharm Biol Sci 5:353–360
203. Uma C, Gomathi D, Muthulakshmi C, Gopalakrishnan VK (2010) Production, purification and characterization of invertase by *Aspergillus flavus* using fruit peel waste as substrate. Adv Bio Res 4:31–36
204. Sudeep KC, Upadhyaya J, Joshi DR, Lekhak B, Chaudhary DK, Pant BR, Bajgai TR, Dhital R, Khanal S, Koirala N, Raghavan V (2020) Production, characterization, and industrial application of pectinase enzyme isolated from fungal strains. Fermentation 6:59

205. El Enshasy HA, Elsayed EA, Suhaimi N et al (2018) Bioprocess optimization for pectinase production using *Aspergillus niger* in a submerged cultivation system. BMC Biotechnol 18:71. https://doi.org/10.1186/s12896-018-0481-7
206. Atalla SMM, El Gamal NG (2020) Production and characterization of xylanase from pomegranate peel by *Chaetomium globosum* and its application on bean under greenhouse condition. Bull Natl Res Cent 44:104. https://doi.org/10.1186/s42269-020-00361-5
207. Atalla SMM, Ahmed NE, Awad HM, El Gamal NG, El-Shamy AR (2020) Statistical optimization of xylanase production, using different agricultural wastes by *Aspergillus oryzae* MN894021, as a biological control of faba bean root diseases. Egypt J Biol Pest Control 30:125
208. Zehra M, Syed MN, Sohail M (2020) Banana peel: a promising substrate for the coproduction of pectinase and xylanase from *Aspergillus fumigatus* MS16. Pol J Microbiol 63(1):19–26
209. Bandikari R, Poondla V, Sarathi V, Reddy O (2014) Enhanced production of xylanase by solid state fermentation using *Trichoderma koeningi* isolate: effect of pretreated agro-residues. 3 Biotech 4:655–664
210. Selvakumar P, Sivashanmugam P (2017) Optimization of lipase production from organic solid waste by anaerobic digestion and its application in biodiesel production. Fuel Process Technol 165:1–8
211. Venkatesagowda B, Ponugupaty E, Barbosa AM, Dekker RF (2015) Solid-state fermentation of coconut kernel-cake as substrate for the production of lipases by the coconut kernel-associated fungus *Lasiodiplodia theobromae* VBE-1. Ann Microbiol 65:129–142
212. Kumar A, Kanwar SS (2012) Lipase production in solid-state fermentation (SSF): recent developments and biotechnological applications. Dyn Biochem Pro Biotech Mol Bio 6:13–27
213. Parihar DK (2012) Production of lipase utilizing linseed oilcake as fermentation substrate. Int J Sci Environ Technol 1:135–143
214. Ramadiyanti M, Djali M, Mardawati E, Andoyo R (2020) Production of laccase enzyme by *Marasmius* sp. from the bark of cocoa beans. Syst Rev Pharm 11(3):405–409
215. Botella C, Diaz A, De Ory I, Webb C, Blandino A (2007) Xylanase and pectinase production by *Aspergillus awamori* on grape pomace in solid state fermentation. Process Biochem 42:98–101
216. Couto SR, Lopez E, Sanromán MA (2006) Utilisation of grape seeds for laccase production in solid-state fermentors. J Food Eng 74:263–267
217. Krishna C, Chandrasekaran M (1995) Economic utilization of cabbage wastes through solid state fermentation by native microflora. J Food Sci Technol 32:199–201
218. Tian M, Wai A, Guha TK, Hausner G, Yuan Q (2018) Production of endoglucanase and xylanase using food waste by solid-state fermentation. Waste Biomass Valorization 9:2391–2398. https://doi.org/10.1007/s12649-017-0192-7
219. Panagiotou G, Kekos G, Macris BJ, Christakopoulos P (2003) Production of cellulolytic and xylanolytic enzymes by *Fusarium oxysporum* grown on corn stover in solid state fermentation. Indian Crop Prod 18:37–45
220. Laufenberg G, Kunz B, Nystroem M (2003) Transformation of vegetable waste into value added products: (A) the upgrading concept; (B) practical implementations. Bioresour Technol 87:167–198
221. Lesage-Meessen L, Stentelaire C, Lomascolo A, Couteau D, Asther M, Moukha S, Record E, Sigoillot JC, Asther M (1999) Fungal transformation of ferulic acid from sugar beet pulp to natural vanillin. J Sci Food Agric 79:487–490
222. Rao GV, Sobha K (2020) Lipases with preferred thermo-tolerance in food industry. Res J Biotechnol 15(7):141–150
223. Torres S, Baigorí MD, Swathy SL, Pandey A, Castro GR (2009) Enzymatic synthesis of banana flavour (isoamyl acetate) by *Bacillus licheniformis* S-86 esterase. Food Res Int 42:454–460
224. Fischbach R, Laufenberg G, Kunz B (2000) Generation of natural flavours by solid-state fermentation of food industry by-products. In: Proceedings of biotechnology, vol 4, Berlin, Frankfurt, 3–8 Sept 2000, pp 266–268
225. Dandavate V, Jinjala J, Keharia H, Madamwar D (2009) Production, partial purification and characterization of organic solvent tolerant lipase from *Burkholderia multivorans* V2 and its application for ester synthesis. Biores Technol 100:3374–3381

226. Laufenberg G, Rosato P, Kunz B (2001) Conversion of vegetable waste into value added products: oil press cake as an exclusive substrate for microbial d-decalactone production. Lipids, fats, and oils: reality and public perception. In: 24th World congress and exhibition of the ISF, 16–20 Sept 2001. AOCS Press, Berlin, p 10ff. ISBN: 1-893997-16-x

227. Malenica D, Bhat R (2020) Review article: current research trends in fruit and vegetables wastes and by-products management-scope and opportunities in the Estonian context. Agron Res 18(S3):1760–1795

228. Vikas OV, Mridul U (2014) Bioconversion of papaya peel waste into vinegar using *Acetobacter aceti*. Int J Sci Res 3:409–411

229. Priscilla D, Gnaneel M (2020) Production of citric acid from banana peel using *Aspergillus niger*. Int J Creative Res Thoughts (IJCRT) 8(2):547–558

230. Prabha MS, Rangaiah GS (2014) Citric acid production using *Ananas comosus* and its waste with the effect of alcohols. Int J Curr Microbiol Appl 3:747–754

231. Acourene S, Ammouche A (2012) Optimization of ethanol, citric acid, and a-amylase production from date wastes by strains of *Saccharomyces cerevisiae*, *Aspergillus niger*, and *Candida guilliermondii*. J Ind Microbiol Biotechnol 39:759–766

232. Dhillon GS, Brar SK, Verma M, Tyagi RD (2011) Enhanced solid-state citric acid bioproduction using apple pomace waste through surface response methodology. J Appl Microbiol 110:1045–1055

233. Ray S (2015) Ferulic acid benefit as antioxidant supplement. www.raysahelian.com/ferulicacid.html

234. Tilay A, Bule M, Krishenkumar J, Annapure U (2008) Preparation of ferulic acid from agricultural wastes: its improved extraction and purification. J Agric Food Chem 56:7644–7648

235. Mora-Villalobos JA, Montero-Zamora J, Barboza N, Rojas-Garbanzo C, Usaga J, Redondo-Solano M, Schroedter L, Olszewska-Widdrat A, López-Gómez JP (2020) Multi-product lactic acid bacteria fermentations: a review. Fermentation 6:23

236. Krishnakumar J (2013) Biological production of succinic acid using a cull peach medium. Masters thesis, Clemson University, Clemson, S.C. Available from: http://tigerprints.clemson.edu/all_theses/1735/

237. Kumar R, Shivakumar S (2014) Production of L-lactic acid from starch and food waste by amylolytic *Rhizopus oryzae* MTCC 8784. Int J ChemTech Res 6:527–537

238. Jawad AH, Alkarkhi AF, Jason OC, Easa AM, Norulaini NN (2013) Production of the lactic acid from mango peel waste—factorial experiment. J King Saud Univ Sci 25:39–45

239. Afifi MM (2011) Enhancement of lactic acid production by utilizing liquid potato wastes. Intl J Biotechnol Biochem 5:91–102

240. Kumar R, Kesavapillai B (2012) Stimulation of extracellular invertase production from spent yeast when sugarcane pressmud used as substrate through solid state fermentation. SpringerPlus 1:81. http://www.springerplus.com/content/1/1/81

241. Krishna C (2005) Solid-state fermentation systems—an overview. Crit Rev Biotechnol 25:1–30

242. Sharma NK, Beniwal V, Kumar N, Kumar S, Pathera AK, Ray A (2014) Production of tannase under solid state fermentation and its application in detannification of guava juice. Prep Biochem Biotechnol 44(3):281–290

243. Prakasham RS, Rao CS, Sarma PN (2006) Green gram husk—an inexpensive substrate for alkaline protease production by *Bacillus* sp. in solid-state fermentation. Bioresour Technol 97:1449–1454

244. Elumalai P, Lim JM, Park YJ et al (2020) Agricultural waste materials enhance protease production by *Bacillus subtilis* B22 in submerged fermentation under blue light-emitting diodes. Bioprocess Biosyst Eng 43:821–830. https://doi.org/10.1007/s00449-019-02277-5

245. Thakrar F, Goswami D, Singh SP (2020) Production of an alkaline protease from Nocardiopsis Alba Om-4, a haloalkaliphilic actinobacteria in solid-state fermentation using agricultural waste products. In: Proceedings of the national conference on innovations in biological sciences (NCIBS). https://doi.org/10.2139/ssrn.3586506

246. Poornima K, Preetha R (2017) Biosynthesis of food flavours and fragrances—a review. Asian J Chem 29(11):2345–2352

247. Vandamme EJ, Soetaert W (2002) Bioflavours and fragrances via fermentation and biocatalysis. J Chem Technol Biotechnol 77(12):1323–1332

248. Dorta E, Sogi DS (2017) Value added processing and utilization of pineapple by-products. In: Lobo MG, Paull RE (edS) Handbook of pineapple technology, production, postharvest science, processing and nutrition. Oxford: John Wiley and Sons. pp. 196–220

249. Ncobela CN, Kanengoni AT, Hlatini VA, Thomas RS, Chimonyo M (2017) A review of the utility of potato by-products asa feed resource for smallholder pig production. Anim Feed Sci Technol 227:107–117. https://doi.org/10.1016/j.anifeedsci.2017.02.008

250. Kumar D, Jain VK, Shanker G, Srivastava A (2003) Utilisation of fruits waste forcitric acid production by solid state fermentation. Process Biochem 38:1725–1729

251. Hang YD (1987) Production of fuels and chemicals from apple AQ11 pomace. Food Technol (March issue):115–117

252. Mehyar GF, Delaimy KS, Ibrahim SA (2005) Citric acid production by *Aspergillus niger* using date-based medium fortified with whey and additives. Food Biotechnol 19:137–144

253. Iralapati V, Kummari S (2015) Population of citric acid from different fruit peels using *A. niger*. Int J Sci Eng Res 3(5):129–130. ID: 15210

254. Mudaliyar P, Sharma L, Kulkarni C (2012) Food waste management—lactic acid production by *Lactobacillus* species. Intl J AdvRes Biol Sci 2:34–8

255. Vuanyuan W, Ma H, Zheng M, Wang K (2015) Lactic acid production from acidogenic fermentation of fruit and vegetable wastes. Bioresour Technol 191:53–58

256. Li T, Shen P, Liu W, Liu C, Liang R, Yan N, Chen J (2014) Major polyphenolics in pineapple peels and their antioxidant interactions. Int J Food Prop 17:1805–1817

257. Jin F, Zhou Z, Moriya T, Kishida H, Higashijima H, Enomoto H (2005) Controlling hydrothermal reaction pathways to improve acetic acid production from carbohydrate biomass. Environ Sci Technol 39:1893–1902

# Applying Circular Economy Principles to Agriculture: Selected Case Studies from the Indian Context

Vaibhav Aggarwal and Ritika Mahajan

**Abstract** This chapter aims to explore circular farming practices adopted by farmers in the states of Rajasthan and Madhya Pradesh and the challenges faced by them. The chapter documents case studies of eight farmers identified through snowball sampling. These farmers have been practicing circular farming for more than two decades and serve as role models for their peers. The chapter provides insights about the struggle stories of these farmers and proposes a framework comprising personal, social and systemic antecedents as well as consequences that drive these farmers to adopt circular farming despite the constraints. The framework can be useful in identifying future directions for research and insights for policy makers.

**Keywords** Circular farming · Sustainable agriculture · Case study · Circular economy · Natural farming

## 1 Introduction

India is an agrarian economy; agriculture is the primary source of livelihood for more than 58 percent of the population. In the 1950s, the agricultural output of food grains in India for a population of 361 million people was merely 50.82 million metric tons [1]. By the 1960s, daunted by natural calamities, increasing population, logistic issues and other challenges, the Government undertook a series of measures and India became the leading importer of food grains. The Green revolution became the Government's most important program to boost agricultural productivity. Vast acres of land was brought under cultivation, hybrid seeds were introduced, natural and organic fertilizers were replaced by chemical fertilizers and locally made pesticides were replaced by chemical pesticides. This boosted productivity manifold in the

V. Aggarwal
Center for Research and Implementation of Sustainable Practices, Jaipur, India

R. Mahajan (✉)
Department of Management Studies, MNIT Jaipur, Jaipur, India
e-mail: ritika.dms@mnit.ac.in
URL: http://OrganicIndiastory.com

© The Author(s), under exclusive license to Springer Nature Singapore Pte Ltd. 2021     227
R. S. Mor et al. (eds.), *Challenges and Opportunities of Circular Economy in Agri-Food Sector*, Environmental Footprints and Eco-design of Products and Processes, https://doi.org/10.1007/978-981-16-3791-9_12

country. According to data provided by the World Economic Forum (2015), the population of India since Independence has tripled but food production has quadrupled recording a substantial increase in food grains available per capita [2]. Nevertheless, unbalanced use of chemical fertilizers and pesticides, improper irrigation techniques, usage of heavy machinery, faulty agricultural practices exacerbated by natural causes led to massive degradation of fertile land and several health hazards [3]. The solution to all these problems lies in adopting principles of circular economy (CE) to agricultural practices. The idea is not new to the Indian context; it is more like going back to the roots and re-adopting practices that will help in implementing the ideals of reducing, reusing, recycling, restoring, designing and rethinking to farming to maintain productivity without externalities harmful for people and the planet.

Although circular farming is not a new concept, preliminary survey and a review of literature suggested that there is limited information available on practical challenges of circular farming. In the Indian context, primary data-based studies as well as qualitative studies in the field are rare. Thus, there exists a gap which needs to be filled by collection of data and knowledge from the field about the pre-requisites and challenges of circular farming. This chapter discusses real-life case studies of farmers in India who are implementing the principles of circular economy (reduce, reuse, rethink, redesign and recycle) in their farms. The case studies were compiled through field visits and semi-structured interviews conducted with a small sample of farmers in different locations in North and Central India including Sagar, Bhopal, Indore, Khandwa, Sikar and Jhunjhunu identified through snowball sampling. Structurally, the chapter provides an overview of circular farming in India, the best practices and associated challenges that may serve as a useful resource for policy makers, researchers, not for profit organizations as well as farmers interested in this domain.

## 2 Background

India has its philosophical and economic roots attached to agriculture. Since its independence in 1947, for many years, the contribution of agricultural and allied activities remained more than 50% in India's gross domestic product (GDP) [4]. Agriculture has been the primary source of livelihood for more than 800 million people of India [5], and it still is the dominant source of livelihood. According to the World Bank (2019), the per capita GDP of India in the 1960s (i.e., post-independence) was US $82.189[1] only [6]. Despite the agrarian nature of the economy, there was a deficit in the quantum of production and productivity. Further, rising population and hunger levels, natural calamities, logistics and other issues led India toward Green Revolution in the mid-1960s as the immediate solution to many problems. It increased production and productivity through the usage of chemical-based fertilizers and pesticides in the farms. The first agricultural university of India was set up in 1960 to carry out research on agriculture; high yielding and hybrid varieties of seeds were

---

[1] Per capita GDP of the UK in 1960 was US $1397.595 and that of the USA was US $3007.123.

developed and adopted. This led to a production of 108.4 MMT of food grains in 1971 which was more than double the production of 55.0 MMT of food grains in 1951. But Green revolution not only increased output, there was also a sharp rise in the input cost to the farmers in the form of capital expenditure. The number of private tube-wells increased from 0.1 million in 1965 to 0.47 million in 1971 and the number of pump sets—diesel and electric increased from 0.88 million to 3.24 million during the same period. Similarly, the consumption of chemical fertilizers per cropped acre increased from 4 to 16 kg, i.e., a whopping 400% in a span of 7 years and the expenditure by agriculture on modern inputs in real terms (1960–1961 prices) increased from INR 207 million to INR 734 million during the first decade ending 1960–1961. In the second decade ending 1970–1971, it went up to INR 4,355 million and further to INR 6,181 million in 1972–1973 [7].

Rising input cost for the farmers, unbalanced usage of chemicals on the field, improper irrigation techniques and many other reasons led to degradation of the soil, health and other serious environmental issues [3]. Some of the regions in India like Malwa (Punjab) have, as a result, been suffering from the major problem of groundwater contamination and serious health issues because of the heavy usage of pesticides on the most fertile land of India. Various healthy and drought resistant varieties of millets, rice, barley and other crops have become non-existent after production of wheat and some varieties of rice as major crops during the Green Revolution [8]. Mono cropping and improper handling of agriculture residues are still posing several environmental problems. Thus, once considered to be a very beneficial movement, the Green Revolution eventually led to serious side effects on the Indian agriculture system.

## 2.1 Problem in Modern Agricultural Practices

Modern agricultural practices have no doubt helped the world in solving the problem of hunger of the rapidly growing world population. The Green revolution helped India in doubling grain production with virtually no increase in net cultivable area. However, overuse of pesticides especially in the production of vegetables and fruits resulted in chemical residues that were above safety levels [9] contaminating soil, groundwater, and river water and creating environmental problems [10]. This also created the problem of agricultural residues. Agricultural residue is waste produced during agricultural operations, agricultural products, agro-industries, animal feed, horticulture, aquaculture, etc. This may include residue from manure, crop, chemical, livestock, processing, etc. As per one of the estimates by the Ministry of New and Renewable Energy (MNRE), India produces around 500 million tons of crop residues annually. Harvesting of various crops generates residues both on and off the farm. This crop residue is used as bedding material for livestock, livestock feeding, soil mulching, biogas generation and other industrial uses. However, majority of the residue is burnt on-farm to clear the bed for sowing the next crop. This burning of residue is a low cost and low labor solution to farmers; resulting in loss of nutrients, change of soil texture, emission of greenhouse gases and air pollution [11]. Thus, modern agriculture practices are more focused on production, usage and wastage giving linearity to the process which was once circular in nature [12]. Further, through its emphasis on high production, the industrial model of agriculture (modern) has affected soil, water and biodiversity and pushed more and more acres in the hands of fewer and fewer farmers crippling rural communities [13]. As per the Agriculture Census in 2015, approximately 67% farmers in India are holding less than 1 acre of land, aggregating just 17% of the total cultivable area [14]. Concerns are also growing about the sustainability of agriculture.

## 2.2   Circularity in Agriculture

Agriculture should connect closely with the natural ecosystem as it is the easiest way for material circulation. Circular economy has been argued to be the most effective path for sustainable development of every country. The development of circular economy requires adoption of closed loop systems which work toward the goals of improved economic and environmental sustainability. The development of such systems is a move away from traditional linear production models that operate via the conversion of natural resources into products and then waste released into the environment. Circular economy is about reduction, reuse, recycling and recovery (4R) [15] and circularity in agriculture is harnessing the concept of circular economy in agriculture.

Circular agriculture (CA) may be defined as the set of agricultural practices based on maximum adoption of 4R's on and off the field for efficient material circulation. Hence, CA involves low-input, high-recyclability, high-efficiency, high-technology and industrialized set of practices. The idea is not new to the Indian context; Vedic references and work of several poets like Gagh, Bhaddari, etc. refer to circularity in agricultural practices in India. The work of Satapatha Brahman and Krishi Parasharis considered to be seminal in the field of CA in India that emphasizes circularity through seed, crop, pest, waste, nutrition, and water management, etc.

## 2.3   Circular Farming Practices

Kirchherr et al. [15] argued that the current definitions of circular economy are focused on reducing, reuse and recycling without considering the importance of behavior and cultural shift. However, they have also agreed that circular economy principles should be adapted to the concept of closed loop material flow. Tang et al. [16] specifically researched about CA and enumerated various factors involved like circular utilization of energy; optimal utilization of biomass energy resources; utilization of every material link in the production process; promotion of clean production and conservation-minded consumption; stringent control of harmful inputs and waste

**Table 1** Practices and impact of circular farming

| Practices | Impact areas |
|---|---|
| Soil preservation | Earth |
| Closing nutrient cycles | |
| Reduction of greenhouse gasses and ammonia | Air |
| Producing sustainable energy | |
| Conserving water and stopping contamination | Water |
| Animal welfare and health | Animal |
| Contribution to regional economy and vitality of the rural area | Societal |
| Maintenance of biodiversity and nature conservation | Overall environment |

*Source* Authors

production; maximum reduction of pollution and ecosystem destruction. The circular farming practices are not standard across the globe and can be decided by farmers, corporations or policy makers based on the circumstances and resources. Further, the role of contemporary technologies is also important for adopting circularity in agriculture [17]. We, however, for simplicity of the study and taking cue from holistic approach of circular economy, have limited ourselves to the practices resulting in following impacts (see Table 1).

## 3 Methodology

The preceding section has explained evolution of farming practices in India and the need and significance of CA in the present context. Based on authors' own experience in this field, and interactions with farmers, it was identified that the knowledge on CA in the Indian context is limited. Conceptual papers and policy-related documents exist, but there is very scanty literature capturing real-life cases of farmers practicing CA and explaining their challenges. This chapter is an attempt to fill this gap. The chapter presents and discusses case studies of a sample of eight farmers engaged in circular and natural farming practices or in other words implementing circular economy principles in their farms across two Indian states of Rajasthan and Madhya Pradesh located in northern and central part of the country. The research design (see Fig. 1) was exploratory for understanding the nature and extent of application of circular economy principles in farming in selected cases. The data was primary in nature, collected by field visits to the respective farms of the individual farmers. The sampling method was snowball sampling. The data collection method was semi-structured interviews and observation. A schedule was prepared prior to data collection based on issues identified from the literature. A snapshot of the schedule is

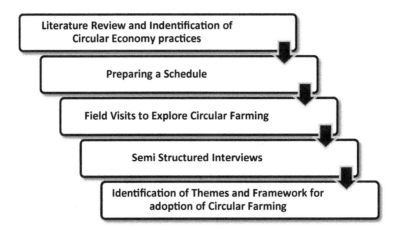

**Fig. 1** Research design. *Source* Authors

presented in Fig. 2. Age of the respondents ranged from 30 to 60 years and average number of years spent in farming was 15 years. During the field visit, detailed notes were prepared along with documentation through photographs. The interviews were audio-recorded and transcribed verbatim. The following section elaborates the cases, followed by key themes emerging from the cases.

| Questions |
|---|
| 1. When, why and how did you start practicing this form of farming? |
| 2. What were the problems in the initial phase? |
| 3. Did you find any change in the quality of your produce after adopting these practices? |
| 4. Explain. |
| 5. How do you provide nutrients to the crops? Please elaborate. |
| 6. How does animal husbandry help in this kind of farming? |
| 7. If this zero budget farming, then what is Organic Farming? |
| 8. What support does the government provide to promote this kind of farming? |

**Fig. 2** Interview schedule. *Source* Authors

# 4  Cases

## 4.1  Case 1: Kan Singh Nirvana, Sikar, Rajasthan

Kan Singh Nirvana is a farmer based in Sikar, Rajasthan, i.e., the largest State in India. He is an extremely progressive farmer practicing circular farming on 35 acres of land and has been awarded many national and international accolades for his work. His ideas are appreciated by people living in and around his farmland and he is a role model for many due to his revolutionary ideas. The field-visit gave the authors a chance to also explore his entrepreneurial skills by virtue of which he plans to develop his native place as a model farm in the near future where people interested in circular farming and nature can come and spend time. He was born in a family of farmers; however, he is the first in his family to engage in circular farming and has been doing so since more than 15 years. He highlighted several key practices as well as issues that impact the application of circular economy principles in agriculture that he understood through his experiential learning. One, nature has immense capacity to generate and rejuvenate on its own, and chemicals applied incessantly in huge volumes are "not only detrimental but simply unnecessary". Basic scientific principle in farming is to understand the need of micro-organisms in the soil that will help in providing the nourishment required for the crops to grow and flourish. Two, in his opinion, the introduction of chemicals and promotion of their usage has a lot to do with a lack of understanding among the farmers about circular economy practices applicable in agriculture as well as lack of government will and policy support. In his opinion, if we stop using chemicals, the associated industries will start losing their business and hence, the problem is not as simple as it seems or in other words, it is not only an awareness issue but also has economic and social implications for various stakeholders. Third, it took him time to convince his family that this kind of farming is possible, sustainable and profit yielding as well. There was little support that he received in terms of advice and information available to carry on circular farming. He eventually met "Subhash Palekar[2]" who helped him in gaining knowledge about circular farming. However, he insisted that it is "only by doing" that he eventually understood the "language of the Earth" and the science behind circular farming. Further, he also investigated and understood how cow dung and cow urine form an integral part of farming and henceforth, the cow became the center of all his farming activities. Fourth, he strongly believes that much of the knowledge about circular farming is simple and easily available, but it is being dominated by certain individuals and organizations who want to make profit by making it sound very difficult and inaccessible. In his words, these people have hardly spent time working in the fields. Last point made by him was that there still is a huge potential in circular farming. It is easily replicable if one understands the fundamentals. Further, there is also the scope for creating organic tourism models for the farmers interested in this domain because eventually given the hustle-bustle of city life and rising levels of

---

[2] https://en.wikipedia.org/wiki/Subhash_Palekar.

pollution in urban areas, the numbers of people looking for countryside experiences are going to increase and the farmers can become entrepreneurs with some training and government support in this regard.

## 4.2  Case 2: Sudhanshu, Bhopal, Madhya Pradesh

This is the case study about Sudhanshu, a farmer based in Vidisha near Bhopal in the state of Madhya Pradesh located in Central India. He has a farmland of 15 acres where he grows crops through circular farming. He is the first-generation farmer in his family and interestingly has completed his formal education in business management from LEEDS University Business School, England. Several key points were highlighted during the visit of the author to his farm. One, he had a very difficult time in gathering knowledge and information about circular farming as he was the first in his family to venture into farming. He sought assistance from many "Gurus" on circular economy but could not gather much support as many of these people were treating it like a business. It was only through hit and trial and research on his own that he learnt the principles of circular farming. Two, he shared his experience about contract farming wherein the farmer gets a pre-determined price for his crop. In his experience, many ill-intentioned people have entered the business of contract farming who end up cheating the farmers. Therefore, background checks and relationship management are very important in this regard. Third, he highlighted the absence of market linkages for selling circular farm produce. After he started growing paddy in his fields through circular farming, he started looking for buyers. Initially, he could only get in touch with a few buyers through his own efforts. His training in business and management domain helped and eventually he created a platform for his own products under the brand name "Jaivik Jeevan" that brings products grown by circular farming practices to consumers at reasonable prices in Bhopal.

## 4.3  Case 3: Rahul, Indore, Madhya Pradesh

Rahul and his sister are farmers from Indore, Madhya Pradesh. He is an engineer by qualification and a first-generation farmer in his family. He quit his job in a multinational company after his mother suffered from cancer to work full-time in the fields along with his sister. In addition to farming, the brother-sister duo run an agro-food company Maa Renuka Foods Private Limited for marketing and selling their products. He considers that lack of access to vegetables grown through circular farming is one of the major causes for cancer today. He discussed several challenges that he encountered during his journey as a farmer at a very young age. One, he said that among a population of 130 crores in India, not more than 10 crore consumers are interested in such products. However, on an optimistic note, the awareness is increasing day by day. Two, he highlighted the role of government to be very critical.

The circular economy ecosystem for agriculture requires policy support because (a) the yield of such crops can be comparatively lower and (b) certification further increases the cost. In the absence of strategic government policies, such farming practices will not scale up the way they should. Third, the farmers are hesitant because of lack of awareness, mental blocks as well as lack of clarity in government initiatives. He shared a personal experience where he tried convincing a farmer to try circular farming only on a part of his land and compare it with the crop yield where urea was used, but he did not even consider trying it. Fourth, in his opinion, creating an ecosystem where the farmers have easy access to the market to sell such products would be very beneficial. He has created a group of more than 400 farmers involved in circular farming diligently and truthfully and they help each other in getting connected to the market. Fifth, he insisted that the mindset is a huge issue in shifting to organic farming. Rather than focusing on quantity, circular farming helps to focus on quality and price. It requires the farmer to have an entrepreneurial mindset to understand this concept.

### 4.4 Case 4: Balbir, Jhunjhunu, Rajasthan

Balbir is engaged in as he suggested "cow-based" farming since 1992 in the Kari village of Nawalgarh sub-district in the Jhunjhunu district of Rajasthan. Balbir comes from the family of farmers but is the first one in his family to apply circular practices to farming. He has a very scientific mindset when it comes to farming and spends a lot of time in experimenting and creating organic fertilizers using plants like neem, satyanashi, banana, amlas, etc. and cow urine. He not only practices circular farming but also cultivates a great deal of zeal in teaching this to others. Apart from the challenges discussed above, i.e., lack of access to knowledge for practicing circular farming, absence of government support and market linkages, and resistance from family members, he shared innovative methods and techniques that he uses in farming. He learnt these methods by experimentation. There was hardly any readymade guidance available and he explored various sources on his own from 1990 to 1992, and his family was clearly unhappy with his decisions. Today, he has taught the principles of circular farming to more than 40 farmers in his own village and has been able to sell wheat and watermelon at fairly profitable prices because of his direct linkage with the consumer base developed by his own efforts. Two main technological interventions described by him were: "cow urine extraction" and "green waste management". He has created a machine-like structure in his farm where cow urine extract is prepared as a nutrient for the plants by mixing it with plant extracts. After boiling these together, the condensed vapor is used for nurturing the plants. Further, he prepares dry and wet manure in his farms for the crops by using cow dung and decomposers, easily available in stores at very affordable prices. He has, thus, created indigenous knowledge which can help many farmers in shifting to circular farming.

## 4.5   Case 5: Akash Chaurasia, Sagar, Madhya Pradesh

Akash Chaurasia is a farmer from Sagar in Bundelkhand district of Madhya Pradesh. He is widely known for multi-level farming in his area and conducts training programs on the same. He strongly supports cow-based and circular farming and propagates about its benefits to the farmers. Several insights were collected by the authors during the visit to his farm. One, like many others, he wishes to see the government playing a more active and significant role in promoting organic farming. According to his experience, the farmers do not get the right prices for their produce because there is lack of systematic planning. It is a challenge to identify the crops to be grown so that neither the farmer bears any loss because of too much quantity nor there is shortage of supply in the market. In both the scenarios, the issue is lack of a systematic approach. Two, he discussed how farming in itself is a very sustainable activity that incorporates waste, water and land management. The issue is not about making farming sustainable but encouraging farmers to stick to a more rooted way of carrying on farming; traditionally whatever they have been doing is sustainable in itself. Third, he highlighted the significance of organic tourism and generating additional income from it. It can also help in positioning India as a favorable tourist destination. Fourth, he shared how farming can be adopted by one and all by taking small steps and cultivating vegetables in their kitchen gardens. He also talked about collaborative and crowd funding models where common people can contribute money to the farmers in helping them raise crops and get vegetables and fruits in return.

## 4.6   Case 6: Om Prakash, Sikar, Rajasthan

Om Prakash is a farmer based in a small village near Kolida in Sikar, Rajasthan. He practices natural and circular farming and is also imparting training about it to the farmers through the Krishi Vigyan Kendra[3] in his area. He started farming at the age of 17 and was using chemicals in his fields for a very long time. When he read that chemicals have contaminated our food to such an extent that even mother's milk is laced with chemicals, he made up his mind to shift to circular farming. He strongly follows natural farming methods in his fields which he has learnt by gathering knowledge on his own and experimenting. He has also spent a lot of time in designing his fields in way that micro-organisms thrive in the soil. His fields have high rise structures and trees at the boundaries for this reason. According to him, by shifting to circular farming, the input cost can be reduced which will in turn benefit the farmer even if the yield is marginally affected. In this kind of farming, the soil starts replenishing good bacteria on its own that provide nutrition to the crops. He insisted that farmers must find their own way out through a curious mindset and experimentation. To him, his field is his laboratory. He uses elements like gajarghaas (carrot grass), green manure water, cow buttermilk and urine in growing crops without

---

[3] https://kvk.icar.gov.in/.

any chemicals in his farms. Most of this knowledge has been created by him through experimentation in his farms. Year after year he also collected seeds and created his own seed bank.

## 4.7   Case 7: Vaishali, Khandwa, Madhya Pradesh

Vaishali from Indore does not come from a family of farmers. After completing her education, she was working as an HR manager in Ethiopia. She came back to India deeply affected by her mother's cancer and pledged to work for the cause of circular farming. She was encouraged by her family. With her determination and hard work, she not only became a successful farmer but today she also heads the Kisan Mandal, conducts training programs and works for women empowerment. Her farmland runs over 10 acres in Khandwa, Madhya Pradesh. Some of the key issues described by her are as follows. One, like all other respondents, she faced the problem of lack of knowledge and access to circular farming practices. She learnt by watching videos and meeting Subhash Palikar. Two, she was living alone 200 km away from her house. She needed assistance and for this purpose, she started training the women around who came to her rescue. She feels that being a woman is not a weakness but a strength as it helped her gather support from rural women. She has not only created her own venture but has provided opportunities to other women. Three, through her experiences she has learnt that "natural farming" is the best way forward where cows and not chemicals form the basis of agriculture. The soil is to be enriched with micro-organisms through cow urine and cow dung; these are "magical" for the crops. Fourth, market linkages are very important for circular economy. She learnt it the hard way as her initial produce got damaged because of lack of access to the market. Fifth, she strongly advocates for the farmers to create their own seed bank. If the farmer is self-sufficient and not dependent on the market, it will improve their financial conditions to the great extent.

## 4.8   Swati, Sagar, Madhya Pradesh

Swati Singhai is based in Sagar district of Madhya Pradesh and is a role model for her villagers who want to learn circular farming. She is not only a farmer but also fulfills the roles of being a mother, a daughter and a wife. She has a work experience of more than six years in the IT sector and has lived in cities like Pune, Delhi, Noida, Kolkata and Mumbai. After her daughter started developing respiratory problems, she shifted to her birthplace Sagar and started working toward becoming a farmer. Although she comes from a family of famers, she is the first-generation circular farmer in her family. She tried to gather knowledge and ended up attending the training workshop of Akash Chaurasia (Case 5) and learnt about circular and multi-layer farming. She particularly highlighted the challenge of gender bias she faced whereby people thought that it will be very difficult for her being a young woman and a mother to venture into an unknown area like circular farming. She has met all the challenges with great conviction and is carving a niche for herself in the domain.

## 5 Discussion

Data collected from the interviews and the field visits was read and re-read to identify key issues based on a reflexive approach [18]. These issues reflect line by line analysis of the interview transcriptions and have been divided into two broad categories—antecedents and consequences (see Fig. 3). Further, the antecedents are of three types—personal, social and systemic. This section provides an explanation of the antecedents and the consequences that form the basis of recommendations and future research directions in this area.

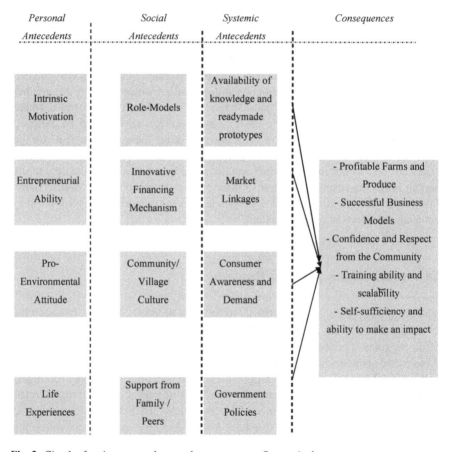

**Fig. 3** Circular farming: antecedents and consequences. *Source* Authors

## 5.1  Antecedents

### 5.1.1  Personal

Antecedents identified at personal level include intrinsic motivation, entrepreneurial ability, pro-environmental attitude and life experiences. It was observed in the case of all the respondents, without exception, that they had great deal of passion and eagerness to learn and practice the art and science of circular farming. Although some of them came from a farming background, all of them are first-generation farmers engaged in circular farming. Further, all of them exhibited an entrepreneurial ability to not only identify sources of learning but also implement innovative and indigenous technologies in their farms and generate market linkages. This was also observed in the methodical planning and usage of farmland by them. Some had developed equipment and machinery to assist circular farming on their own. It also emerged during the discussions, that the respondents took great interest in pro-environmental activities and their own life experiences that contributed to their association with circular farming.

### 5.1.2  Social

At a social level, antecedents have been divided into four categories-availability of role models, innovative finance mechanism, village or community culture and peer support. All the respondents made regular visits to successful farmers/role models practicing circular farming across the country. It came out of the discussions that unlike conventional farming, there are hardly any easily accessible role models in the sphere of circular farming. There are only few big names scattered all over the country. Thus, even with a genuine intention to learn and practice circular farming, this societal issue acts as an inhibitor. Another important aspect is the availability of finance. Majority of the respondents had to face hardship in the initial years of farming. As the knowledge of circular farming is still not widely available, there are possible chances of losses in the initial years because farmers have to adopt hit and trial methods. Thus, the farmers struggling for survival may not enter this zone. Innovative financing mechanisms like crowd funding, self-help groups at community level, profit sharing ventures, and value co-creation models with consumers, etc. can be helpful. Culture at the community level also emerged as a significant social factor. Demonstration effect works effectively in rural areas. In case of the most of the respondents, each circular farmer in the village inspired many others like him or her. The last social factor is family/peer support, as in villages family unity as well as community feeling is very strong. Both these factors, if present can make the life of farmers engaged in circular farming comparatively easy.

### 5.1.3 Systemic

Availability of knowledge and readymade prototypes, market linkages, consumer awareness and demand, and government policies have been identified as the systemic factors. As observed during the field visits, each respondent faced a lot of difficulty in reaching the right source to learn about circular farming. Further, there is satisfactory knowledge base available for understanding problems in chemical-based traditional farming but no platform or database is available for exploring solutions for problems in circular farming. Eventually, all the respondents learnt by doing themselves. As the farmer is not getting the right price for the produce; thus, farmer community as a whole is in dire need of proper market linkages in India. This aspect becomes very essential in case of circular farming as the farmer is going an extra mile for producing environmental friendly products. All of the respondents have marketed themselves to develop the linkages in their nearby areas directly with the consumers. This gives rise to another antecedent which is consumer awareness. Without consumer awareness that generates demand for such products, circular economy cannot flourish. The consumers must demand and must be ready to pay for products produced organically, without chemicals. This requires an ecosystem created by the government through right kind of policies to promote circular farming. Policies which support farmers, create platforms, logistic support and infrastructure for them are essential. Consumer awareness programs and financial support can also be helpful.

## 5.2 Consequences

If the enablers or antecedents exist, the circular farming can create several benefits for the farmers and society. First and foremost, as shared by the respondents, circular farming has the potential to create profitable opportunities for the farmers unlike the common perception that it does not yield profits. Successful circular farmers have created business out of their farmlands by reducing input cost, and generating market linkages. It has helped them position themselves not only as pro-environmental activists but also enabled them to become role models and entrepreneurs for their co-villagers. Many of them have become trainers who now provide knowledge and skills of circular farming to many of their fellow farmers. They have earned their community's confidence and respect. Finally, their entrepreneurial spirit has made them self-sufficient and has provided them with ample opportunities to make social and environmental impacts.

# 6 Conclusion, Implications, Limitations and Future Directions

This chapter discusses circular farming practices and the associated challenges through case studies of a sample of eight farmers from north India. These practices are circular because they ingrain the principles of reducing, reusing, recycling and restoring nutrients, water and energy in cultivating crops. Optimization of resources, waste and water management, reduced carbon emissions and providing healthy food options to a growing population are some of the main benefits of such farming practices. However, several challenges come in the way of adoption of circular farming which have been identified and explained in this chapter.

## 6.1 Implications

The study provides useful insights for stakeholder interested in the area of circular farming. Specifically, it can guide future researchers on identifying challenges for farmers in adoption of the circular farming practices in India. This may result in robust practical solutions and provide basis for policy recommendations. In this context, some important future research and policy-related questions emerging from the analysis are as follows:

- What is the relative significance/weightage of the antecedents in circular farming?
- What are the best business models developed by circular farmers?
- How can technology enable market linkages for circular farmers?
- What are the best policy recommendations for making the existing farmers shift to circular farming?
- How can we connect circular farming with the sustainable development goals?
- How can circular farming be extended to sustainable agri-business supply chains?
- How can circular farmers be connected to eco-labeling and sustainable packaging practices?
- How can large-scale industry further the cause of circular farming?
- How can MSMEs engaged in circular farming become scalable and more profitable?
- How can circular farmers contribute to green product designing and development?
- How can latest concepts like blockchain help in certification of goods produced through circular farming?

## 6.2 Limitations and Future Research Directions

The sample size was very small and the research was exploratory. Further studies based on larger sample sizes and ethnographic studies or descriptive surveys can

**Table 2** Limitations and future research directions

| Limitation | Direction for future research |
| --- | --- |
| Small samples size | Studies with larger sample sizes |
| Exploratory study | Descriptive studies to test hypothesis formed on the basis of the findings emerging from this study |
| Limited to few areas of Rajasthan and Madhya Pradesh | Studies in different cultural, geographical, and economic contexts can be conducted |
| Based on cross-sectional data | Longitudinal studies to measure impact over a time-span |
| Based on only case studies | Ethnographic studies could help in understanding ground-level issues in further depth |

*Source* Authors

provide more insights for generalizability. The research was also limited to few areas in Rajasthan and Madhya Pradesh. Different cultural, economic, and geographical contexts can be explored in future studies. Data collected was cross-sectional. Longitudinal studies can help in measuring impact over time. The limitations of case study methodology also apply. In-depth exploration of issues can be conducted in future through qualitative methods like ethnography or phenomenology. These limitations and future research directions have also been mapped in Table 2.

# References

1. World Development Indicators (2015) The World Bank
2. Ways to boost Indian agriculture (2015) World Economic Forum. https://www.weforum.org/agenda/2015/11/3-ways-to-boost-indian-agriculture
3. Reversing land degradation in India (2020) TERI. https://www.teriin.org/article/reversing-land-degradation-india
4. Agriculture, forestry, and fishing, value added (% of GDP)—India|Data (2021) The World Bank. https://data.worldbank.org/indicator/NV.AGR.TOTL.ZS?end=2019&locations=IN&start=1960&view=chart
5. Census (2011)
6. GDP per capita (current US$)—India|Data (2019) The World Bank. https://data.worldbank.org/indicator/NY.GDP.PCAP.CD?locations=IN
7. Dantwala ML (1976) Agricultural policy in india since independence. Agricult Food Policy Product Econ 276–295
8. Nelson ARLE, Ravichandran K, Antony U (2019) The impact of the Green Revolution on indigenous crops of India. J Ethnic Food 6(1). https://doi.org/10.1186/s42779-019-0011-9
9. Chourasiya S, Khillare PS, Jyethi DS (2015) Health risk assessment of organochlorine pesticide exposure through dietary intake of vegetables grown in the periurban sites of Delhi, India. Environ Sci Pollut Res 22(8):5793–5806. https://doi.org/10.1007/s11356-014-3791-x
10. Prasad R (2005) Organic farming vis-à-vis modern agriculture. Curr Sci 89(2):252–254
11. National Policy for Management of Crop Residues (2014)

12. Toop TA, Ward S, Oldfield T, Hull M, Kirby ME, Theodorou MK (2017) AgroCycle—developing a circular economy in agriculture. Energy Procedia 123:76–80. https://doi.org/10.1016/j.egypro.2017.07.269
13. Earles R (2005) Sustainable agriculture: an introduction. ATTRA, the National Sustainable Agriculture Information Service
14. Agriculture Census 2015 (2016)
15. Kirchherr J, Reike D, Hekkert M (2017) Conceptualizing the circular economy: an analysis of 114 definitions. Resour Conserv Recycl 127:221–232. https://doi.org/10.1016/j.resconrec.2017.09.005
16. Tang H, Qiu J, Ranst EV, Li C (2006) Estimations of soil organic carbon storage in cropland of China based on DNDC model. Geoderma 134:1–2. https://doi.org/10.1016/j.geoderma.2005.10.005
17. Rosemarin A, Macura B, Carolus J, Barquet K, Ek F, Järnberg L, Lorick D, Johannesdottir S, Pedersen SM, Koskiaho J, Haddaway NR, Okruszko T (2020) Circular nutrient solutions for agriculture and wastewater—a review of technologies and practices. Curr Opin Environ Sustainability 45:78–91. https://doi.org/10.1016/j.cosust.2020.09.007
18. Alvesson M, Skoldberg K (2000) Reflexive methodology: new vistas for qualitative research. Sage, London. https://doi.org/10.1177/002194360203900107

# Transition Toward a Circular Economy Through Surplus Food Management

**Anuj Mittal, Caroline C. Krejci, Michi Lopez, Jake Kundert, Carter Oswood, and Jason Grimm**

**Abstract** Most surplus food goes into landfills, which is a major environmental concern. Reducing waste at the source and rescuing edible surplus food and delivering it to food-insecure people can help reduce the environmental impact of surplus food. Reduction at the source refers to reducing the amount of surplus food generated by the food supply system. Through food rescue, edible surplus food is collected and delivered to food-insecure people, thereby reducing both the amount of food being diverted to landfills and food insecurity. This chapter describes four case studies that demonstrate the importance of local food systems in reducing food waste at the source and the role of non-profit organizations in rescuing surplus food from farms, grocery stores, restaurants, and institutional food service, all of which contribute a large volume of food waste. The contribution of local food systems and food rescue organizations in transitioning toward a circular economy and promoting U.N. Sustainable Development Goals for 2030 is discussed. Challenges faced by local food systems and non-profits in rescuing food and potential solutions to overcome

A. Mittal (✉)
Department of Industrial Engineering Technology, School of Engineering, Dunwoody College of Technology, 818 Dunwoody Blvd, Minneapolis, MN 55403, USA
e-mail: amittal@dunwoody.edu

C. C. Krejci
Department of Industrial, Manufacturing, and Systems Engineering, The University of Texas at Arlington, Box 19017, Arlington, TX 76019, USA
e-mail: caroline.krejci@uta.edu

M. Lopez · J. Kundert · J. Grimm
Iowa Valley Resource Conservation and Development, 920 48th Avenue, Amana, IA 52203, USA
e-mail: michi@ivrcd.org

J. Kundert
e-mail: jake@ivrcd.org

J. Grimm
e-mail: jason@ivrcd.org

C. Oswood
Feed Iowa First, PO Box 1190, Cedar Rapids, IA 52406, USA
e-mail: carter@feediowa1st.com

© The Author(s), under exclusive license to Springer Nature Singapore Pte Ltd. 2021       245
R. S. Mor et al. (eds.), *Challenges and Opportunities of Circular Economy in Agri-Food Sector*, Environmental Footprints and Eco-design of Products and Processes,
https://doi.org/10.1007/978-981-16-3791-9_13

them with the help of information and communication technology (ICT) are also discussed.

**Keywords** Circular economy · Food rescue · Food waste · Decentralized supply chains · Crowd logistics · Information and communication technology · ICT · U.N. Sustainable Development Goals

# 1 Introduction

The US food supply chain exemplifies the linear economy, in which unused products and products at the end of their useful life are thrown away. This is a major concern for the environment. The US Department of Agriculture (USDA) estimates that 31% of the 430 billion pounds of available food supply at retail and consumer levels in the USA goes uneaten each year [1]. This results in 30.3 million tons of food waste sent to landfills annually, which accounts for nearly 18% percent of anthropogenic methane emissions in the USA [2], making food waste a major contributor to climate change. Furthermore, 18–28% of cropland in the USA is used to grow food that is ultimately not eaten [3], which is a needless and enormous environmental burden, since agricultural activities in the USA tend to be ecologically intensive. For example, runoff from agriculture lands is a major cause of water pollution in inland and coastal waters [4]. Likewise, more than 25% of total freshwater in the USA is used to produce food that is eventually wasted [5].

Food waste occurs at every level of the food supply chain. In the USA, the majority of food waste occurs at the household level (43%), followed by restaurants (18%), farms (16%), grocery and distribution (13%), institutional food service (8%), and manufacturers (2%) [3]. Crops may be wasted due to inaccurate demand forecasts, labor shortages, and products that are edible but fail to meet strict color, size, and quality standards for major markets [6]. In addition, many farmers intentionally and understandably overproduce to mitigate the risk of low or imperfect yields. Grocery stores generate 10.5 million tons of surplus food each year in the U.S [7]. They carry an enormous variety of products so that they can offer customers choice and convenience, but this presents a challenge for accurate forecasting and inventory management, and subsequently leads to food waste [7]. Perishable products, including produce, meat, seafood, ready-made food, and dairy, constitute the largest percentage of grocery food waste, with grocery stores typically discarding packaged foods two to three days before their expiry dates [3]. Restaurants are another major source of food waste in the USA, generating 11.4 million tons each year, of which 390,000 tons could be recovered to yield 643 million meals [8]. However, fewer than 5% of the more than 1 million restaurants in the USA donate surplus food [9]. A major barrier is logistics: because the restaurant sector consists of many independent locations with relatively small volumes of reusable food per location, efficient collection and distribution of surplus food is challenging [10].

Transitioning food supply chains to reflect a circular economy would reduce the environmental impacts of food waste. Accordingly, the US Environmental Protection Agency (EPA) has recommended various approaches to prevent and divert food waste, such as reducing the volume of surplus food generated, donating edible surplus food to hungry people, feeding animals, industrial processing of food to generate energy, and composting [11]. Among these, the US EPA considers reducing food waste at the source and rescuing edible surplus food and delivering it to food-insecure people as the most preferred approaches to managing surplus food. Preventing food from becoming waste in the first place, i.e., at the source, offers the greatest financial and environmental benefits. From the financial perspective, it reduces the cost of purchasing, handling, and ultimately disposing food that is not consumed. From the environmental perspective, it helps avoid the use of water, energy, fertilizers, and other resources used to produce, process, transport, package, and dispose of food that is not eaten, as well as reduce the greenhouse gas emissions resulting from sending that food to the landfills. One approach to reducing food waste at the source is to align food production with consumer demand.

However, surplus food can also be used to address food insecurity. Food insecurity in the USA is a serious humanitarian concern, with 10.5% of US households lacking consistent access to sufficient nutritious food [12]. To address this problem, the USDA supports multiple initiatives, including food distribution programs, child nutrition programs, the Supplemental Nutrition Assistance Program (SNAP), and a special SNAP for women, infants, and children (WIC). However, one-third of food-insecure individuals do not qualify for federal assistance because they do not meet the income requirements [13]. Much of the surplus food generated by farms, grocery stores, and restaurants is still suitable for human consumption. Rescuing this surplus edible food and diverting it to food-insecure people will not only help mitigate environmental impact of food waste but will also help address the problem of food insecurity. Non-profits play a pivotal role in managing surplus food to be rescued from donors and delivering it to hunger relief agencies with the help of volunteers.

This chapter describes four case studies that demonstrate the importance of local food systems in reducing food waste at the source and non-profit organizations in rescuing surplus food from farms, grocery stores, restaurants, and institutional food services. Through these case studies, the advantages of decentralized supply chains in managing surplus food, with the help of information and communication technology (ICT), is discussed. In addition, contribution of local food systems and food rescue organizations in promoting U.N. Sustainable Development Goals (SDGs) for 2030, especially of zero hunger (SDG 2), decent work and economic growth (SDG 8), responsible consumption and production (SDG 12), and climate action (SDG 13), is also discussed.

## 2   Case Studies

### 2.1   Grow: Johnson County

Grow: Johnson County (hereinafter referred to as Grow), is a hunger relief and educational vegetable farm operating on five acres of land located at the Johnson County Historic Poor Farm in Iowa City, Iowa. The farm is managed by Iowa Valley Resource Conservation and Development (IVRCD), a non-profit located in Amana, Iowa. Grow's mission is to improve healthy food access through charitable food production and hands-on education. To combat food insecurity in the Johnson County, Grow freely provides fresh, organic produce to hunger relief agencies such as shelters, neighborhood centers, food pantries, and free meal sites. The farm also leads an educational program to empower community members of all ages and backgrounds to grow their own food.

The Grow farm began distributing produce in 2016 on two acres and distributed 12,000 pounds of produce to 9 partner agencies. In 2016, the selection of crops to be grown and distributed was decided based on staff experience, the land available, and equipment access. All the harvested food was distributed by Table To Table (a non-profit delivery partner) to the partner agencies. Evaluation was conducted through informal conversations with partner agency staff to identify what were the crops and quantities that met the need of the community. Overall, the consensus was that each agency wanted more produce: the clientele loved being able to freely access organic produce.

By 2018, the farm expanded its operations to four acres and distributed 20,000 pounds of produce with a market value of $45,000 to 15 partner agencies. However, in 2018, partner agencies began informing the Grow staff that some of the distributed food was going to waste. At that point, distributions were not tailored to each receiving agency, but instead partner agencies were able to pick the items and quantities they wanted from whatever had been sent out each day. This meant that the first agency on each route got first pick, the second site got second pick, and so on, until the last site was left with the less desirable items. These foods might have been less desirable to an agency because of limited refrigerated space, limited processing capacity, or clientele preference. For example, one day the last agency on a delivery route was hoping to get kale and cutting greens that they could use for green smoothies to give the children at their free meal site. When Table To Table delivery van finally arrived, the only items left were turnips. The agency, which had a pre-approved menu that they must stick to, was forced to turn away the turnips. On another occasion in 2018, Napa cabbage was distributed to a partner agency with the understanding that all fresh produce is desirable. After a couple of days, a significant quantity of the cabbage remained on the shelves and ended up getting composted.

Farm staff knew they had to implement a more rigorous data collection process regarding what food items to grow. Input from the consumers of Grow's produce would not only aid the farm in planning the next growing year, but also potentially reduce food waste. Staff at partner agencies began asking their clientele directly or by providing surveys to determine what types of food and in what quantity Grow could provide. This direct approach did not bring forth the best results, as clientele were wary of providing any "negative" feedback so as to not seem ungrateful for the food they were receiving. However, in less formal discussions with agency staff, clientele would very openly express their preferences. To collect the desired feedback, the collection process must address the barrier of direct communication.

IVRCD and Running Robots, a local technology company located in Iowa City, Iowa, developed a survey tool in 2019, named Voice Your Choice (VYC), that would collect clientele preferences without having to give "negative" feedback. It was developed with an objective to allow Grow to embrace "food sovereignty" within their operations. Food sovereignty is defined as the right of people to healthy and culturally appropriate food produced through ecologically sound and sustainable methods, and their right to define their own food and agriculture systems [14]. It not only makes sure that people have access to food but emphasizes building a relationship between consumers, producers, and the land.

VYC is an iPad-based survey tool that allows agencies to gain valuable food preference information from their clients in a non-intrusive way. The iPad with VYC survey is placed in a secure, standing kiosk. The kiosk is then left at participating partner sites and strategically placed in high-traffic areas. The survey is then filled out by the recipients whose responses were recorded and stored on Google Sheets. The survey is imaged-based and uses limited text; however, instructions are available in seven languages. Language options were selected to cater to the diversity of the recipient population and informed by partner agencies. The main component of the survey asks the recipients to simply select the image of produce they enjoy and submit their responses. Figure 1 shows a snapshot of the survey tool in which consumers can tap on the different vegetable pictures to enter their preferences.

The survey was piloted in January 2020 at five agency partner sites for 2–4 weeks and 160 responses in five languages were recorded and used to inform the crop planning decisions for the 2020 season. For example, onions were among the top ten most requested items of the survey, so from 2019 to 2020, onion production increased by 177.5%, produce such as Napa cabbage and Brussels sprouts were eliminated, and new crops like cauliflower were planted for the first time.

Not only were the Grow farmers able to collect overall data to inform the master growing plan, but because the iPad had been connected to Wi-Fi at the partner agencies, data was able to be sorted by IP address, providing each site's preferences which allowed for more tailored distribution. Using the VYC data, and information about the partner sites' capacity from partner agency staff, predetermined quantities of a variety of foods are sent out to each site. So, if Site One on a particular route only has a strong preference for cabbages and little preference for carrots and Site Two has less preference for cabbage and more for carrots, then the Grow staff can direct the appropriate amounts of each crop to match those preferences.

VYC has helped in reducing food waste at partner agency sites. In 2019, before the VYC tool was implemented, three partner agencies reported to IVRCD about composting excess produce. In 2020, no agencies reported composting excess produce. With such success in such a short window of time, VYC was continued

## Step 2 of 3 - We Value Your Feedback

**Fig. 1** Snapshot of VYC tool showing how consumers can tap on the different vegetables pictures to enter their preferences. *Source* Author

and is currently still being used by the Grow farm, as well as a partner organization, Feed Iowa First. Both farming programs want to know what the consumers want to eat, so that each organization can produce food that matches the preferences of the communities being served. While this tool was developed to allow these two organizations to grow the food that consumers want to eat, it also is translatable to inform food assistance agencies' decisions on what items to make available for clients. A similar survey could also include dairy and meat options, dry goods, or canned products.

## 2.2  Feed Iowa First

Feed Iowa First (FIF), a non-profit organization located in Cedar Rapids, Iowa, rescues produce from the farmers that does not reach the supermarket shelves through two different methods: gleaning, which is defined as extraction of economically non-viable material, and, by aggregating already harvested cosmetically imperfect food from the farmers. All the food rescued is Non-GMO and free of chemical fertilizers. Through their program called "Don't Waste Donate", FIF works with partner farmers and hunger relief agencies to redirect excess produce to Linn County residents in Iowa that lack access to fresh locally grown and chemical-free produce.

Excess crops at the farm are typically either left to die or are tilled back in because of poor demand, high yielding seasons, and/or poor crop planning. Farmers choose to abandon these crops in the field to save the expense and labor involved in harvesting less than perfect produce, or due to lack of storage space available at the farm. FIF coordinates with a network of volunteers that glean this crop from the farm. Upon gleaning, the crop is brought to the FIF warehouse, where it is weighed, washed, packed, stored and distributed directly to the families in need, or through their partner hunger relief agencies. For example, FIF captured over 12,000 pounds of onions from a farm in Iowa in 2020. FIF is now a part of the Iowa Gleaning Network, which was established as part of the Feeding Iowans Task Force during the COVID-19 crisis to help Iowa's hunger relief agencies fill gaps in existing gleaning programs, and launch new gleaning efforts across the state, with the help of AmeriCorps members as gleaning coordinators. The network has been established to help coordinate gleaning efforts across the state of Iowa by connecting volunteers who are willing to donate time to farmers who are willing to donate food.

Another method of rescuing food by FIF is through aggregating extra harvested produce by farmers. For example, if a farmer grows only a particular crop for a grocery chain, they have to meet particular standards with respect to size, shape and color. The grocery chain generally does not accept anything that is a little too big, too small, or too colorful. In this method of rescuing food from farms, partner farms on their own bring this cosmetically imperfect food meant for the supermarket shelves to FIF. FIF aggregates, wash, pack and distributes this food direct to families, or through their partner hunger relief agencies. Many farmers actively donate cosmetically imperfect produce to FIF. In one such instance, one of the partner farms of FIF, Jupiter Ridge Farm, went around at the end of the Dubuque Farmers Market to ask farmers if they were left with any unsold produce and want to donate. After aggregating produce from the farmers in the market, Jupiter Ridge Farm brought the produce to FIF, where it was sorted, washed, packed, and distributed direct to families and through their partner agencies.

FIF distributes food in the neighborhoods on Wednesday, Thursday, and Friday of each week. At each of their distribution points, FIF sets up a market and encourage recipients to choose the type and quantity of produce they want at no cost with no questions asked. They do not distribute food in a box, which might lead to more food waste, as recipients will only consume what they need and know how to prepare. The food which is not picked up by the recipients is further directed to be donated to social service agencies (e.g., Green Square Meals, located in Cedar Rapids, Iowa), which process the produce to prepare meals and further help people in the need of fresh and healthy food.

Over the years, FIF has established a strong network of local farmers from where they glean produce from the farm, as well as collaborate with them to donate cosmetically imperfect food. Establishing this network of farmers is critical and FIF employs different strategies to collaborate with more local farms. For example, one of the farmers uses FIF cold storage facility and in return donates a part of his farm's produce. Another farmer uses the FIF facility for her CSA stand and donates a couple of CSA boxes through FIF each week. In 2019, FIF donated a total of 7000 pounds of food through gleaning and aggregating produce from farmers' donations. However, this number went to 25,710 pounds in 2020.

The major challenge for gleaning is getting volunteers to a farm, which can sometimes be located far from the city. Before the COVID-19 pandemic, FIF worked with volunteers to meet at a common location, from where it coordinated the transportation to the farm. However, during the pandemic, volunteers reach the farm by themselves where gleaning operations need to be performed. Thus, ensuring appropriate number of volunteers available at farms located far away from the city is sometimes challenging for FIF.

FIF is also deploying the VYC survey tool as used by the Grow farm (discussed in the above case study), to understand the needs of their pantries and recipients in the neighborhoods. This will help them direct the foods to different neighborhoods and pantries based on their individual needs and preferences. Currently, FIF is also using different GIS datasets (e.g., food insecurity rate in a region, availability of grocery stores, and bus routes) to generate a map of most food-insecure region in Cedar Rapids, Iowa. This will help determine locations of new access points, where a greater number of food-insecure people can get fresh and healthy food through FIF.

## 2.3 Food Rescue US

Food Rescue US is a non-profit organization, offering collection of surplus food from donors and delivery to local hunger relief agencies. Food Rescue US partners with restaurants, grocery stores, hospitals, corporate cafeterias, and any other food establishment motivated to participate in food donation. Privately owned restaurants tend to donate food after larger-scale events, such as Sunday brunches, that would otherwise be wasted and will end up in landfills. Food donations help restaurants meet their sustainability goals by reducing the amount of food that they send to landfills, reducing their waste disposal fees as well as getting financial benefits in the form of tax credits [15]. Sole proprietorships, partnerships, corporations, and limited liability companies can all benefit from federal tax credits by donating surplus food to charitable organizations, as long as it is "apparently wholesome" [16]. The Bill Emerson Good Samaritan Act defines "apparently wholesome" donations as "food that meets all quality and labeling standards imposed by federal, state, and local laws and regulations even though the food may not be readily marketable due to appearance, age, freshness, grade, size, surplus, or other conditions" [17]. In addition, the Emerson Act protects donors participating in food donation with good faith from civil and criminal liability.

The recipients of food donation include but not limited to food pantries, soup kitchens, homeless shelters, and senior centers. The highest demand is on perishable food [18]. Cooked and ready-to-eat foods are sent to homeless shelters and senior centers rather than pantries. When there is a new establishment motivated for donation, pickup is scheduled for one day in a week [18]. As the establishment gets familiar with the donation process, Food Rescue US tries to gradually increase the frequency of donations.

Food Rescue US provides the technology platform (a mobile application) for food rescuers to directly transfer excess food from donors to local social service agencies that feed the food insecure. Donors use cardboard cartons or aluminum trays to pack food. The mobile application lists food rescue opportunities with a daily, weekly, and monthly schedule. Volunteers simply log on to the mobile application and sign up for a rescue. The application provides all details of every rescue including how much food there will be, where to pick up and deliver and who to contact, among others. Food rescue trips are designed such that the total time from pickup to delivery is around 30 minutes and the time food spends in non-refrigerated vehicles is minimized due to food safety concerns. In case of a large amount of food donation, volunteers may make more than one trip, or they deliver as much as they can. Occasionally, more than one volunteer can be assigned to one pickup location.

The model adopted by Food Rescue US to rescue food is also usually referred to as a grocery-to-home model where the time it takes for a volunteer to pick up and deliver the donation is approximately equal to the time one will spend for grocery shopping [18]. Food Rescue US does not hire employees for pickups and deliveries and depends on volunteers to use their personal vehicles to pick up the deliveries from the donation sites and deliver it to the recipients. Thus, Food Rescue US heavily depends on volunteer participation to sustain its operations. Although large group of people participate, regular volunteers generally include senior citizens, high income non-working women, and stay-at-home mothers. Couples with young kids, members of the churches and people with an obligation to do public service are among other group of volunteers. Food Rescue US does not provide any type of incentives to initiate or maintain the participation of volunteers.

For-profit organizations like Door Dash are also leveraging their existing logistics network to deliver food from restaurants to hunger relief agencies [19]. Door Dash launched a program called Project Dash in partnership with a non-profit organization, Feeding America. Restaurants who have surplus edible food available to donate could use Meal Connect, a mobile application developed by Feeding America to provide information about the food. The mobile application pairs the donation with a nearby hunger relief agency and Door Dash uses their existing drivers to fulfill the delivery.

## 2.4 Second Harvest Heartland

Second Harvest Heartland (SHH), a food bank located in the greater Minneapolis-Saint Paul region in Minnesota, is a member of Feeding America—the nationwide network of more than 200 food banks in the USA. Food banks act as food storage and distribution depots for smaller front line agencies; and usually do not distribute food directly to people struggling with hunger. SHH partners with nearly 1,000 food shelves, group meal centers, shelters, and other programs to help distribute food directly to food-insecure people [20]. These agency partners provide meals for food-insecure people across 59 counties in Minnesota and western Wisconsin [20]. SHH rescue food from donors including retail stores, manufacturers, distributors, and farmers. In addition, SHH also receives food donation through government programs as well as purchases it to address food insecurity in their region.

Food rescue through retail grocery stores constituted highest percentage of food donated by SHH in 2019. SHH rescues food from around 276 national retailers in Minnesota and Wisconsin which are in the network of Feeding America. SHH also partners with regional retailers in the region to rescue food. However, as of 2019, SHH rescued food from only about six percent of the total regional retailers. Logistics is one of the major challenges for SHH to rescue food and deliver it to their agency partners. SHH has a fleet of refrigerated trucks, which currently delivers approximately 40% of the food to their agency partners. For example, these trucks travel from the SHH warehouse to pick up surplus food at the grocery stores and deliver it to their agency partners. The remaining 60% of the distributed food is picked up by the agency partners themselves in coordination with SHH. Figure 2 shows the location of the two SHH's warehouses, their agency partners, and retail grocery stores from which they rescue food.

In the fall of 2019, SHH partnered with Dunwoody College of Technology to use data for developing effective strategies to improve their current distribution methods that would ultimately help SHH to rescue greater amount of food from both national and regional grocery stores within their region [21]. The partnership was established as a result of a capstone project that a group of students within the Industrial Engineering Technology program at Dunwoody College of Technology did with SHH. In particular, SHH aimed to meet the demand of food within their serving region locally by their agency partners themselves and/or by developing new distribution infrastructure. This would also help agency partners to increase utilization of their existing distribution infrastructure by rescuing food from donors within few miles from their facility, thereby, also reducing dependence on SHH fleet of trucks to deliver food to them.

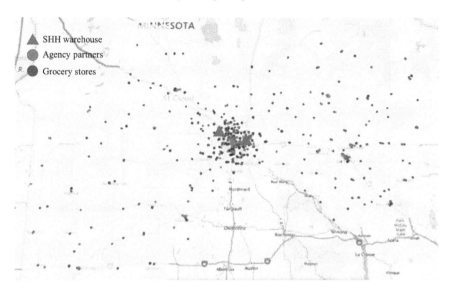

**Fig. 2** Map showing the SHH serving region (in blue), location of their two warehouses, agency partners, and grocery stores from which they rescue food. *Source* Author

Three different datasets were used to better understand the geographic relationship between demand, supply, and existing distribution capacity of food within the SHH service region. Specifically, this included analysis of demand of food (food insecurity rate), food potentially available for donation by national and regional retailers, and the existing distribution capacities of their agency partners. Population and food insecurity rate within the SHH region as estimated by Feeding America were used to estimate pounds of food needed within each census tract in the SHH service region [22]. Potential food available for donation by national retailers was provided through Feeding America for the SHH region. However, SHH also had data on total annual food sales (in pounds) and the amount of food donated (in pounds) by the regional stores they were already working with. A relationship between store size, based on total pounds of food sold, and their subsequent food donation percentage was identified. It was found that smaller stores tend to donate a larger percentage of their overall sales, while larger stores donate a much smaller percentage. This relationship, upon verification from SHH, was applied to the regional stores in SHH's region from which they were not receiving donations, to estimate the total potential food supply from national and regional stores within each census tract in their region. Pounds of food distributed each year by the agency partners was used as a proxy for assessing their current distribution capacities.

After aggregating the three datasets, key regions were identified for making recommendations to SHH with respect to distribution capacity improvements. First region consisted of five counties (Big Stone, Swift, Lac qui Parle, Chippewa, and Yellow Medicine) in the Minnesota region (Fig. 3) which have excess of food available to donate but only few agency partners in the region to be able to rescue and distribute

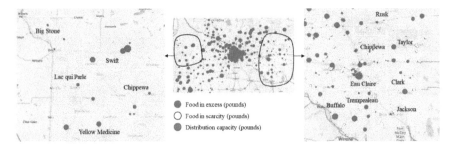

**Fig. 3** Regions in Minnesota and Wisconsin identified for distribution capacity improvements within the SHH serving region. *Source* Author

food. Red and white bubbles in the figure represent pounds of food available in excess and scarcity, respectively at each location (bigger the size of the bubble, greater the amount of food in pounds). The size of the blue bubbles represents the distribution capacity (in pounds) of agency partners. A similar observation was made for the second region consisting of Renville and Sibley County in Minnesota. The third region represented a group of eight counties in Wisconsin (Rusk, Taylor, Chippewa, Clark, Eau Claire, Jackson, Trempealeau, and Buffalo) as shown in Fig. 3. This region had excess supply in some places, and excess demand in others, however, the capacity of agency partners was not found to be enough to meet the gap.

As the first two regions identified in Minnesota lack availability of agency partners, distribution capacity improvements could be achieved either by partnering with new agency partners in the regions or by building new warehouses. For the region in Wisconsin, increasing the capacity of existing agency partners was recommended by adding refrigerated space, creating extra shelving and being open longer hours or more days in the week. A GIS-based tool was developed for SHH which can be used to add other datasets (e.g., availability of refrigerated trucks by their agency partners) for further improved decision-making with respect to distribution capacity improvements [21]. This tool can also be used by other food banks within the Feeding America network to understand and improve their current distribution systems. Tools such as the one discussed above could help food banks and similar non-profit organizations make data-driven decisions and can ultimately help better manage surplus food by diverting it from donors to food-insecure people.

# 3   Discussion

The case studies presented in this chapter suggest how decentralized food supply systems can improve the management of surplus food, especially when they are able to leverage scale-appropriate ICT solutions.

## 3.1  Decentralized Food Supply Systems

The four case studies discussed in this chapter demonstrate the advantages of decentralized food supply systems (i.e., systems with no centralized control and few intermediaries between consumers and providers) in supporting a circular economy through better management of surplus food. In particular, local food systems, in which food is produced and distributed locally to consumers by small and mid-sized farmers [23], are decentralized. For example, Grow: Johnson County and FIF are members of a local food system in Iowa. Owing to their geographic proximity and social connections, local producers are better able to understand their consumers' preferences and requirements, which can help them to balance supply and demand and reduce wastage caused by overproduction [24]. Reducing food waste on farms by planning production in alignment with consumer demand will help reduce surplus food from the farms going into the landfills, thereby supporting the U.N. Sustainable Development Goal of combating climate change and its impacts (SDG 13).

In addition, reducing the number of intermediaries between producers and consumers helps farmers retain greater percentage of their revenues. This supports the U.N. Sustainable Development Goal of decent work and economic growth for all (SDG 8). Further, when customers buy directly from farmers, it reduces food waste due to damage caused by handling and prolonged exposure to ambient temperatures [3]. Local food systems can also support a closed loop food supply chain [25], in which soil nutrients can be retained and replenished by composting local food scraps. This supports the U.N. Sustainable Development Goal of responsible consumption and production (SDG 12).

The approaches to surplus food rescue taken by SHH and Food Rescue US are also examples of decentralized supply systems. Centralized supply systems rely on large centrally located warehouses to aggregate products and then distribute them to the retailers using large trucks. By contrast, SHH intends to move to a more decentralized distribution structure, in which agency partners directly rescue food for their clients from nearby grocery stores, rather than relying on SHH to deliver food to them from their central warehouse. This would help in increased vehicle capacity utilization of agency partners. Similarly, Food Rescue US follows a direct rescue model in which food is transferred directly from donors to recipients, using the concept of crowd logistics. Crowd logistics is a term used to describe crowd sourced transport and delivery of goods and freight, taking advantage of the underutilized capacity of individuals' personal vehicles. Senders use an online platform to source logistics services from a carrier crowd to deliver items to recipients, with a platform management system acting as a mediator. For example, Grub hub, Uber Eats, and Door Dash deliver restaurant meals to consumers using crowd logistics. Food Rescue US's crowd logistics-based distribution model increases clients' access to highly perishable foods from restaurants, which would be difficult to rescue through traditional food recovery models that rely on warehousing. The crowd logistics approach is also more economically sustainable for non-profits, as it does not require investments in infrastructure and instead empowers communities to use their existing resources. Thus, decentralized distribution can improve the operational efficiency of food rescue organizations, thereby rescuing more surplus food for food-insecure people and promoting U.N. Sustainable Development Goals of zero hunger (SDG 2) and climate action (SDG 13).

## 3.2 ICT Solutions to Manage Decentralized Supply Systems

ICT solutions play an important role in facilitating coordination between actors in decentralized supply systems. For example, the Grow farm adopted an ICT solution that allowed them to efficiently and effectively collect data to understand their consumers' needs. Food Rescue US relies on an ICT platform (a mobile application) to facilitate coordination among donors, volunteer delivery drivers, and hunger relief agencies. Non-profits can also use ICT solutions to track the volume of donations by food type from various donors. Providing this information to donors could encourage them to order less food, thereby reducing food waste. This information can also help donors evaluate amount of eligible federal tax credits through donation of surplus food to charitable organizations [16].

Although ICT solutions enable effective communication and better management of resources in a decentralized food supply system, deployment can be challenging, especially among local food systems and non-profits. Local food systems consist of small and mid-sized farmers and agricultural enterprises (e.g., food hubs and farmers' markets), which typically have constrained budgets [26]. In addition, most existing ICT solutions are designed for large entities and include features not affordable or needed by small agricultural businesses. Similarly, non-profits often do not have access to the technical resources that many for-profit companies have. However, lack of access to ICT solutions due to financial and/or technical constraints could be mitigated by leveraging resources from degree granting institutions through student engagement. Universities and technical degree granting institutes, through educational and research projects, can provide low-cost technological skills to the local community and offer continued support as operations evolve [27]. For example, collaboration between Dunwoody College of Technology and SHH demonstrates a win–win partnership. Students get the opportunity to be involved in real-world projects that allow them to put their technical knowledge into practice, while also connecting them more closely to their community.

## 4   Conclusion

Current strategies for managing surplus food through food rescue benefit everyone involved: donors get financial incentives in terms of tax credits, surplus food is prevented from being sent to landfills, and food-insecure people have increased access to food. However, the long-term goal should be to reduce the size of the population that is food insecure. Furthermore, food-insecure people deserve the same high-quality food as other people who are buying directly from the grocery stores. This will help build an equitable society, thus, ultimately promoting the U.N. Sustainable Development Goal of reduced inequalities (SDG 10).

# References

1. Buzby JC, Wells HF, Hyman J (2014) The estimated amount, value, and calories of postharvest food losses at the retail and consumer levels in the United States. USDA Econ Res Serv. https://doi.org/10.2139/ssrn.2501659
2. US EPA (2017) Inventory of U.S. Greenhouse Gas Emissions and Sinks, pp 1990–2015
3. Gunders D, Bloom J (2017) Wasted: how America is losing up to 40 percent of its food from farm to fork to landfill. Natural Resources Defense Council, vol 1. Retrieved from https://www.nrdc.org/sites/default/files/wasted-2017-report.pdf
4. Wu J (2008) The magazine of food, farm, and resource issues. In: Land Use Changes: economic, social, and environmental impacts. CHOICES 4th Quarter, vol 23(4)
5. Hall KD, Guo J, Dore M, Chow CC (2009) The progressive increase of food waste in America and its environmental impact. PLoS ONE 4(11):7940. https://doi.org/10.1371/journal.pone.0007940
6. Gunders D (2012) Wasted: how America is losing up to 40 percent of its food from farm to fork to landfill. NRDC Issue Paper, 1–26. Retrieved from https://www.nrdc.org/sites/default/files/wasted-food-IP.pdf
7. ReFED (2021) Stakeholder retailers. Retrieved from https://refed.com/stakeholders/retailers/
8. ReFED (2018) Restaurant Food Waste Action Guide. Retrieved from https://refed.com/downloads/Restaurant_Guide_Web.pdf
9. Berkenkamp J, Phillips C (2017) Modeling the potential to increase food rescue, Denver, New York and Nashville, NRDC. Retrieved from https://www.nrdc.org/sites/default/files/modeling-potential-increase-food-rescue-report.pdf
10. Mittal A, Gibson NO, Krejci CC, Marusak AA (2021) Crowd-shipping for urban food rescue logistics. Int J Phys Distrib Logist Manag. https://doi.org/10.1108/IJPDLM-01-2020-0001
11. US EPA (2020) Food recovery hierarchy. Retrieved from https://www.epa.gov/sustainable-management-food/food-recovery-hierarchy
12. USDA (2019) Key statistics and graphics. Retrieved from https://www.ers.usda.gov/topics/food-nutrition-assistance/food-security-in-the-us/key-statistics-graphics.aspx
13. Feeding America (2020) Study shows state and local food insecurity reached lowest levels since its inception 10 years ago. Retrieved from https://www.feedingamerica.org/about-us/press-room/study-shows-state-and-local-food-insecurity-reached-lowest-levels-its-inception
14. Sélingué M (2007) Declaration of Nyéléni. Retrieved from https://nyeleni.org/spip.php?article290
15. Mittal A, Gibson NO, Krejci CC (2019) An agent-based model of surplus food rescue using crowd-shipping. In: Proceedings—Winter Simulation Conference. https://doi.org/10.1109/WSC40007.2019.9004732
16. NRDC (2016) A farmer's guide to the enhanced federal tax deduction for food donation. Retrieved from https://www.nrdc.org/sites/default/files/farmers-federal-tax-deduction-food-donation.pdf
17. USDA (2021) Frequently asked questions about the Bill Emerson Good Samaritan Food Donation Act. Retrieved from https://www.usda.gov/sites/default/files/documents/usda-good-samaritan-faqs.pdf

18. Krejci CC, Gibson NO (2019) Interview with Nicole Straight. Fairfield County Site Director of Food Rescue US
19. Peters A (2018) DoorDash is now using its algorithm to deliver extra food from restaurants to food banks. Retrieved from https://www.fastcompany.com/40517038/doordash-is-now-using-its-algorithm-to-deliver-extra-food-from-restaurants-to-food-banks
20. Second Harvest Heartland (2021) Agency partners the direct link to people experiencing hunger. Retrieved from https://www.2harvest.org/who--how-we-help/services-and-programs/services/agency-partners.html
21. Mittal A, Adolfson D, Matson M, Hiniker W, Patterson G, Coleman P (2020) Community-based capstone projects in industrial engineering. In: Institute of Industrial and Systems Engineering Annual Conference, IISE
22. Gundersen C, Dewey A, Kato M, Crumbaugh A, Strayer M (2019) Map the meal gap 2019: a report on county and congressional district food insecurity and county food cost in the United States in 2017. Feeding America
23. Mittal A, Krejci CC, Craven TJ (2018) Logistics best practices for regional food systems: A review. Sustainability (Switzerland). https://doi.org/10.3390/su10010168
24. Ogrodnick A (2020) Earth day: how buying local reduces food waste. Retrieved from https://blog.freshharvestga.com/reduce-food-waste-in-georgia-farming/
25. Jurgilevich A, Birge T, Kentala-Lehtonen J, Korhonen-Kurki K, Pietikäinen J, Saikku L, Schösler H (2016) Transition towards circular economy in the food system. Sustainability (Switzerland). https://doi.org/10.3390/su8010069
26. Mittal A, Grimm J (2020) ICT solutions to support local food supply chains during the COVID-19 pandemic. J Agr Food Syst Community Dev. https://doi.org/10.5304/jafscd.2020.101.015
27. Marusak A, Sadeghiamirshahidi N, Krejci CC, Mittal A, Beckwith S, Cantu J, Grimm J (2021) Resilient regional food supply chains and rethinking the way forward: key takeaways from the COVID-19 pandemic. Agr Syst 190:103101